Monographs on Theoretical and Applied Genetics 19

Monographs on Theoretical and Applied Genetics

Volume 1 **Meiotic Configurations**
A Source of Information for Estimating Genetic Parameters
By J. Sybenga (1975)

Volume 2 **Pollination Mechanisms, Reproduction and Plant Breeding**
By R. Frankel and E. Galun (1977)

Volume 3 **Incompatibility in Angiosperms**
By D. de Nettancourt (1977)

Volume 4 **Gene Interactions in Development**
By L.I. Korochkin (1981)

Volume 5 **The Molecular Theory of Radiation Biology**
By K.H. Chadwick and H.P. Leenhouts (1981)

Volume 6 **Heterosis**
Reappraisal of Theory and Practice
Editor: R. Frankel (1983)

Volume 7 **Induced Mutations in Plant Breeding**
By W. Gottschalk and G. Wolff (1983)

Volume 8 **Protoplast Fusion**
Genetic Engineering in Higher Plants
By Y.Y. Gleba and K.M. Sytnik (1984)

Volume 9 **Petunia**
Editor: K.C. Sink (1984)

Volume 10 **Male Sterility in Higher Plants**
By M.L.H. Kaul (1988)

Volume 11 **Tree Breeding: Principles and Strategies**
By G. Namkoong, H.C. Kang, and J.S. Brouard (1988)

Volume 12 **The Wheat Rusts – Breeding for Resistance**
By D.R. Knott (1989)

Volume 13 **Population Dynamics and the Tribolium Model**
Genetics and Demography
By R.F. Costantino and R.A. Desharnais (1991)

Volume 14 **Genetic Improvement of Tomato**
Editor: G. Kalloo (1991)

Volume 15 **Evolution and Function of Heterostyly**
Editor: S.C.H. Barrett (1992)

Volume 16 **Distant Hybridization of Crop Plants**
Editors: G. Kalloo and J.B. Chowdhury (1992)

Volume 17 **Cytogenetics in Genetics and Plant Breeding**
By J. Sybenga (1992)

Volume 18 **Cytogenetics of the Festuca – Lolium Complex**
By P.P. Jauhar (1993)

Volume 19 **Breeding Oilseed Brassicas**
Editors: K.S. Labana, S.S. Banga, and S.K. Banga (1993)

K.S. Labana S.S. Banga S.K. Banga (Eds.)

Breeding Oilseed Brassicas

With 17 Illustrations

Springer-Verlag
Berlin Heidelberg GmbH

Editors
Dr. Kuldeep S. Labana (Deceased)
Ex-Head, Department of Plant Breeding
Punjab Agricultural University
Ludhiana 141 004, India

Dr. Surinder S. Banga
Dr. Shashi K. Banga
Department of Plant Breeding
Punjab Agricultural University
Ludhiana 141 004, India

ISBN 978-3-662-06168-8 ISBN 978-3-662-06166-4 (eBook)
DOI 10.1007/978-3-662-06166-4

Originally published by Springer-Verlag Berlin Heidelberg New York in 1993
Softcover reprint of the hardcover 1st edition 1993

31 3140 - 543210

Typesetting: Indira Printers, New Delhi-110020, India.

To Mr. S.S. Banga Mrs. Satpal Kaur Banga and Mrs. Nirmal Labana

Preface

From ugly duckling of agricultural crops to the Cinderella crop, the significance of oilseed brassicas in international trade has increased tremendously with global production having almost trebled in the past three decades. Over 13.2% of the world's edible oil requirement is now met from this source. The stimulus for this dramatic reversal was provided by the identification of rapeseed genotypes with greatly reduced levels of nutritionally undesirable erucic acid and meal glucosinolates during the early 1960s. Since then, rapeseed breeding has been oriented towards the need to improve oil quality. Though past achievements have been impressive, greater strides are expected in upgrading productivity through the development of F_1 hybrids and better mobilization of genetic diversity available in wild allies. Recent emphasis on cellular and molecular biology will help to bypass sexual constraints for genetic enrichment of crop brassicas, and to reduce the time frame for cultivar development. The present book was conceived to bring together critical and comprehensive reviews on research approaches, achievements and limitations to breeding better brassicas. As the book has a strong practical bias, certain topics like taxonomy, cytogenetics, etc. are dealt with in relatively lesser detail. Thus, '*Brassica* crops and wild allies', edited by S. Tsunoda, K. Hinata and C. Gõmez-Campo, will serve as an ideal companion book.

It is a matter of immense sorrow that the senior editor, Prof. K. S. Labana, passed away before the completion of the book project. He died on 17.10.1990 at the age of 60. A doyen of oilseed breeders in India, Prof. Labana had 20 varieties and more than 160 research publications to his credit.

The completion of this volume has depended much on the cooperation of all the contributors, our gratitude to them and to Dr. R. Frankel for useful comments. We are grateful to Padma Bhushan Prof. Khem Singh Gill, Vice Chancellor, Punjab Agricultural University, Ludhiana, for encouragement and permission to edit the book. Appreciation is also expressed for the support provided by Mr. N. K. Mehra and Mr. M. S. Sejwal, Narosa Publishing House, New Delhi, and to the members of the Biology Editorial Department, Springer-Verlag. Technical help by Ms. Rosy is also acknowledged.

Regrets to the children, Winnie and Richie, for the period of neglect.

SURINDER S. BANGA
SHASHI K. BANGA

List of Contributors

1. ÅHMAN, INGER, Svalöf AB, S-26881, Svalöv, Sweden.
2. AHUJA, K.L., Department of Plant Breeding, Punjab Agricultural University, Ludhiana, India.
3. BANGA, SHASHI K., Department of Plant Breeding, Punjab Agricultural University, Ludhiana, India.
4. BANGA, S.S., Department of Plant Breeding, Punjab Agricultural University, Ludhiana, India.
5. BHATIA, C.R., Nuclear Agricultural Division, Bhabha Atomic Research Centre, Bombay, India.
6. CHOPRA, V.L., Biotechnology Centre, Indian Agricultural Research Institute, New Delhi, India.
7. DHAWAN, A.K., Regional Research Station, Haryana Agricultural University, Karnal, India.
8. DHILLON, S.S., Regional Research Station, Punjab Agricultural University, Bathinda, India.
9. DUKE, L.H., Ontario Agricultural College, Department of Crop Science, University of Guelph, Guelph, Ontario, NIG2W1, Canada.
10. GUPTA, M.L., Department of Plant Breeding, Punjab Agricultural University, Ludhiana, India.
11. GUPTA, NEENA, Department of Plant Breeding, Punjab Agricultural University, Ludhiana, India.
12. HOLBROOK, L.A., Department of Biological Sciences, Faculty of Science, University of Calgary, Alberta, Canada.
13. INOMATA, N., Department of Biology, College of Liberal Arts and Sciences, Okayama University, Tsushima, Okayama 700, Japan.
14. KUMAR, P.R., Directorate of Research, Rajendra Agricultural University, Samastipur, India.
15. LABANA, K.S. (deceased), Department of Plant Breeding, Punjab Agricultural University, Ludhiana, India.
16. MITRA, R., Nuclear Agricultural Division, Bhabha Atomic Research Centre, Bombay, India.

17. MOLONEY, M.M., Department of Biological Sciences, University of Calgary, 2500 University Drive N.W., Alberta, T2N 1N4 Canada.

18. PRAKASH, S., Biotechnology Centre, Indian Agricultural Research Institute, New Delhi, India.

19. SAHARAN, G.S., Department of Plant Pathology, Haryana Agricultural University, Hissar, India.

20. SEKHON, B.S., Regional Research Station, Punjab Agricultural University, Bathinda, India.

21. THORPE, M.L., Ontario Agricultural College, Department of Crop Science, University of Guelph, Guelph, Ontario NIG2W1, Canada.

22. THURLING, N., School of Agriculture, Crop and Pasture Sciences, University of Western Australia, Nedlands, W.A. 6009, Australia.

23. YARROW, S., Allelix Crop Technologies, 6850, Goreway Drive, Mississauga, Ontario, Canada.

Contents

CHAPTER 1

Importance and Origin

K.S. Labana and *M.L. Gupta*

1.1 Introduction

The oilseed brassicas comprise four species, namely *B. campestris*, *B. juncea*, *B. napus* and *B. carinata*. Together they occupy over 14 million hectares with an annual production of 24.61 million metric tonnes (Anonymous 1991). This provides about 12% of the world's edible vegetable oil supplies. These crops have wide adaptation and are grown under varied agroclimatic conditions. Both winter and spring forms of rape (*B. napus*) and turnip rape (*B. campestris*) are cultivated in Europe, while spring-grown cultivars of rape and turnip rape account for most of the rapeseed production in Canada. Winter rape is also the major oilseed crop in China, whereas Indian mustard (*B. juncea*) is the predominant *Brassica* of the Indian subcontinent. Ethiopian mustard (*B. carinata*) is grown mainly in Ethiopia. Considerable potential exists for extending the cultivation of *Brassica* oilseeds to the USA and Australia. The importance of these crops in international trade can be judged from the tremendous increases in both cultivated acreage and yield per hectare during recent years. Production almost trebled between 1971-74 and 1989-90. This can be attributed to the development and large-scale adaptation of zero erucic and low glucosinolate varieties. Quality improvement is likely to remain an important factor in the furtherance of the status of oilseed brassicas in international trade through the development of cultivars with varied fatty acid composition to suit specific nutritional and industrial needs. The toxicant-free rapeseed-mustard meal is being increasingly used for animal feeding and as a human protein supplement.

1.2 Nomenclature and Uses

B. campestris, *B. oleracea* and *B. juncea* occur as a range of morphotypes which represent a parallel evolution of different vegetable and oilseed forms (Table 1). Secondary and tertiary evolution has resulted in adaptation of species to different regions of the world. Each of these species now has heading forms, open leafy rosettes, swollen stems, enlarged roots and prominent floral or oilseed-producing forms. *B. nigra*, *B. carinata* and *B. napus* are less varied.

1.3 Species Relationship and Origin

Initial cytogenetic researches demonstrated that crop brassicas comprise three elementary diploid species, namely *B. campestris* ($2n = 20$; AA), *B. nigra*

Table 1. Nomenclature, chromosome number and genomic descriptors of crop brassicas

Species	Subspecies or variety	2*n* chromosome no.; genome descriptor	Common name	Use
B. nigra	—	16; BB	Black mustard	Condiment
B. oleracea	*acephala*	18; CC.a	Kale	Vegetable/Fodder
	alboglabra	18; CC.al	Chinese kale	Vegetable
	botrytis	18; CC.b	Cauliflower	Vegetable
	capitata	18; CC.c	Cabbage	Vegetable
	costata	18; CC.co	Portuguese cabbage	Vegetable
	gemmifera	18; CC.g	Brussels sprouts	Vegetable
	gongylodes	18; CC.go	Kohl rabi	Vegetable/Fodder
	italica	18; CC.i	Broccoli, Calabrese	Vegetable
	medullosa	18; CC.m	Marrow stem kale	Vegetable
	palmifolia	18; CC.p	Kale	Vegetable
	ramosa	18; CC.ra	Thousand-head kale	Vegetable
	sabauda	18; CC.s	Savoy cabbage	Vegetable
	sabellica	18; CC.sa	Collards	—
	selensia	18; CC.se	Borecole	—
B. campestris syn. *rapa*	*chinensis*	20; AA.c	Pak choi	Vegetable
	narinosa	20; AA.na	—	Vegetable
	nipposinica	20; AA.n	Curled mustard	Vegetable
	oleifera	20; AA.o	Turnip rape, *toria*	Oilseed
	pekinensis	20; AA.p	Chinese cabbage, Petsai	Vegetable
	parachinensis	20; AA.pa	Choy sum	Vegetable
	perviridis	20; AA.pe	Komatsuna, Mustard spinach	Vegetable
	rapifera	20; AA.r	Turnip	Vegetable
	trilocularis	20; AA.t	*Sarson*	Oilseed
	utilis	20; AA.u	—	—
B. carinata		34; BBCC	Ethiopian mustard	Vegetable/Oilseed
B. juncea	*capitata*	36; AABB.c	Head mustard	Vegetable
	crispifolia	36; AABB.cr	Cut leaf mustard	Vegetable/Fodder
	faciliflora	36; AABB.f	Broccoli mustard	Vegetable
	lapitata	36; AABB.l	Large petiole mustard	Vegetable
	multiceps	36; AABB.m	Multishoot mustard	Vegetable/Fodder
	oleifera	36; AABB.o	Indian mustard	Oilseed
	rapifera	36; AABB.r	Root mustard	Vegetable
	rugosa	36; AABB.ru	Leaf mustard	Vegetable
	spicea	36; AABB.sp	Mustard	—
	tsa-tsai	36; AABB.t	Big stem mustard	Pickle/Vegetable
B. napus	*oleifera*	38; AACC.o	Oilseed rape	Oilseed
	rapifera	38; AACC.r	Swede, rutabaga	Vegetable/Fodder

($2n$ = 16; BB) and *B. oleracea* ($2n$ = 18; CC) and three amphidiploids, which originated through interspecific hybridization between any two of the three elementary species. These are *B. juncea* ($2n$ = 36; AABB), *B. napus* ($2n$ =38; AACC) and *B. carinata* ($2n$ = 34; BBCC). The cytogenetic relatedness between these species was established by U (1935). This was later confirmed by chromosome pairing and artificial synthesis of amphiploids (Prakash and Hinata 1980), nuclear DNA content (Verma and Rees 1974), DNA analysis (Erickson et al. 1983) and the use of genome-specific chromosome markers (Hosaka et al. 1990). The close homoelogy between three elementary species was established by homoelogous pairing (Attia and Röbbelen 1986) and the presence of shared fragments between A, B and C genomes (Hosaka et al. 1990). The available evidence suggests that all three elementary species evolved from a common ancestor, probably having a chromosome number of n = 6 (Röbbelen 1960) through an increase of some chromosomes. Analysis of cotyledons of various taxa led to the proposal that *Brassica* evolved from the *Sinapidendron* archetype through the *Diplotaxis* and *Erucastrum* complexes (Gõmez-Campo and Tortosa 1974).

Based on nuclear restriction fragment length polymorphisms (RFLPs), Song et al. (1990) proposed a new hypothesis, according to which species with chromosome numbers of n = 7 constituted the most ancient group. This probably arose from the prototype species of n = 6. Two evolutionary pathways were suggested for various accessions included in the study. *B. adpressa, B. fruticulosa, B. nigra* and *Sinapis arvensis* compose one group, having ascending chromosome number with *B. adpressa* or a close relative as the primary ancestor. *Diplotaxis erucoides, B. oleracea* and *B. rapa* compose another ascending group with *D. erucoides* or a close relative as the primary progenitor.

1.3.1 *B. nigra*

Although little information is available, the occurrence of land races in Europe, the Mediterranean and the Ethiopian plateau (Schulz 1919; Bailey 1922; Mizushima and Tsunoda 1967) indicates that *B. nigra* probably originated in central and south Europe (Bailey 1922; Zeven and Zhukovsky 1975). It is presumed that *nigra* was introduced into India comparatively recently, even though it found mention in the Sanskrit book *Upanisadas* as *Sarshap* (Prakash 1961).

1.3.2 *B. oleracea*

There is great diversity of morphotypes in *B. oleracea* (Table 1), and all are thought to have originated from a wild west European *B. oleracea*, which was domesticated during the first millennium B.C. (De Candolle 1886). The hypothesis of monophyletic origin was further supported by the RFLP analysis of nine cultivated morphotypes of *B. oleracea* (Song et al. 1990). On the other hand, triple (Helm 1963) or even multiple origin (Lizgunova 1959; Mithen et al. 1987) was also proposed. Gates (1953) believed that cabbage, Brussels sprouts and kale originated from *oleracea* in western Europe, while cauliflower and broccoli evolved from a related species in the eastern Mediterranean region. Syria was also proposed as the centre of domestication of cauliflower (Hyams 1971). Helm (1963) suggested a scheme indicating the origin of various *B. oleracea* morphotypes from a wild cabbage (var. *sylvestris*) which gradually gave rise to various forms through mutation or chromosomal structural changes (Harberd 1972, 1975; Chiang and Grant 1975). Selective introgressions from wild species such as *B. incana* from the west and south coasts of Italy, Sicily and Yugoslavia, *B. montana* from Europe, northeast Spain and southern Italy, *B. rupestris* and *B. villosa* from Sicily, *B. cretica* from the Aegean sea region and *B. insularis* from Corsica and Sardinia substantially increased the variability and adaptability (Prakash and Hinata 1980). This was further confirmed by the occurrence of specific markers (e.g. Clone PB 177) found in certain kale, as well as in some accessions of *B. insularis* and *B. incana* (Hosaka et al. 1990). According to the hypothesis of Song et al. (1990), the cultivated morphotypes of *B. oleracea* originated from a single progenitor that was similar to wild *B. oleracea* and *B. alboglabra*. The earliest cultivated *B. oleracea* was probably a leafy kale from which a variety of kales originated. Ancient broccoli might have originated from a type of kale grown in Italy that had specialized inflorescence. Cauliflower later evolved from broccoli. Ancient cabbage perhaps evolved from a leafy kale, and Portuguese cabbage may be an intermediate type from which common cabbage developed.

1.3.3 *B. campestris*

In *B. campestris*, a large number of important subspecies are recognized (Table 1). All these subspecies have a wide range of variation and were cultivated in China even before the time of Christ. The earliest written reference pertains to *yellow sarson* in ancient Sanskrit literature *Upanisadas* and *Brahamanas* (Ca. 1500 B.C.), where it was mentioned as *Siddhartha* (Prakash 1961). *B. campestris* has the widest distribution, with secondary centres of diversity in Europe, western Russia, central Asia and the near East (Vavilov 1949; Mizushima and Tsunoda 1967; Zeven and Zhukovsky 1975). Burkill (cf. Prakash and Hinata 1980) regarded Europe as the place where *campestris* first originated. The wide range of subspecies in *B. campestris* is believed to have resulted from varied selection pressures (including human selection) in different geographic regions. Turnip (ssp. *rapifera*) is probably of European origin. In the east, the selection for leafy vegetables resulted in great diversity of Chinese cabbages. In the Indian subcontinent, human selection for oil content led to three morphologically distinct oleiferous races, namely *yellow sarson, brown sarson* (*lotni* and *tora* groups) and *toria*. The possible centre of origin being eastern Afghanistan and adjoining areas of Pakistan and northwest India. *Brown sarson* is perhaps the oldest in origin, from which *yellow sarson* and *toria* differentiated. The oleiferous forms probably originated separately in the Mediterranean region and Asia. The existence of two races was also confirmed by protein analysis (Denford 1975). Song et al. (1990) conducted nuclear RFLP analysis of various morphotypes of *B. campestris* (syn. *rapa*). Based on these studies, they proposed that wild *B. rapa* originated in Europe from a common ancestor of *B. rapa* and *B. oleracea*. Primitive *B. rapa* then spread through the Middle East to

India and south China to form different populations. In Europe, wild *B. rapa* was domesticated as turnip rape. Pak choi and *narinosa* were domesticated independently in south China; Chinese cabbage evolved from pak choi, heading cabbage evolved later on. Various Japanese vegetables differentiated from pak choi.

1.3.4 *B. juncea*

Indian mustard is grown widely in Asia. Due to ecogeographic variation and human selection, a number of morphologically distinct forms are available (Table 1). These include oleiferous, semi-oleiferous, rapiferous and leafy types. Due to the sympatric range of distribution of the two diploid progenitors, *B. nigra* and *B. campestris*, the allopolyploid *B. juncea* may have originated anywhere between eastern Europe and China. Prakash and Hinata (1980) hold the view that *B. juncea* first evolved in the Middle East, where its putative diploid progenitors, *B. campestris* and *B. nigra*, had geographic sympatry. This is further confirmed by the fact that the various forms of *B. campestris* and *B. juncea* can still be observed growing wild in the Middle East especially on the plateaus of Asia Minor and Southern Iran (Olsson 1960; Mizushima and Tsunoda 1967; Tsunoda and Nishi 1968). The occurrence of parallel variation in forms between 20-chromosome species and *B. juncea* indicates that the latter may have originated due to natural hybridization between different subspecies of *B. campestris* with *B. nigra* as a common genome, thus accounting for the various centres of diversity observed in this species (Hemingway 1976; Prakash and Hinata 1980).

1.3.5 *B. napus*

This crop is of comparatively recent origin (Olsson 1960) and wild populations of *B. napus* have not been found (Prakash and Hinata 1980). Sinskaia (1928) and Schiemann (1932) believed it to be of southwest European or Mediterranean origin. It is generally accepted that *B. napus* originated in southern Europe, where there is overlapping in the geographic distribution of wild *B. campestris* and *B. oleracea*. Winter rape first appeared as a crop around the year 1600 in Flanders and rapidly replaced the existing winter turnip rape throughout northwestern Europe. The Dutch name is Koolzad, from which the French name colza is derived. The German name is Ölraps. As considerable variability exists in *B. napus*, it is possible that it arose at different locations involving different constitutive genomes (Rudorf 1950). Recent reports provide evidence for the polyphyletic origin of *B. napus* (Palmer et al. 1983; Song et al. 1988; Hosaka et al. 1990).

1.3.6 *B. carinata*

This is used as both oilseed and vegetable crop in Ethiopia, with little differentiation into various crop types. While Mizushima and Tsunoda (1967) failed to locate any wild types in the Ethiopian plateau, they observed wild-growing *B. nigra* and cultivated kale-like forms of *B. oleracea*. Thus it is possible that these are the ancestral forms of *B. carinata*.

1.4 Conclusions

Various *Brassica* species occur as a range of morphotypes which represent a parallel evolution of oilseed and vegetable forms. Available evidence indicates that all the present day crop brassicas arose from a single ancient prototype species with a chromosome number of n=6.

References

Anonymous (1991) Cereal and oilseeds review, February 1991, p 37

Attia T, Röbbelen G (1986) Cytogenetic relationship within cultivated *Brassica* analyzed in amphiploids from the three diploid ancestors. Can J Genet Cytol 28: 323–329

Bailey LH (1922) The cultivated *Brassicas*. I. Genetes Herb 1: 53–108

Chiang BY, Grant WF (1975) A putative heterozygous interchange in cabbage (*Brassica oleracea* var. *capitata*) cultivar Badger Shipper. Euphytica 24: 581–584

De Candolle A (1886) Origin of cultivated plants. Reprint, 2nd Ed Facsim ed Hafner, New York 1967 p 468

Denford KE (1975) Isozyme studies in members of the genus *Brassica*. Bot Not 128: 455–462

Erickson LR, Straus NA, Beversdorf WD (1983) Restriction patterns reveal origins of chloroplast genomes in *Brassica* amphidiploids. Theor Appl Genet 65: 201–206

Gates RR (1953) Wild cabbages and effects of cultivation. J Genet 51: 363–372

Gõmez-Campo C, Tortosa ME (1974) The taxonomic and evolutionary significance of some juvenile characters in Brassiceae. Bot J Linn Soc 69: 105–124

Harberd DJ (1972) A contribution to the cytotaxonomy of *Brassica* (cruciferae) and its allies. Bot J Linn Soc 65: 1–23

Harberd DJ (1975) Cytotaxonomic studies of *Brassica* and related genera. In: Vaughan JG, Macleod AJ, Jones BM (eds) The biology and chemistry of Cruciferae. London pp 47–68

Helm J (1963) Morphologisch taxonomische Gliederung der Kultursippen von *Brassica oleracea* L. Die Kulturpflanze 11: 92–110

Hemingway JS (1976) *Brassica* ssp. and *Sinapis alba* (Cruciferae) In: Simmonds NW (ed) Evolution of crop plants. Longmans, Green, New York pp 19–21

Hosaka K, Kianian SF, McGarlh JM, Quiros CF (1990) Development and chromosomal localization of genome specific DNA markers of *Brassica* and the evolution of amphidiploids and n = 9 diploid species. Genome 33: 131–142

Hyams E (1971) Cabbages and Kings. In: Plants in the service of man. London. pp. 33–61

Lizgunova TV (1959) The history of botanical studies of the cabbage, *Brassica oleracea* L. Bull Appl Bot Genet Pl. Breed 32: 37–70

Mithen RF, Lewis BG, Heaney RK, Fenwik GR (1987) Glucosinolates of wild and cultivated *Brassica* species. Phytochem. 26: 1969–1973

Mizushima U, Tsunoda S (1967) A plant exploration in *Brassica* and allied genera. Tohoku J Agri Res 17: 249–276

Olsson G (1960) Species crosses within the genus *Brassica*. I. Artificial *Brassica juncea* Coss. Hereditas 46: 171–222

Palmer JD, Shield CR, Cohen DB, Orton TJ (1983) Chloroplast DNA evolution and origin of amphidiploid *Brassica* species. Theor Appl Genet 65: 181–189

Prakash O (1961) Food and drinks in Ancient India. Delhi pp 265–266

Prakash S, Hinata K (1980) Taxonomy, cytogenetics and origin of crop brassicas, a review. Opera Bot 55: 1–57

Röbbelen G (1960) Beitrage zur Analyse des *Brassica*-Genomes. Chromosoma 11: 205–228

Rudorf W (1950) Über die Erzeugung und die Eigenschaften synthetischer Rapsformen. Zur Pflanzenzüchtung 29: 35–54

Schiemann E (1932) Entstehung der Kulturpflanzen Handb vererbwis Lfg 15

Schulz OE (1919) Cruciferae-Brassicaceae, Part 1: Brassicinae and Raphaninae. In Engler A (ed), Das Pflanzenreich IV: 1–290 Leipzig

Sinskaia EN (1928) The oleiferous plants and root crops of the family Cruciferae. Bull Appl Bot Genet Pl Breed 17: 3–166

Song KM, Osborn TC, Williams PH (1988) *Brassica* taxonomy based on nuclear restriction fragment length polymorphisms (RFLP). I. Genome evolution of diploid and amphidiploid species. Theor Appl Genet 75: 784–794

Song KM, Osborn TC, Williams PH (1990) *Brassica* taxonomy based on nuclear restriction fragment length polymorphisms (RFLPs) 3. Genome relationships in *Brassica* and related genera and origin of *B. oleracea* and *B. rapa* (syn. *campestris*). Theor Appl Genet 79: 497–506

Tsunoda S, Nishi S (1968) Origin, differentiation and breeding of cultivated *Brassica*. In: Proc XII Int Congr Genet 2: 77–78

UN (1935) Genome analysis in *Brassica* with a special reference to the experimental formation of *B. napus* and peculiar mode of fertilization. Jpn J Bot 7: 389–452

Vavilov NI (1949) The origin, variation, immunity and breeding of cultivated plants. Chron Bot 13: 1–364

Verma SC, Rees H (1974) Nuclear DNA and the evolution of allotetraploid *Brassicae*. Heredity 33: 61–68

Zeven AC, Zhukovsky PM (1975) Dictionary of cultivated plants and their centres of diversity. Pudoc pp 29–30

CHAPTER 2

Breeding Objectives and Methodologies

S.S. Dhillon, P.R. Kumar and *Neena Gupta*

2.1 Introduction

The cultivar development in rapeseed-mustard aims at satisfying the requirements of the producer, miller and consumer. During the last two decades, the need for modification in the fatty acid composition of the oil and the elimination of the glucosinolates from the seed meal attracted great attention from *Brassica* breeders in Canada and the European countries. This led to the release and widespread adoption of '00' varieties of rape and turnip rape. In Asian countries, greater seed yield and stability were the primary objectives. Attempts are now underway to extend the advantages of quality breeding to Indian rape and Indian mustard. Overall, the major breeding goals are to increase yield, improve quality and reduce production costs.

The type of desired cultivar and the mode of reproduction determine the methods of breeding. For the cross-pollinating ecotypes of *B. campestris*, the choice is mainly between composites, synthetics, open-pollinating populations and hybrids. Development of F_1 hybrids is difficult due to the occurrence of sporophytic self-incompatibility. In *B. juncea* and *B. napus*, the option was previously restricted only to the production of pure line cultivars. With the availability of cytoplasmic male sterility-fertility restoration systems (see Banga, Chap. 3, this Vol.), the development of commercial hybrid cultivars is now a real possibility. In view of high outcrossing rates (20–30%), the partial exploitation of heterosis is also possible through the development of synthetic cultivars (Becker 1989). Breeding objectives and methodologies of varietal development of *Brassica* oilseeds will be discussed in this chapter.

2.2 Breeding Objectives

2.2.1 Seed Yield

Higher seed yield and its stability are the prime objectives. Yield is the end product of many biological processes, which are under the control of complex polygenic systems. Since the response to selection for this complex trait is normally slow, it may, therefore, be profitable to select for any yield-contributing factor like branch number, pods per plant, seeds per pod and seed size. Further yield increases could result from increase in biomass and/or harvest index. Increased biomass can result from reduced photorespiration, and increased light saturated rate of photosynthesis.

2.2.2 Early Maturity

Early maturing varieties are needed for use in various multiple cropping sequences in China and India. In western Canada, *B. campestris* varieties maturing in less than 100 days are required to escape frost damage. Short-duration rapeseed varieties are suitable in southern Australia to avoid late-season drought stress.

2.2.3 Resistance/Tolerance to Abiotic Factors

Brassica yields suffer because of many abiotic stresses associated with low temperatures, frost, drought, alkalinity and salinity. Winter hardiness is very important for winter types of *B. campestris* and *B. napus*. In the northwest region of the Indian subcontinent, frost resistance is needed to prevent substantial yield losses.

2.2.4 Resistance to Biotic Stresses

A large number of fungi and viruses are pathogenic to oilseed brassicas. Blackleg (*Phoma lingam*) and *Sclerotinia* rot (*Sclerotinia sclerotiorum*) are the important diseases afflicting *B. campestris* and *B. napus* in the majority of rapeseed-mustard growing countries, except for the Indian subcontinent. White rust [*Albugo candida* (Lev.) Kunze] also causes substantial yield losses in *B. campestris* and *B. juncea*. All European and Canadian *B. napus* varieties are resistant to white rust. However, staghead infection was observed during 1989 in many *B. napus* cultivars bred in India. White rust incidence on oilseed rape has also been reported from China. *B. juncea* genotypes, with few exceptions, are generally susceptible. *Alternaria* blight is the most destructive disease, especially in India, and no well-characterized source of resistance is available. Many other diseases are also known to cause substantial economic losses (see Saharan, Chap. 12, this Vol.).

Among the insects, mustard aphid (*Lipaphis erysimi*) is the most devastating insect pest in India. Extensive screening efforts have failed to identify any source of resistance. Both winter and spring rape forms of *B. napus* suffer considerable damage by flower bud and seed-pod pests (see Sekhon and Åhman, Chap. 13, this Vol.).

2.2.5 Herbicide Resistance

Incorporation of herbicide resistance (e.g. atrazine, simazine etc.) in all oilseed brassicas is an important crop management strategy to overcome weed problems. Few sources of resistance are available.

2.2.6 Intercropping

In India, intercropping of *B. napus* and *B. campestris* has proved lucrative. There is a need to breed cultivars of both *B. campestris* and *B. napus* which are especially suitable for this practice.

2.2.7 Shattering Resistance

Pod shattering in *B. napus* is a major problem and no known source of shattering resistance is available in world germplasm of *B. napus*. Recently, Prakash and Chopra (1988) were able to introgress resistance to pod shattering in *B. napus* from *B. juncea*. This can prove to be a very important source for shattering resistance.

2.2.8 Oil Content and Quality

Increased oil content is an important breeding objective as, apart from its economic value, in the majority of European countries the seed is purchased on the basis of its oil percentage. A large number of genetic stocks with very high oil content (45%) are available. However, there is a metabolic limit to increasing the oil content without the yield suffering. A simple switch-over to the yellow-seeded forms would mean a bonus increase in oil content of about 2%. Development of low erucic acid cultivars is an essential nutritional objective, which has been accomplished in both rape and turnip rape but is still to be achieved at the commercial level in Indian rape and Indian mustard. For industrial purposes, increased erucic acid is desirable. Genotypes with about 60% erucic acid level are available while the theoretical limit is 66.6%. Reduced linolenic acid content (< 3%) and increased linoleic acid (> 30%) are also desirable. Higher levels of palmitic or palmitoleic acid are needed for improved margarine quality, while increased oleic acid is required for ideal cooking oil.

2.2.9 Meal Quality

The fixed goal for glucosinolate content is <10 μmol/g defatted meal. In *B. napus* and *B. campestris*, cultivars with glucosinolate levels less than 15 μmol/g and 30 μmol/g respectively of defatted meal have been developed. This objective is now receiving the attention of Indian breeders.

2.3 Breeding Methodologies

The choice of breeding procedure depends on the immediate and long-term objectives, the extent of genetic variability and available resources. In the earlier stages of varietal development, selection from a heterogeneous population is the obvious first step. This is followed by planned matings or induced mutagenesis to supplement the existing variability. Various procedures have been devised over the years to evaluate and channelize the variability into improved cultivars.

2.3.1 Simple Selection

Selection in a variable population is the simplest and most widely followed method of crop improvement. In the earlier times, mass selection was the method of choice and was followed both in self-pollinating types like *B. juncea* and *B. napus* as well as in cross-pollinating ecotypes of *B. campestris*. In the self-pollinating crops, mass selection coupled with self-pollination provides an inexpensive procedure for increasing the frequency of desirable genotypes in the population during inbreeding. In cross-pollinating crops mass selection aims at developing a superior open-pollinating population. The best use of this technique is in varietal purification. In the self-pollinators, the procedure involves the selection of individual plants from a variable material. The individual plant to progeny rows are then grown and undesirable progenies are discarded. The selected progenies are put in initial yield trials or observation rows followed by replicated yield trials. A number of mustard varieties in India, namely Durgamaani, Seeta, Shekhar, Krishna and Kranti, were developed through selection. The first winter rape cultivar Lembkes was also a selection from local landraces in Germany.

2.3.2 Hybridization and Handling of Segregating Populations

Creation of a segregating population with characters of interest is a major step in varietal development. This is normally done by hybridization of genetically diverse parents. The simplest and most frequently used population is formed by mating the two parents. Sometimes, however, it is imperative to develop three-parent, four-parent or even complex populations. Selection of the parents of a cross depends upon their complementary characters and the extent of genetic diversity among the parents. The advantage of mating genetically diverse parents is that each may contribute desirable alleles, which, when combined, may result in superior individuals. Many mating designs are available which aid in forming such populations and also in deciphering the genetic make-up of the population being handled (Mather and Jinks 1982).

In the self-pollinating brassicas, a number of methods are available for inbreeding a segregating population until practical homozygosity is realized. These include the bulk method, the pedigree method, and the single-seed descent method. The pedigree method predominates, though only few breeders follow it strictly. The actual pedigree method involves the selection on a single plant basis in each generation, till family characters start appearing. In the subsequent generations, desirable families are identified first, then the desirable lines from within the families, and lastly the desirable plants within the selected lines. The process is continued until homozygous progeny is obtained. This is then multiplied to provide sufficient seed for the multilocation yield trials. A major limitation of this system is the unnecessary record keeping. The majority of the pure line varieties bred in *B. juncea* and *B. napus* owe their origin to pedigree selection. In *B. napus*, when the breeding goal is to combine higher yield, winter hardiness and low erucic acid, pedigree selection is carried out in combination with the backcross method. The selection is normally initiated in the BC_2 or BC_3 generation. This modification is also used for combining disease resistance (e.g. blackleg resistance) with other agronomic characters.

Studies in *B. juncea* have shown that the visual selection on single-plant basis for seed yield and oil content is ineffective in the F_2 generation (Chopra 1981). This is in conformity with the results obtained in cereals (McKenzie and Lambert 1961; Knott 1972). Thus the breeders normally prefer to delay any selection for yield until F_4 or even later generations. This, however, invites the risk of losing desirable recombinants. Partial alleviation of the problem can be achieved through the appropriate choice of parents and use of alternatives like the bulk selection or single-seed descent method. The single-seed descent method can be integrated very easily into the usual pedigree method.

2.3.3 Backcross Method

The highly developed varieties are likely to have the best alleles at many loci. These linkages can be regarded as an advantage, because they tend to hold together the existing favourable combinations (Allard 1960). The backcross method is used to introduce specific alterations without disturbing the background genotype, as backcrossing is known to decrease the yield of recombinants. Ever since the technique was first proposed (Harlan and Pope 1922), the backcross method has proved very effective for incorporating disease resistance into susceptible but otherwise agronomically superior varieties (Briggs and Allard 1953). The term backcross refers to the repeated crossing of hybrid progeny to one of the parents. The parent which is repeatedly crossed to the hybrid progeny is known as the recurrent parent, while the parent contributing the genes for the characters under consideration is called the donor or the non-recurrent parent. The recurrent parent should be a highly acceptable genotype, except for the character for which a modification is desired. If the recovery of the cytoplasm of the non-recurrent parent is desirable, then this parent must be used as female in the initial cross. In the subsequent backcrosses also, the recurrent parent should be used as female.

In the absence of linkage and selection, the average percentage of genes of the recurrent parent increased by one half of the percentage of the genes that were present during the previous generation. The average recovery of the genotype of the recurrent parent is $1 - (1/2)^{n+1}$, where n is the number of backcrosses to the recurrent parent. Thus in about nine backcross generations the genotype of the recurrent parent is recovered up to 99.902%. The backcross approach has been widely used in rapeseed-mustard crops for the transfer of simply inherited traits like low erucic acid and glucosinolate content into agronomically superior types (Harvey and Downey 1964; Lein 1972; Morice 1974; Kirk and Hurlstone 1983).

Apart from its use in the selective improvement of specific characters, the utility of the backcross technique in the improvement of several quantitative characters was also demonstrated (Duvick 1974; Grafius et al. 1976). In *B. napus*, Thurling (1982) developed a breeding procedure in which two successive backcrosses followed by two successive generations of selfing gave rise to a number of lines in each of the two crosses studied which significantly outyielded the high-yielding recurrent parent.

2.3.4 Recurrent Selection

The economic potential of a variety or a hybrid is determined by a large number of genes that constitute the genetic make-up of an individual. The plant breeders aim to increase the frequency of the desirable alleles in a population through selection of superior recombinants derived from crossing between the complementary individuals. Intercrossing is known to increase the yield of recombinants. At least one or more, and preferably four, intermating cycles should precede the selfing generations to ensure a degree of breakup of the linkage groups and to increase the genetic recombinations within the linkage group. This is well achieved by recurrent selection with a low selection intensity. In a comprehensive strategy of varietal development, it is necessary to combine population improvement and varietal development (Gallais 1986). The improved population with an increased frequency of desirable alleles can be used as a cultivar per se, as a source of superior individuals that can be used as inbred lines, pure line cultivars or parents of a hybrid.

A number of intra- and inter-population improvement techniques suited mainly to the cross-pollinating crops, are available (Sprague and Eberhart 1977; Hallauer and Miranda 1981; Moll and Hanson 1984). These techniques can be easily extended to the improvement of the cross-pollinating *B. campestris*. Apart from a few attempts, not many results are available to demonstrate the efficacy of these systems in brassicas. In the related crop of white mustard (*Sinapis alba*), long-term experiments (12 years) on recurrent selection have led to the improvement in the number of seeds per pod, oil content, fatty acid composition and nematode resistance (Olsson and Persson 1986). After completion of selection, the number of seeds per pod was increased by 67%. Similarly, the oil content could be increased up to 37.4% in the positive selection against 30.0% of the standard. The total difference between the positive and negative selections was 14.78 ± 0.63% (Olsson and Persson 1986). Simultaneous selection for seed yield and oil content was also effective.

The mating system of plants is not a biological constant as a large amount of genetic variability can exist within the same species (Becker 1989). Both *B. juncea* and *B. napus* are largely self-pollinating crops, but substantial outcrossing (up to 30%) is known to occur (Labana and Banga 1984; Rackow and Woods 1987; Dhillon and Labana 1988). Recently self-incompatible genotypes have also been developed (see Banga, Chap. 3, this Vol.). In addition, the available sources of genetic male sterility can be used to facilitate cross-pollination in *B. juncea* and *B. napus*.

A number of methods of recurrent selection can be used with genetic male sterility to improve a population (Burton and Brim 1981). The general principal is to harvest seeds from the male sterile segregants. This can be continued for many cycles of recurrent mating, following which the seed is taken only from fertile plants. The male sterility allele is then bred out of the material by progeny testing. Doggett (1972) utilized the recurrent phenotypic selection based on genetic male sterility in sorghum. Selection only among the male sterile plants resulted in an average yield increase of 7% after three cycles of recurrent selection. Burton and Brim (1981) increased the oil percentage of population from 18.8 to 19.7% after three cycles of selection among male sterile individuals in soybean.

2.3.5 Introduction of Alien Variation

Interspecific or intergeneric hybridization has been employed quite often in oilseed brassicas to transfer desirable genes from wild species to their domesticated relatives. These include incorporation of resistance to blackleg in *B. napus* (Roy 1984), to white rust and leaf miner in *B. juncea* (Banga, pers. commun.) and to pod shattering in *B. napus* (Prakash and Chopra 1988). Many studies have been undertaken to resynthesize the alloploid *Brassica* species utilizing a wide range of parents from each of the respective elementary species. Intergeneric or interspecific cytoplasm substitutions have proved useful in developing new sources of cytoplasmic male sterility. The techniques involved (see Prakash and Chopra, Chap. 8, this Vol.), also enable the breeders to analyze and exploit the effects of whole genome, chromosomes or chromosome segments introduced into the cultivated types from related wild species or genera.

2.3.6 Heterosis Breeding

With the availability of cytoplasmic male sterility and fertility restoration systems, the way is now clear for the development of commercial F_1 hybrids in *B. napus* and *B. juncea*. The current level of research on this aspect is discussed in detail in the chapter on heterosis and its utilization.

2.3.7 Mutation Breeding

Genetic variability is an important component of crop improvement based on recombinational breeding. The induction of mutations through physical or chemical mutagens can contribute substantially towards widening the genetic base of variability for the characters of economic significance. This is mainly because of the fact that the mutation frequency can be augmented remarkably with mutagen treatment, as spontaneous mutation frequency is only 10^{-5} to 10^{-7} on the average (Fujii 1981). Practical results of induced mutagenesis are significant, since the number of known cultivars derived from induced mutants increased from 77 in 1969 to 1163 in 1988. This figure includes 916 mutant cultivars in the narrow sense and 247 cultivars developed by using induced mutations in cross-breeding (Micke et al. 1989).

Major changes in the plants in terms of morphology or reaction to various pests and diseases have been reported (Fowler and Stefansson 1975; Ahmed and Fatma 1979; Balicka et al. 1979; Labana 1980; Walters et al. 1987). A maximum number of mutations have been reported for seed colour in *B. campestris* and *B. juncea*. A number of useful mutations could be induced including early floral initiation and maturity (Verma and Rai 1980; Nagamani et al. 1981; Labana et al. 1985), genetic male sterility (Takagi 1970), self-compatible or self-incompatible *B. campestris* variants (Swamy Rao 1979), frost resistance (Ohlsson 1983) and resistance to various biotic stresses (Srinivasachar and Verma 1971; Labana et al. 1979; Das and Rahman 1988). Reduction of linolenic acid (18 : 3) content in *B. napus* through induced mutagenesis (Röbbelen and Rakow 1970; Röbbelen and Nitsch 1975; Röbbelen 1982) is a classical example of the efficacy of this technique.

Induced genetic variation for several quantitative traits like seed yield and its components has been reported in *B. juncea* (Labana et al. 1979) and *B. campestris* (Kumar and Das 1977; Swamy Rao 1979). Irrespective of the fact that in the majority of the studies, increased variability was almost always accompanied by a negative shift in the mean values, a number of mutants with higher yielding property were identified. Two mutant varieties in *B. napus* (Regina Varraps elite A and F) were released in Sweden. Three Indian varieties (RLM 198, RLM 514, RLM 619) of *B. juncea* are also induced mutants.

Despite valuable results which have been obtained by a number of researchers, the use of induced mutagenesis is limited due to diplontic selection, adverse linkages and the undesirable pleiotropic effects normally associated with mutants.

2.3.8 Induction of Variability and Selection at Cell Level

Cultured cells or protoplasts provide excellent material for mutagenic treatment and isolation of desirable mutants. Sacristan (1982, 1985) treated cell cultures of *B. napus* with a mutagen (EMS or N-methyl-N-nitro-N-nitrosoguanidine) for inducing genetic variability for reaction against *Phoma lingam*. The medium containing phytotoxin produced by *Phoma lingam* was used as a selection medium. About 22% of the plants regenerated from resistant cell colonies gave resistance or tolerance response to the infection. The acquired resistance was heritable. Contrarily Macdonald and Ingram (1986) could obtain only transient and non-genetic resistance to *alternaria* species. A higher level of resistance was also reported from EMS-treated microspores (Macdonald et al. 1989). Apart from screening for disease resistance, it has been possible to induce morphological mutants, in the homozygous state, in regenerants from mutagen treated haploid cell cultures (Hoffman et al. 1982). In *B. juncea*, selection at cell level for salt tolerance has proved effective (Sarfraj, pers. commun.). Efficient selection of mutants from cell cultures requires the application of a number of procedures (Wersuhn 1989):

1. Haploidization of stock material.
2. Development and subculture of genetically stable cell cultures.
3. Mutagenesis in cell culture.

4. Selection in cell culture.
5. Regeneration (including reploidization).
6. Selection on level of regenerants.
7. Testing of both the genetic basis and stability of each new trait.

2.3.9 Somaclonal Variation

It is now well established that heritable variation occurs when plants are regenerated from protoplast, whole cell or callus cultures. This variation can be a useful supplement to normal breeding procedures. Somaclonal variants have been observed in Indian mustard for dwarfness (Jain et al. 1989) and yellow seed colour (George and Rao 1983); and for many agronomically useful characters in *B. carinata* (Kirti, pers. commun.). Somaclonal variation affects cytoplasmic genome as readily as the nuclear genome, and provides a possibility of adding the desired trait to an already acceptable agronomic base without disturbing the existing favourable gene complexes.

2.3.10 Haploid Induction and Utilization

Development of homozygous lines by conventional methods involves repeated selfing and selection for several generations within a population. Early-generation selection of the desirable genotypes is also limited by the heterozygosity of the material handled. Haploidization followed by chromosome doubling of early-generation material can expedite the approach to homozygosity and shorten the breeding cycle. For the advantageous use of haploidy in breeding programmes, efficient procedures must be available for producing haploids in large numbers and for doubling their chromosome number. Additive genetic variation is more readily expressed in doubled haploids than in conventional breeding lines, because of the masking of variation due to dominance. Apart from their use in the development of instant homozygous lines, haploids are also needed for the development of S allele homozygotes in *B. campestris*.

In oilseed brassicas, haploids can be obtained through anther/microspore culture (Keller et al. 1982). The induction of haploids through the anther culture technique was first reported in *B. oleracea* by Kameya and Hinata (1970). Since then, excellent progress has been achieved in producing and characterizing anther-derived haploids in *B. napus* (Thomas and Wenzel 1975; Keller and Armstrong 1978; Renard and Dosba 1980; Lichter 1981; Mathias 1988), *B. campestris* (Keller and Armstrong 1979) and *B. juncea* (George and Rao 1980; George et al. 1987). In *Brassica*, the developmental pathway for haploid induction normally involves the induction of microspore embryogenesis; however, the proliferation of the microspore-derived callus and subsequent regeneration of plants through embryogenesis has been observed. In *B. napus* it is now possible to recover up to 1300 embryos per anther (Kott et al. 1988). This was possible due to specific selection of buds with microspores in the very late uninucleate stage of cell cycle. Despite these achievements, the frequency of embryogenesis is influenced by a number of factors like pollen development phase, composition of culture medium, physiological condition of the anther donor, and anther pre-treatments etc. Use of microspore-derived haploids is now a routine practice in *B. napus*, especially the winter types. This has resulted in a substantial reduction in the time required to develop new varieties. A number of homozygous lines obtained from the spontaneously occurring haploids have demonstrated yields superior to those of the parental cultivar (e.g. Maris Haplona). Haploidy might also aid in rapid development of multiple recessive stocks in *B. napus* (Henderson and Pauls 1992).

Apart from their use in the production of instant pure lines, the microspore selection system has several advantages over the somatic tissue selection system (Polsoni et al. 1988). These are: (1) rapid generation of large number of microspore-derived embryos, (2) cost effectiveness in processing of large microspore populations, (3) direct conversion of microspore to embryo reduces the occurrence of cytogenetic anomalies common in callus systems, and (4) no masking effect of dominance.

2.4 Conclusions

Crop improvement is a continuous pursuit of greater productivity. Excellent breeding gains have been made in the past. The efficiency of varietal development can be further enhanced by the judicious choice of genetic material and selection criteria. There is a need to develop alternative selection criteria based on physiological parameters as direct selection for yield in early generations has not proved effective. Future breeding methods will have the advantage of increased use of the newer tools of plant biotechnology. That the input is already important is evident from the routine use of doubled haploids in expediting the breeding process in *B. napus*. The development of F_1 hybrids in *B. juncea* and *B. napus* is an immediate objective. A different approach for disease resistance will be the synthesis by genetic means of resistance mechanisms that are analogues of systemic fungicides.

References

Ahmed SU, Fatma A (1979) Effects of irradiation in *Brassica*. J Sci Univ Kuwait, 6:153-157
Allard RW (1960) Principles of plant breeding. John Wiley, New York
Balicka MB, Barcikowska WM, Mlyniec W, Szyld L (1979) Changes of chemical composition in some *Brassica* species. Cruciferae Newslett 4: 16–17
Becker HC (1989) Breeding synthetic varieties in partially allogamous crops. Proc XII Congr Eucarpia, Göttingen, pp 81–90
Briggs FN, Allard RW (1953) The current status of the backcross method of plant breeding. Agron J 45: 131–138
Burton JW, Brim CA (1981) Recurrent selection in soybeans III. Selection of increased percent oil in seeds. Crop Sci 21: 31–34
Chopra S (1981) Early generation selection for high oil content and seed yield in Indian mustard (*Brassica juncea* (L.) Czern and Coss). MSc Theses Punjab Agril Univ Ludhiana, India
Das ML, Rahman A (1988) Induced mutagenesis for the development of high yielding varieties in mustard. J Nuclear Agric Biol 17: 1–4

Dhillon SS, Labana KS (1988) Outcrossing in Indian mustard *Brassica juncea* (L.) Coss. Ann Biol 4: 100–102

Doggett H (1972) Recurrent selection in sorghum populations. Heredity 28: 9–29

Duvick DN (1974) Continuous backcrossing to transfer prolificacy to a single-eared inbred line of maize. Crop Sci 14: 69–71

Fowler DB, Stefansson BR (1975) Ethyl methane sulphonate-induced mutations in rape (*Brassica napus*). Can J Plant Sci 55: 817–821

Fujii T (1981) Future prospects of mutation breeding with genetic engineering. In: Gamma field Symp No 20, Prog Mutat breeding, pp 41–54

Gallais A (1986) A general strategy of plant breeding for varietal development. Proc Plant Breed Symp Agric Soc NZ Publ. 5, pp 42–47

George L, Abraham V, Surya-Vanshi DR, Sipahilani AT, Srinivasan VT (1987) Yield, oil content and fatty acid composition—evaluation in androgenetic plants in *Brassica juncea*. Plant Breed 98: 72–74

George L, Rao DS (1980) In vitro regeneration of mustard plants (*Brassica juncea* var. Rai 5) on cotyledon explants from non-irradiated, irradiated and mutagen-treated seeds. Ann Bot 46: 107–112

George L, Rao DS (1983) Yellow-seeded variants in in vitro regenerants of mustard (*Brassica juncea* Coss var Rai 5). Plant Sci Lett 30: 327–330

Grafius JE, Thomas RL, Barnard J (1976) Effect of parental complementation on yield and components of yield in barley. Crop Sci 26: 673–677

Hallauer AR, Miranda JB (1981) Quantitative genetics in maize breeding. Iowa State Univ Press, Ames

Harlan HV, Pope MN (1922) The use and value of backcross in small grain breeding. J Hered 13: 319–322

Harvey BL, Downey RK (1964) The inheritance of erucic acid content in rapeseed (*Brassica napus*). Can J Plant Sci 44: 104–111

Henderson CAP, Pauls KP (1992) The use of haploidy to develop plants that express several recessive traits using light seeded canola (*Brassica napus*) as an example. Theor Appl Genet 83: 476–479

Hoffman F, Thomas E, Wen G (1982) Anther culture as a breeding tool in rape II. Progeny analysis of androgenetic lines and induced mutants from haploid cultures. Theor Appl Genet 61: 225–232

Jain RK, Sharma DR, Chowdhary JB (1989) High frequency regeneration and heritable somaclonal variation in *Brassica juncea*. Euphytica 40: 75–81

Kameya T, Hinata K (1970) Induction of haploid plants from pollen grains of *Brassica*. Jpn J Breed 20: 82–87

Keller WA, Armstrong KC (1978) High frequency production of microspore-derived plants from *Brassica napus* anther cultures. Z Pflanzenzuecht 80: 100–108

Keller WA, Armstrong KC (1979) Simulation of embryogenesis and haploid production in *Brassica campestris* anther cultures by elevated temperature treatments. Theor Appl Genet 55: 65–67

Keller WA, Armstrong KC, de la Roche AI (1982) The production and utilization of microspore-derived haploids in *Brassica* crops. In: Giles KL, Sen SK (eds) Plant Cell Culture in Crop Improvement. Plenum New York, pp 169–183

Kirk JTO, Hurlstone CJ (1983) Variation and inheritance of erucic acid content in *Brassica juncea*. Z Pflanzenzuecht 90: 331–338

Knott DR (1972) Effects of selection for F_2 plant yield on subsequent generations in wheat. Can J Plant Sci 52: 721–726

Kott LS, Polson L, Ellis B, Beversdorf WD (1988) Autotoxicity in isolated microspore cultures of *Brassica napus*. Can J Bot 66: 1665–1670

Kumar PR, Das K (1977) Induced quantitative variation in self-compatible and self-incompatible forms in *Brassica*. Indian J Genet Plant Breed 37: 5–11

Labana KS (1980) Objectives, breeding approaches and achievements in *raya* (*Brassica juncea* L. Coss) In: Gill KS (ed) Breeding oilseed crops. Punjab Agril Univ, Ludhiana India, pp 253–266

Labana KS, Badwal SS, Chaurassia BD (1979) Induced mutations in breeding of *B. juncea* (L.) Czern and Coss. Proc Symp Role ind mutat Crop Improv. Osmania Univ, Hyderabad, pp 211–218

Labana KS, Badwal SS, Gupta ML (1985) RLM 619—an early maturity dwarf variety of *Raya*. Prog Fmg 22: 9–10

Labana KS, Banga SS (1984) Floral biology in Indian mustard (*Brassica juncea* (L.) Coss). Genet Agr 38: 131–138

Lein KA (1972) Genetische und physiologische Untersuchungen zur Bildung von Glucosinolaten in Rapssmen I. Zur Verenbung der Glucosinolatarmut. Z Pflanzenzuecht 67: 243–256

Lichter R (1981) Anther culture of *Brassica napus* in a liquid culture medium. Z Pflanzenphysiol 103: 229–237

Macdonald MV, Ingram DS (1986) Towards selection in vitro for resistance to *Alternaria brassicicola* (Schw) Wilts in *Brassica napus* ssp. *Oleifera* (Metzg) Sinsk winter oilseed rape. New Phytol 104: 621–629

Macdonald MV, Ahmed I, Ingram DS (1989) Mutagenesis and haploid culture for disease resistance in *Brassica napus*. XII Eucarpia Congr Sci Plant Breed 25: 22–27 (Abstr)

Mather K, Jinks JL (1982) Biometrical Genetics, 3rd Edn. Chapman and Hall, London

McKenzie RTH, Lambert J (1961) Comparison of F_3 lines and their related F_6 lines in two barley crosses. Crop Sci 1: 246–248

Mathias R (1988) An improved in vitro culture procedure for embryoids derived from isolated microspores of Rape (*Brassica napus* L.) Plant Breed 100: 320–322

Micke A, Maluszynski M, Sigurbjorsson B (1989) International achievements on plant mutation breeding. XII Eucarpia Congr Sci Plant Breed 15: 22–1 (Abstr)

Moll RH, Hanson WD (1984) Comparisons of effects of intrapopulation versus interpopulation selection in maize. Crop Sci 24: 1047–1052

Morice J (1974) Selection d'une variété de colza sans acide érucique et sans glucosinolates. In: Proc 4th Int Rapeseed Conf, Giessen, pp 31–47

Nagamani B, Kamla T, Rao TC, Lakshminarayana GL (1981) Effect of irradiation and chemical treatment on yield and quality of *B. juncea* (L.) Seed and oil. In: Proc Indian Acad Sci (Plant Sci) 90: 281–284

Ohlsson I (1983) Indian *Brassica* material tested for cold tolerance in Sweden. In: Proc 6th Int Rapeseed Conf, Paris, 1: 504–511

Olsson G, Persson C (1986) Recurrent selection in white mustard (*Sinapis alba* L.) In: Olsson G (ed) Svalof 1886–1986: Research and Results in Plant Breeding, pp 53–56

Polsoni L, Kott LS, Beversdorf WD (1988) Large-scale microspore culture technique for mutation selection studies in *Brassica napus*. Can J Bot 66: 1681–1685

Prakash S, Chopra VL (1988) Introgression of resistance to shattering in *Brassica napus* from *Brassica juncea* through non-homologous recombination. Plant Breed 101: 167–168

Rackow G, Woods DL (1987) Outcrossing in rape and mustard under Saskatchewan prairie conditions. Can J Plant Sci 67: 147–151

Renard M, Dosba F (1980) Etude de l' haploidie chez le colza (*Brassica napus* L. var. *oleifera* Metzger). Ann Amelior Plant 30: 191–209

Röbbelen G (1982) Changes and limitations of breeding for improved polyenoic fatty acids content in rapeseed (Abstr). J Am Oil Chem Soc 59: 305A

Röbbelen G, Nitsch A (1975) Genetical and physiological investigations on the mutants for polyenoic fatty acids in rapeseed, *B. napus* L. Selection and description of new mutants. Z Pflanzenzuecht 75: 73–105

Röbbelen G, Rakow G (1970) Selection for fatty acids in rapeseed. In: Proc 3rd Int Rapeseed Conf Ste-Adele, Quebec, pp 476–490

Roy NN (1984) Interspecific transfer of *Brassica juncea* type high blackleg resistance to *Brassica napus*. Euphytica 33: 295–303

Sacristan MD (1982) Resistance responses to *Phoma lingam* of plants regenerated from selected cell and embryogenic cultures of haploid *Brassica napus*. Theor Appl Genet 61: 193–200

Sacristan MD (1985) Selection for disease resistance in *Brassica* cultures. Hereditas Suppl 3: 57–63

Sprague GF, Eberhart SA (1977) Corn breeding. In: Sprague GF (ed) Corn and corn improvement. Am Soc Agron, Madison Wisconsin

Srinivasachar D, Verma RK (1971) Induced aphid resistance in *Brassica juncea* L Coss. Curr Sci 40: 311

Swamy Rao T (1979) Induced polygenic variations in *brown sarson* In: Symp role ind mutat Crop Improv Osmania Univ Hyderabad pp 77 (Abstr)

Takagi Y (1970) Monogenic recessive sterility in oilseed rape *Brassica napus* L induced by gamma irradiation. Z Pflanzenzuecht 64: 242–247

Thomas E, Wenzel G (1975) Embryogenesis from microspores of *Brassica napus*. Z Pflanzenzuecht 74: 77–81

Thurling N (1982) The utilization of backcrossing in improving the seed yield of spring rape (*Brassica napus* L). Z Pflanzenzuecht 88: 43–53

Verma VD, Rai B (1980) Note on induced mutagenesis for spotting out the source of resistance to *alternaria* leaf spot in Indian mustard. Indian J Agric Sci 50: 545–548

Walters T, Moynihan MR, Mutschler MA, Earle ED (1987) Seed mutagenesis of *Brassica campestris* for generation of cytoplasmic mutants. Cruciferae Newslett 12: 16–17

Wersuhn G (1989) Obtaining mutants from cell cultures. Plant Breed 102: 1–9

CHAPTER 3

Heterosis and Its Utilization

S.S. Banga

3.1 Introduction

Traditionally, pure line cultivars have been bred in *B. juncea* and *B. napus*, while improved populations are the mainstay of the breeding efforts in oleiferous forms of *B. campestris*. Lately, it has become apparent that the new techniques of plant breeding are imperative to sustain the generation of new high-yielding varieties. The need arises from the fact that the traditional methods of breeding do not mobilize a sufficient amount of genetic diversity. The success of hybrid rice is an impressive example of what aggressive application of basic genetic concepts to specific problems can accomplish. It is almost 50 years since Owen (1942) and Jones and Clarke (1943) proposed a system of hybrid seed production, the former in sugarbeet and the latter in onion. The major requirement of this system is the genetic or chemical circumvention of self-fertilization in the female line, which is an endeavour to change the reproductive system from autogamy to exogamy. Prerequisites to such a programme include sufficient yield heterosis, genetic emasculation and effective pollen shedders. Investigations on the occurrence of heterosis in rapeseed-mustard are encouraging, with reports of yield heterosis up to 150%. Hybrid advantage is not simply a function of heterosis. Three factors influence the end result: (1) breeding method efficiency (rate of progress factor), (2) the negative or positive effects of cytoplasmic male sterility-fertility restoration system used to produce the hybrid, and (3) the inherent heterosis (Virmani and Edwards 1983). The present chapter describes the extent of heterosis, different male sterility systems and associated problems. The prospects of commercial hybrids in rapeseed-mustard are also discussed.

3.2 Yield Heterosis

The magnitude of the heterotic effect will define the success of a hybrid breeding programme. For considerations of commercialization, the important criterion is the improved yield of hybrids from the best commercial pure line variety. A survey of the literature revealed that heterosis has been obtained by a large number of researchers in rapeseed-mustard under a range of test conditions. Unfortunately, the majority of these studies report heterosis over mean parental value or better parent value; data based on superiority to commercial variety are rare. Yield heterosis of more than 200% has been recorded in Indian mustard (Labana et al. 1975; Gupta 1976; Banga and Labana 1984a; Dhillon et al. 1990). A major observation was that the hybrids between genotypes of different geographic origins were more productive than the crosses involving lines of the same origin. Anand (1987) also emphasized the use of artificially synthesized allopolyploids of *B. napus* and *B. juncea* in obtaining heterotic combinations. Increased yield in a good majority of heterotic combinations resulted from increased branch number (Singh and Singh 1972; Asthana and Singh 1973; Labana et al. 1975; Banga and Labana 1984a; Dhillon et al. 1990). Heterosis for days to flowering was generally low. This is expected, as heterosis for this trait has mostly been reported in either photoinsensitive or qualitatively photosensitive crops (Singh 1973). Indian mustard, on the other hand, is a quantitatively photosensitive crop which flowers only when total photoperiodic requirements are met. Most of the hybrids evaluated (Banga and Labana 1984a) showed no significant improvement in harvest index as well as in oil content. This could be attributed to the fact that, physiologically, a specific metabolic system of plant genus sets an absolute limit to the directional partitioning of available photosynthates into more valuable compounds like oil at the cost of carbohydrates (see Bhatia and Mitra, Chap. 5, this Vol.).

A number of workers have studied the yield potential of hand-made single-cross hybrids in oilseed rape (*B. napus*). Heterosis for seed yield in F_1 hybrids up to 40-70% has been frequently reported (Schuster 1969; Shelkudenko 1968; 1972; Schuster and Michael 1976; Shiga 1976; Lefort-Buson and Dattee 1982; 1985a; Sernyk 1983; Grant and Beversdorf 1985; Lefort-Buson et al. 1987a). In the majority of these cases, heterosis was evaluated either over mid-parental values or over the better-performing parent in the cross. As was expected, negative or no heterosis was observed for the percent protein and oil content (Grant and Beversdorf 1985). Similar to the results in *B. juncea*, heterosis in oilseed rape was greater in hybrids between the parents of diverse origin. For example, maximum heterosis was manifested in Canadian × European crosses (Grant and Beversdorf 1985) or between European × Asiatic lines (Lefort-Buson et al. 1987b). The hybrids in almost every investigation were intermediate in maturity.

Patnaik and Murty (1978) reported 42.5% heterosis in *B. campestris* var. *brown sarson*. Heterosis was marginal for maturity and development traits, mainly due to internal cancellation of components. High better-parent heterosis for seed yield (63%) was also recorded in *B. campestris* var. *yellow sarson*. Higher heterosis values were attributed to the siliqua number per plant (Labana et al. 1978). Hybrids between the parents selected on the basis of harvest index and plant height proved highly heterotic (106%) for seed yield in *B. campestris* var. *toria*.

The results of the various investigations demonstrate that sufficient heterosis is available in single-cross hybrids in oilseed brassicas. Attention must, however, be paid to the fact that the branch number which constitutes the prime component of yield heterosis is a character with poor heritability, and is severely affected by population density.

3.3 Genetic Basis of Heterosis

The utilization of heterosis in plant breeding depends on its genetic basis. In cross-fertilized plants, where heterosis is more frequent, it can generally be explained by the mutational load. However, in both autogamous or allogamous plants it is difficult to exclude a role of marginal overdominance which could justify the hybrid variety production (Gallais 1988). Different hypotheses have been proposed to explain heterosis (Hayes 1968). These are:

1. Combined action of favourable dominant or partially dominant factors with linkage of multiple factors,
2. non-allelic interaction or epistasis, or
3. complementary action of alleles at the same locus resulting in overdominance.

If heterosis is due to dominance, it can theoretically be fixed in a homozygous line. Based on various biometrical analyses, Jinks (1983) concluded that heterosis is fixable for a great part. In oilseed brassicas, the majority of studies present a complex picture for yield inheritance, with strong evidence of dominance and epistasis with occasional overdominance. The main yield components generally exhibit a simpler pattern of inheritance, suggesting additive gene action with some dominance and occasional epistasis (Singh and Singh 1972; Lefort-Buson and Dattee 1985a; Dhillon et al. 1990). The importance of both additive as well as non-additive gene effects has also been emphasized (Labana et al. 1978; 1980; Grant and Beversdorf 1985). Non-fixable heterosis depends solely on non-additive factors that find their importance in the specific combining ability (SCA) of inbred lines (Dorsman 1976). Experiments with different brassicas have indicated that general combining ability (GCA) effects were highly significant for the yield components studied in F_1 populations; SCA effects, though significant, were on a much lower level. The ratio GCA : SCA was generally larger for the components of yield than for complex character yield (Singh 1973; Amrithadevarathinam et al. 1976; Ram et al. 1976; Labana et al. 1978; 1980; Grant and Beversdorf 1985; Dhillon et al. 1990). It is also possible to develop some predictors of heterosis or specific combining ability between the two lines, using a criterion of genetic distance (Lefort-Buson et al. 1987a).

Based on the available genetic information, it can be concluded that the importance of additive or dominant factors on crop productivity varies with the experimental material used and the environment. F_1 hybrids are not always the one and only possibility to achieve greater productivity. Hybrids can be more justified in the earlier stages, and lines can be justified later if all heterosis is fixable (Gallais 1988). Agronomically and physiologically, hybrids will need to have a well-balanced combination of yield components, a high harvest index, stability and resistance to prevailing diseases (Hayes and Foster 1976).

3.4 Sources of Cytoplasmic Male Sterility (CMS)

3.4.1 *Raphanus*-Based CMS System

Ogura (1968) observed male sterility in a Japanese radish variety. This sterility was controlled by a pair of nuclear genes (ms ms) and a sterility-inducing cytoplasm.

Bannerot et al. (1974) successfully introduced cabbage (*B. oleracea*) nucleus into CMS Japanese radish by backcrossing. CMS *B. napus* was also prepared in a similar fashion (Bannerot et al. 1977). This sterility was subsequently transferred to *B. campestris* (Heyn 1978) and *B. juncea* (Labana and Banga 1989). Resultant male sterility is highly thermostable (Polowick and Sawhney 1987) but three major problems arose in the utilization of this system in all these species:

1. Chlorophyll deficiency especially at low temperature (< 13°C).
2. Low nectar production.
3. Lack of fertility restoration.

The first two problems have been solved in *B. napus* by protoplast fusion (Pelletier et al. 1983; Jarl et al. 1989). Cybrids could be regenerated which combine the CMS trait from *B. napus* with chloroplasts of *B. napus* or of a triazine-resistant *B. campestris*. Male sterile and chlorophyll-non-deficient plants with well-developed nectaries are now available (Pelletier et al. 1987). The cybrid 27 developed in France has excellent nectar production which ensures good seed setting on the male sterile lines (Mesquida et al. 1987).

Fertility Restoration. Test crosses with a wide array of genotypes belonging to the crop brassicas did not throw up any fertility restorer genotype. Restorer genes were, however, identified in many European radish varieties (Bonnet 1975). Crosses between CMS rapeseed with Ogura type cytoplasms and *Raphanobrassica* strains showed that the restorers were present in *Raphanobrassica* (Heyn 1976). Occurrence of restorers in *R. sativus* × *B. oleracea* amphiploid was also confirmed by Rousselle and Dosba (1985). Introgression of the restorer genes from radish to rapeseed was achieved from the progeny of the material developed by Heyn (1978; 1979). The genes for fertility restoration seemed to be linked to flower colour because plants with intermediate to complete fertility level were always associated with yellow-flowered plants. The inheritance of male fertility restoration is still not clear because of abnormal segregation ratios (Pellan-Delourme 1986; Pellan-Delourme et al. 1987).

The introgressed restorer genes were associated with reduced female fertility and consequently a much reduced seed set. The low female fertility was primarily due to a very high percentage of ovule abortion (Pellan-Delourme and Renard 1987). The cytogenetic study of restorer material indicated meiotic irregularities despite a balanced ($2n = 38$) chromosome number. Up to three multivalents are expected in the presence of the chromosome(s) with translocations belonging to the radish genome. Thus, elimination of unfavourable radish genetic material is needed before using this material in hybrid seed production.

3.4.2 Bronowski-Shiga-Thompson System

Cytoplasmic male sterility in *B. napus* was first reported by Thompson (1972). Male sterile plants occurred in the F_2 generation of the crosses of some European spring and winter rape collections that had Bronowski spring rape as the male parent. The reciprocal cross did not reveal any segregation for male sterility. Thus it was suggested that Bronowski had recessive genes for sterility but fertilizing cytoplasm. The pollinator parents used in the crosses, on the other hand, had genes for fertility restoration but sterilizing cytoplasm. The combination of sterilizing cytoplasm with the recessive genes for sterility resulted in male sterility. Later, Shiga and Baba (1971) independently identified male sterility in the intervarietal crosses of Chisaya-natane and Hokuriku 23. Both these systems were later shown to be similar, as Thompson's CMS line was maintained by its maintainer Bronowski as well as by Shiga's maintainers, Isuzu-natane and Murasaki-natane. Shiga's CMS was also maintained by Bronowski. Both these lines had common restorers and thus both these systems were considered identical (Shiga et al. 1983). This was further confirmed by the fact that with every enzyme used to date, the chloroplast (cp) restriction patterns of Bronowski are the same as those of Isuzu-natane (Erickson et al. 1983; 1986). Male sterile plants are also found in the progeny of three-way crosses involving one oilseed rape breeding line and two resynthesized *B. napus* lines (Sv 84-28053 × No 7076) × No 7406 (Chen et al. 1990). Test crosses with *nap* maintainer and restorer lines support the conclusion that the observed male sterility is of *nap* type.

The restriction patterns of chloroplast (cp) and mitochondrial (mt) DNA in *B. napus* reveal the alloplasmic nature of this cytoplasmic male sterility. This system probably exploits cytoplasmic diversity from *B. campestris* (Erickson et al. 1986). Nuclear genes specific to this system do not cause sterility in maintainers (Bronowski and Isuzu-natane) because they have *campestris* cytoplasm, but give rise to sterility in *B. napus* cytoplasm. The breakdown of sterility at higher temperatures is a major limitation of this system (Fan and Stefansson 1986; Banga et al. 1988).

Fertility Restoration. The majority of the European and Japanese cultivars have restorer genes (Shiga 1980; Fan et al. 1986); while Thompson (1972) found that restoration was controlled by a single dominant gene, up to four dominant genes have also been implicated (Sernyk 1982; Shiga et al. 1983; Banga et al. 1988).

3.4.3 *nigra*-Based System

Pearson (1972) transferred *B. oleracea* genome into *B. nigra* cytoplasm and developed fairly stable male sterility. Initially, the male sterile flowers lacked nectaries and were malformed. Dickson (1975) improved the male sterile line, which was then released. The new line had almost complete female fertility and moderately sized nectaries. Male sterility was later transferred into *B. napus* (Pellan-Delourme et al. 1987). Observations on the male sterile plants indicated that the male sterility was very stable, with petaloid or pistilloid stamens. Male sterile flowers had no nectaries and had varied seed set.

Fertility Restoration. Almost all the *B. napus* lines evaluated proved to be restorers of this system. No maintainer has yet been found (Pellan-Delourme et al. 1987).

3.4.4 *Diplotaxis muralis* System

Hinata and Konno (1976) transferred the genome of *B. campestris* cv. Yukina to the cytoplasm of *Diplotaxis muralis* through backcross substitution. The male sterile plants were characterized by indehiscent anthers and linear narrow petals. The female sterility was marginally reduced. This male sterility was highly influenced by atmospheric temperature (above 20 °C). The sterility was subsequently transferred into *B. napus* cultivars Mangun and Hinchu through backcrossing (Pellan-Delourme and Renard 1987).

Fertility Restoration. The majority of *B. campestris* lines acted as the maintainers of this sterility. The fertility restoration was characterized by two Fr genes, which were present only in a few *B. campestris* varieties. Restoration behaviour, like the sterility expression, was influenced by temperature and photoperiod (Hinata and Konno 1976). On the other hand, the restorer genes are readily available in *B. napus*.

3.4.5 Rawat and Anand System

Male sterile plants were originally discovered in an unknown alloplasmic population of *B. juncea* (Rawat and Anand 1979). The plants had rudimentary or petaloid anthers completely devoid of pollen grains. The male sterility was stable over a wide range of environmental conditions. Subsequently, the male sterility inducing cytoplasm was transferred to *B. napus* (Mathias 1985), *B. carinata* (Anand 1987) and *B. campestris* (Banga, unpubl.). A limitation to this system is much reduced nectaries and the frequent occurrence of malformed flowers. Female fertility is normal. While the exact origin of the sterilizing cytoplasm in this system remains obscure, the restriction patterns of mitochondrial DNA indicated that *B. tournefortii* is the most probable donor of this cytoplasm (Pental, pers. commun.).

Fertility Restoration. Donors for partial fertility restoration could be identified in some accessions of *B. campestris*, *B. nigra*, *B. napus*, *B. juncea* and *B. carinata*. Unfortunately, however, the restoration is incomplete (24-60%) and strongly influenced by environmental conditions.

3.4.6 Polima System

Male sterile plants (*pol*) were initially discovered in a *B. napus* cultivar Polima from Poland (Fu 1981). The male sterile flower had small petals, short filaments and degenerative anthers. The analysis of F_2, BC_1 and BC_2 generations indicated that sterility in Polima resulted from the combination of the sterilizing cytoplasm with one or two pairs of the recessive nuclear genes (Liu et al. 1987). A major limitation of this system is that this cytoplasm is influenced by environmental conditions. The sterility is maintained up to a day-night temperature regime of 26/20 °C but there is partial restoration at 30/24 °C (Fan and Stefansson 1986). It is possible that this reversion resulted from the presence of certain temperature-sensitive genes for fertility restoration. If this is so, then it should be possible to eliminate the latent restorer genes by simple breeding manipulations. Based on temperature sensitivity, the Polima CMS line has been divided into three groups: (1) high temperature line, (2) low temperature line and (3) stable CMS line (Fu et al. 1990). These studies indicate that the temperature sensitivity of this system depends on the nuclear background rather than the cytoplasm. The studies conducted at Punjab Agricultural University, India, showed that two oilseed rape genotypes, ISN 126 and GSL 8909, acted as perfect maintainers for this system, and sterility was relatively stable even at a higher temperature regime. Significantly, both ISN 126 and GSL 8909 are synthetic alloploids. Recently, protoplast fusion was used to combine cytoplasmic triazine resistance (ctr) and Polima-type cytoplasmic male sterility (Barsby et al. 1987). Such cybrids constitute a major biological input required for the production of commercial single-cross hybrids (Beversdorf et al. 1985).

The origin of Polima cytoplasm is unknown, but the Pst 1 pattern of Polima mt-DNA is most similar to *B. campestris* (Erickson et al. 1986). This may indicate that the Polima cytoplasm is from *B. juncea*, an amphiploid thought to have acquired its cytoplasm from *B. campestris* (Erickson et al. 1983; Banga et al. 1983).

Fertility Restoration. Most of the *B. napus* cultivars evaluated were maintainers or partial maintainers of sterility (Fan et al. 1986). However, the F_1 hybrid between *pol* CMS and *B. juncea* cv. Zem revealed fertility restoration, as the F_1 plants produced normal anthers and were fertile. This F_1 was backcrossed six times to *B. napus* cv. Regent, followed by three selfings of BC_6 to generate a pure breeding population of the *pol*-Regent R line (McVetty et al. 1990). A single BC_6F_4 family having $2n = 40$, i.e. a pure breeding addition line genotype, has been developed and can be used as R-line in hybrid production. In another study (Fang and McVetty 1987), two winter-type *B. napus* lines namely Italy and UM 2353 were identified as restorers for this CMS system. The analysis of the segregating generations indicated that both Italy and UM 2353 contained a single dominant Mendelian gene for fertility restoration. Tests for allelism (Fang and McVetty 1989) of the restorer genes indicated that the fertility restorer gene from Italy (designated Rfp1) was different (i.e. non-allelic) from the restorer gene possessed by UM 2353 (designated Rfp2). Two moderately good restorers have also been identified from spring rape strains GSL 8807 and GSL 9002. Several new restorers have also been discovered in China (Yang and Fu 1990). Restoration was controlled by one pair of restorer genes (Rf) and the Rf genes in all restorers were alleles at the same locus.

3.4.7 *juncea* System

Male sterile plants of Indian mustard were observed in the F_2 generation of the cross RLM 198 × EJ 33. Genetic analysis indicated that male sterility occurred when the cytoplasm of RLM 198 interacted with recessive nuclear genes of EJ 33 (Banga and Labana 1984b). The genetic constitution of RLM 198 was postulated to be (S) RFRF, EJ 33 as (F) rfrf, and the male sterile plants as (S) rfrf. Male sterile flowers had reduced filaments with vestigeal and indehiscent anthers (Fig. 1).

Fertility Restoration. Almost all the Indian cultivars carried restorer genes with good penetrance, while the European lines acted more like maintainers, showing varying degrees of fertility restoration. Can-81 was an almost perfect maintainer. The majority of Indian *B. juncea* lines seem to have a sterility-inducing cytoplasm and, as

Fig. 1. a, b Male sterile flowers in *carinata* CMS system in *B. juncea*. **c, d** Vestigeal and petaloid anthers in *Ogura* CMS system in *B. juncea*. **e, f** Male sterile flowers in Anand and Ogura CMS systems respectively in *B. napus*. **g, h** Polima CMS in *B. napus* showing partially restored anthers at high temperature (**g**) and vestigeal anthers at low temperature (**h**).

a consequence, carry fertility restorer genes. Therefore, there exists a greater chance of isolating recessive maintainer genes from exotic accessions with fertile cytoplasm.

3.4.8 *campestris* CMS System

Male sterile plants were detected in the progeny of the intraspecific hybrids between a female plant of the New Zealand wild turnip (I 4 line) *B. campestris* ssp. *rapifera* and some Japanese cultivars of *B. campestris*. Sterility was manifested by a decrease in the number of pollen grains (Ohkawa 1984). Study of the reciprocal backcrosses confirmed the cytoplasmic nature of the male sterility. Male sterility was stable even under high temperature conditions.

Fertility Restoration. Eight cultivars, consisting of *rapifera* (six cultivars) and *oleifera* (two) had fertility restorer genes. On the other hand, six cultivars, including *chinensis* (one), *parachinensis* (one), *nipposinica* (one) and *rapifera* (three), had no fertility restorer gene(s). Thus male sterility can be easily established only in these cultivars (Ohkawa 1984).

3.4.9 *oxyrrhina* System

Male sterility resulted from the alloplasmic combination of *B. oxyrrhina* cytoplasm with the nuclear genomes of *B. campestris* (Prakash and Chopra 1988) and *B. juncea* (Prakash and Chopra 1990). The synthetic alloploid between *B. oxyrrhina* (OO; $2n = 18$) and *B. campestris* spp. *oleifera* var. *brown sarson* cv. pusa kalyani (AA; $2n = 20$) was obtained by rescueing the hybrid embryo and doubling the chromosomes. For developing alloplasmic lines, synthetic alloploid *B. oxyrrhina-campestris* was repeatedly backcrossed as female separately with *B. campestris* and *B. juncea* (Fig. 2). Completely male sterile plants resembling the genome donor parent were identified in BC_5 for *B. campestris* and in BC_3 for *B. juncea*.

In both instances, male sterile *B. campestris* and *B. juncea* plants resembled normal fertile plants in growth and morphology. Flowering was delayed by about 10 days in both alloplasmics. Male sterile flowers could be easily identified by slender anthers with sterile pollen (Fig. 3). Plants remained male sterile even at high temperature normally recorded late in the season. The leaves, however, were mildly chlorophyll-deficient in the earlier growth stages but turned completely green at a later development phase. Chlorosis was attributed to the *B. oxyrrhina* plastids. Attempts are already underway to overcome leaf yellowing through plastid substitution. It was also observed that, while the synthetic alloploid *B. oxyrrhina-campestris* and its hybrid with *B. juncea* ($2n = 37$, AABO) were normal green, chlorotic segregants appeared following the elimination of the *B. oxyrrhina* genome. It was proposed that *B. oxyrrhina* possessed gene(s) that can rectify the cytoplasmically induced chlorosis. This was confirmed by isolation of certain green but completely male sterile *B. juncea* plants ($2n = 36$). These plants probably resulted from introgression of a segment of the *B. oxyrrhina* chromosome, containing gene(s) for overcoming chlorosis, into the *B. juncea* genome. Cytological support for this came from the allosyndetic pairing

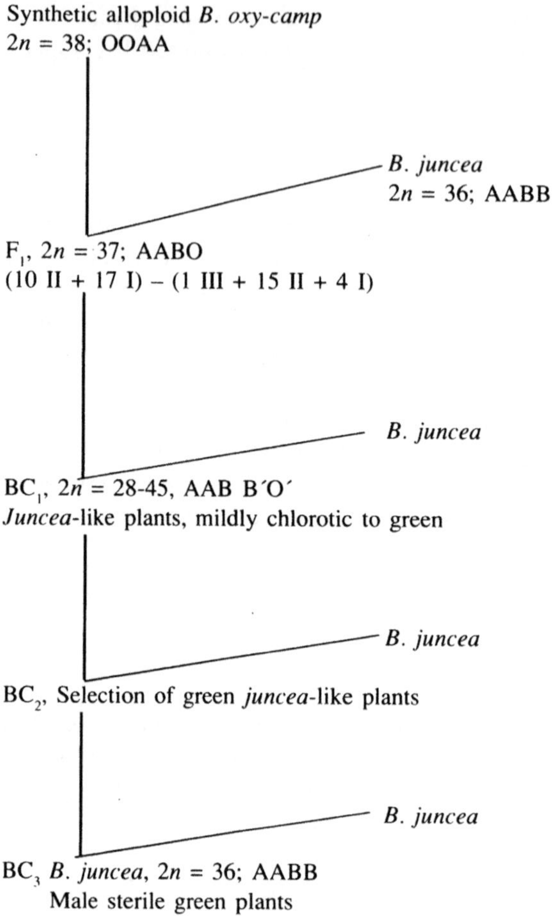

Fig. 2. Scheme employed for the development of alloplasmic male sterile *B. juncea*. *B′* and *O′* are hypodiploid genomes. (Prakash and Chopra 1990).

observed between *B. juncea* and *B. oxyrrhina* chromosomes during F_1 and BC_1 generations (Prakash and Chopra 1990). A major feature of this CMS system is the presence of normal flowers with excellent nectaries and female fertility. The male sterility has now been transferred to *B. napus*.

Fertility Restoration. Hybridization of the male sterile line with a large number of *B. campestris* and *B. juncea* accessions resulted in a male sterile progeny only. Thus it can be concluded that the euplasmic forms of *B. campestris* and *B. juncea* are functional maintainers for this CMS system (Prakash, pers. commun.). Introgression of restorer genes from *B. oxyrrhina* to *B. campestris* and *B. juncea* is already underway. Some fertile plants, showing monosomic addition ($2n + 1$) for a single *B. oxyrrhina* chromosome have been obtained.

3.4.10 Other CMS Systems

The backcross substitution of the *B. juncea* genome into the cytoplasmic background

Fig. 3. *Oxyrrhina* CMS system **a**, **b** fertile and sterile flowers of *B. campestris* **c**, **d** fertile and sterile flowers of *B. juncea*

of *B. carinata* resulted in a stable male sterile line (Banga et al. 1983). The male sterile flowers had reduced anthers with sterile pollen grains and reduced nectaries. Experimental evidence so far available suggests that *B. nigra* is the cytoplasm donor species to *B. carinata*, and hence this male sterility can also be attributed to *B. nigra* cytoplasm. Out of 126 *B. juncea* lines tested so far, only two maintainers could be identified. This is a major problem, as it puts a serious limit on the number of female lines that can be used in hybrid combinations. Genetic studies indicated the involvement of two or three dominant genes for fertility restoration.

B. oleracea cytoplasm was also found to induce stable male sterility in *B. campestris* and *B. juncea* (Prakash, pers. commun.). The male sterility in this system showed no visible symptom of any adverse effect on plant morphology.

3.5 Genetic Male Sterility

A number of sources of genetic male sterility are available in *Brassica* (Table 1). The exploitation of genetic male sterility is not economically viable, as extra labour is required to rouge out the fertile plants from the ms line before anthesis. No linked seedling marker or pleiotropic effect of the male sterility gene has been found which could make it possible to identify the male fertile plants before the initiation of flowering. The micropropogation of genetic male sterile plants of *B. napus* and their subsequent transplanting in hybrid seed production plots can be an attractive proposition.

Table 1. Sources of genetic male sterility

Crop	Inheritance	Origin	Reference
B. napus	Monogenic recessive	Induced mutation	Takagi (1970)
	Monogenic recessive	Induced mutation	Koch and Peters (1953)
	Digenic	Spontaneous	Heyn (1973)
	—	Spontaneous	Lee and Zhang (1983)
B. juncea	Monogenic	Spontaneous	Banga (unpubl.)
	Digenic epistatic	Intervarietal cross	Banga and Labana (1988)
B. campestris			
Brown sarson	Monogenic	Spontaneous	Das and Pandey (1961) Chowdhury and Das (1966; 1967a, b; 1968)
Yellow sarson	Monogenic	Spontaneous	Chowdhury and Das (1966)

3.6 Self-Incompatibility

Self-incompatibility is an excellent mechanism to enforce outcrossing in angiosperms. This system has been widely used for the F_1 hybrid breeding in cruciferous vegetables. A major advantage of self-incompatibility over male sterility is that the F_1 hybrid seed can be harvested from both the parents, thereby imparting greater economy in the hybrid seed production. Strongly self-incompatible S alleles are needed to obtain F_1 seed without self-contamination. Its use in the crops where seed is of economic consequence is, however, limited, as self-incompatibility is dominant over self-compatibility and self-incompatible × self-compatible hybrids are self-incompatible. It should, however, be possible to produce three-way or double-cross hybrids based on this system in *B. campestris*. As the background genotype and S alleles take part in the final expression of self-incompatibility, the parents should be selected not only for their genetic background but also characterized for their S allele composition. For catalogueing S alleles in the parents, a S allele tester set is needed. Such a tester set is available in *B. oleracea* and vegetable types of *B. campestris*. As a first step in this direction, a S allele tester set consisting of 15 S alleles (S_1 to S_{15}) has been developed in *B. campestris* var. *toria* (Banga, unpubl.). These S alleles were isolated from 27 varieties of Indian and exotic origin. S_4 was the most common allele, while S_{15} was the rarest. Some of the S alleles present in Canadian cultivars (Torch, Tobin and Candle) are similar to the S alleles present in the Indian cultivars. In spite of the inherent problems, attempts have been/are being made to utilize this phenomenon in hybrid seed production.

3.6.1 Use of Recessive Self-Incompatibility

Self-incompatibility is dominant to self compatibility in *B. napus* (Mackay 1976, 1977a), but Olsson (1960a) reported segregation for the self-incompatible plants after the selfing of self-compatible plants, thus indicating the possibility of obtaining recessive self-incompatibility. Thompson (1975) identified a self-incompatible, homozygous diploid line, obtained by chromosome doubling of a naturally occurring haploid. Genetic analysis indicated the control of a single recessive gene. This type of self-incompatibility can be profitably used for the production of F_1 hybrids.

Self-Compatible F_1 Hybrids. The self-incompatible line can be multiplied by inducing self seed set through CO_2 treatment in adequately ventilated polythene tunnels. Self-incomptible lines can be sown alternating with a self-compatible pollinator to produce large quantities of F_1 hybrid seed.

Self-Compatible Three-Way Hybrids. The technique, originally suggested by Thompson (1978), consists of the use of two recessive self-incompatible (RSI) lines. In the first step, these incompatible lines are hybridized under natural conditions to produce a single cross self-incompatible hybrid. The single cross hybrid is then sown along with a dominant self-compatible pollinator in alternating blocks (Fig. 4). The three-way self-compatible hybrid seed is marketed.

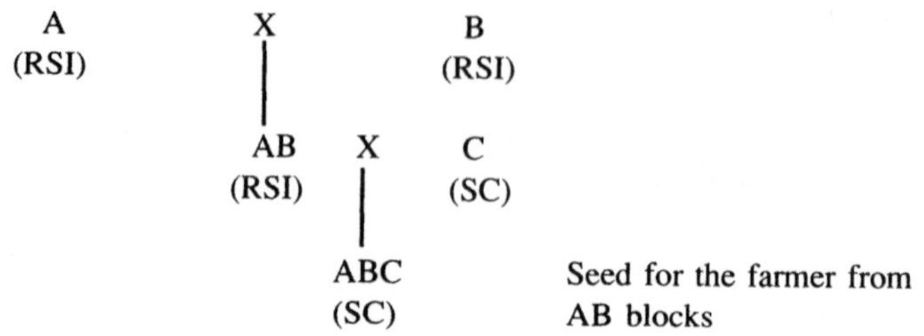

Fig. 4. Scheme for developing three-way hybrid in oilseed rape, RSI = recessive self-incompatible; SC = self-compatible

Thompson (1983) suggested a modification of this technique by the use of atrazine-resistant RSI lines and an atrazine-susceptible SC line. Instead of growing RSI and SC lines in alternating blocks, it is possible to sow mixed seed of RSI F_1 hybrid and SC parent in a ratio which ensures satisfactory seed set of the RSI F_1 hybrid. After the completion of flowering, the whole seed production block is sprayed with the herbicide atrazine, which ensures the survival of only RSI plants . This permits bulk harvesting of the hybrid seed.

A limitation to the use of recessive self-incompatibility may be that the lines active only for S alleles low in the dominance series can be used, and these may be less self-incompatible, which may cause self-contamination at higher temperatures (>20 °C) towards the end of flowering (Johnson 1971). The self-incompatibility reaction of Thompson's SI line in India was found to be very low and almost normal seed setting was observed on selfing (Banga, unpubl.).

3.6.2 Use of Sporophytic Self-Incompatibility

Brassica campestris. Once the lines homozygous for a large number of different S alleles become available, it is possible to produce double-cross hybrids based on normal sporophytic self-incompatibility. As was mentioned earlier, such lines are now available in *B. campestris* var. *toria*. These are currently being used for producing double-cross hybrids at Punjab Agricultural University, India. Care must be taken to exclude S alleles very high up in the dominance series. Partial exploitation of heterosis is also feasible by producing inbred (homozygous for S alleles) × variety cross. F_1 hybrid seed can be harvested from the inbred lines sown in alternating blocks with the pollinator variety. The resultant hybrid will be uniform for the morphological characters but much varied for S allele composition. This will ensure normal seed setting on F_1 plants. The S allele homozygote can be maintained by inducing self seed setting through CO_2 treatment. The concept is currently being evaluated.

Brassica napus and Brassica juncea. Both of these species are self-fertile natural allotetraploids. The parental species in each case possess an effective sporophytic self-incompatible system controlled by a multiple series of S alleles at a single locus (Thompson 1957; Richards and Thurling 1973). Naturally occurring self-incompatible plants have been reported in both fodder and oilseed types of *B. napus* (Gowers 1981; Olsson 1960a). Self-incompatible plants were also obtained following artificial synthesis of *B. napus* (Davey 1958; Olsson 1960b; Gowers 1973; Gowers 1979). Functional S alleles have been successfully introgressed into *B. napus* ssp. *oleifera* from *B. campestris* ssp. *rapifera* (Mackay 1977b). In *B. juncea* three different S alleles could be transferred from *B. campestris* cvs. Candle and Tobin (Banga, unpubl.). A new scheme was proposed (Gowers 1980) to overcome cross-incompatibility controlled by two S genes by the use of modified double-cross using self-incompatible and self-compatible isogenic lines. In *B. napus*, 19 lines were proved to be self-incompatible, having nine different S alleles (Gemmell et al. 1989). The background genotype could modify the strength of the self-incompatibility reaction (Gowers 1989). Theoretically, it is possible to produce three-way or double-cross hybrids based on self-incompatibility.

3.7 Chemical Hybridizing Agents

Basic deficiencies in the exploitation of cytosterility systems in hybrid seed production include the time required to transfer desirable genotypes into male sterile cytoplasm, and the lack of good fertility restorers with stable expression. A chemical with the ability to induce pollen sterility or check pollen shed will not only overcome these problems, but will also impart greater flexibility in the choice of genotype as a potential female parent in the development of hybrids. An ideal chemical hybridizing agent (CHA) or pollen suppressant (PS) should fulfil the following criteria (Chopra et al. 1960).

1. The treatment should cause only pollen abortion and not affect ovule fertility.
2. It should have no mutagenic effect.
3. The method of application should be easy and economical.
4. The precise dose when applied at a certain stage of growth in the life cycle of the plant should give consistent results, i.e. the effects should be reproducible.
5. There should be no undue hazards either to man or plant.

Very few researchers have evaluated the use of chemical hybridizing agents in *Brassica*. In Indian mustard, Banga and Labana (1983b, 1984b) and Banga et al. (1986) have reported some work on this aspect. Three chemicals, namely Ethephon, GA_3 and DPX 3778 were evaluated. DPX 3778 and GA_3 failed to induce any androcidal response. Foliar application of 2000-3000 ppm Ethrel led to anther retardation when applied two to three times before flowering shoot emergence. Treated anthers were up to 4 days late in the pollen release. Delayed treatment reduced the period of anther retardation. Pollen viability remained unchanged, but their number was reduced substantially. Compared to the check, treated variants achieved up to 90% self sterility. Female fertility was affected at higher doses but cross-pollination was possible. A genetical check revealed up to 54% total seed set in an Ethrel-treated population to be of hybrid origin (Banga and Labana 1984b). A major factor limiting the use of Ethrel as CHA is that the gametocidal effect is present for only up to 20-25 days after treatment, while flowering in Indian mustard is continuous for 60-75 days (Banga et al. 1986). Higher dose and repeated application resulted in a longer duration of gametocidal response, but this caused reduced female fertility and other abnormalities, like phytotoxic effects.

Since 1977, a number of experiments with CHAs have been undertaken in China (Gaun and Wang 1987). As many as 30 chemicals have been evaluated. MG1 (zinc methyl arsenate), MG2 and MG3 were found to be very effective in rapeseed (*B. napus*). Spraying at the bud stage with suitable concentration (MG1 = 0.03%, MG2 = 0.15% and MG3 = 0.4%) resulted in 60-80% male sterile plants. Chemical treatment resulted in reduced anthers with no or few pollen grains. When present, pollen grains were irregular and had no vigour. Pollen sterility was attributed to irregular meiosis. Chemical application also resulted in chlorotic leaves, reduced growth and even death of the plant. Male sterility could also be caused with the application of cyanide-sensitive respiratory pathway inhibitors like salicylhydroxamic acid, prophyl gallate and disulfiran in *B. oleracea* (Sachie and Noborn 1990). Sterility was complete when applications were made to the flower buds having anthers less than about 2 mm in length. Male sterile flowers were characterized by reduced filaments and pollen numbers. Female fertility was also affected.

The most important shortcomings observed in the evaluated CHAs include erratic or incomplete pollen control, transient expression, impaired female fertility and phytotoxic effects. Thus the CHAs do not seem to have immediate practical utility but they can aid cytosterility-based hybrid breeding programmes for screening heterotic combinations and evaluation of genotypes for combining ability.

3.8 Hybrid Seed Production

The production of hybrid seed by cross-pollination is the most important factor affecting the bio-economics of seed production. Success depends upon maximizing the seed set on female lines using various production techniques. For commercial purposes, the additional benefits of cultivating hybrid seed must exceed the cost of hybrid seed and the need to replace the seed every year.

3.8.1 Extent of Outcrossing

Both Indian mustard and oilseed rape are essentially self-pollinating crops, where selfing accounts for 80-90% of the total seed set. *B. campestris* var. *toria* and *brown sarson* (*lotni* group) are largely cross-pollinated. Although many experiments have been conducted, the outcrossing rates and role of potent pollinating agents is still not clear. In Indian mustard, the outcrossing varied from 7.6 to 18% (Labana and Banga 1984; Dhillon and Labana 1988). In the USSR, the extent of outcrossing ranged from 4.1 to 32.5% (Hodyrev 1967). In *B. napus*, outcrossing up to 16% was observed (Banga, unpubl.). Olsson (1955) has reported wind as the main agent of pollination. The second school which favours insect pollination is represented by Williams (1978) and Eisikowitch (1981). It has been emphasized that only those pollen grains which are initially disturbed by insects may become truly air-borne. Mesquida et al. (1987)

observed a decrease in seed production on the male sterile lines due to decreasing bee density. The crucial role of nectar production for bee foraging and seed set on male sterile plants was established. Insect pollination is also an important component in *B. juncea* (Labana and Banga 1984).

3.8.2 Seed Set on Male Sterile Plants

The effect of field layout and the orientation of the seed production blocks on outcrossing rates was investigated in Indian mustard male sterile plants (Banga and Labana 1983b). The study incorporated three male : female combinations, i.e. 1 : 2, 1 : 3 and 2 : 4, laid in east-west, north-south directions and ciruclar design. The production of the seed set was maximum in the 2 : 4 combination. The outcrossing varied from 19 to 79%. To evaluate the seed set on the male sterile (*ogu*) plants of *B. napus* (Renard et al. 1987), four isolated plots of spring plants of rapeseed were established with 5, 10, 20, and 30% male fertile parent, the rest being with cytoplasmic male sterile parent. pod set (95%) was comparable to that of male fertile control, from 10% mixtures and above. The number of seeds per pod, seed size, and seed weight per plant were all higher in male sterile cybrids than for the related male fertile donors. Thus the extent of seed set on female rows can be maximized by manipulating insect activity and planting design (Renard and Mesquida 1983). The extent of hybrid seed setting on Polima CMS plants was also investigated in *B. napus* (Pinnish and McVetty 1990). A 10 : 1 ratio of male sterile and pollen parent was used. Leaf cutter bees (*Megachile rotundata*), when used as pollen vector, were found to be effective pollinators. Seed production was low and hybrid seed yield and percent hybridity declined linearly as distance from pollen source increased. Hybrid seed production by selective elimination of pollinator plants from bulk mixtures with herbicide application (Beversdorf et al. 1985) has also been proposed.

3.9 Evaluation of F_1 Hybrids

Only a few reports are available regarding the expected yield increase in hybrids based on male sterility. In Indian mustard, nine hybrids developed by using chemical sterilants were evaluated in a small-scale trial (Banga and Labana 1983). The heterosis over the best commercial variety, Varuna, ranged up to 164%; however, the F_1 hybrids showed a carry-over effect due to Ethrel treatment (Banga and Labana 1983). In *B. juncea*, two hybrids, namely PHR 2 and PHR 7, have done remarkably well during some large-scale yield evaluation trials (1989-90 and 1990-91). The yield superiority of these hybrids over the best national check, Varuna, varied from 23 to 58%.

In *B. napus*, a study was made to compare the performance of the male fertility—restored intervarietal F_1 hybrids in *nap* and *pol* cytoplasms. F_1 hybrids in both the cytoplasms exhibited superior relative performance compared to the conventional cultivar Regent for seed yield, harvest index, total oil yield and total protein yield (McVetty et al. 1990). Calculations based on the estimates made by Penning de

Vries et al. (1974) of the bio-energetic cost of various seed components indicate that the *nap* F_1 hybrid groups produced about 23% more energy than did the *pol* F_1 hybrid group. The loss of seed yield potential (23%) was roughly divided between decrease in total dry matter production and harvest index of F_1 hybrids in the *pol* cytoplasm as compared to the F_1 hybrids in the *nap* cytoplasm (McVetty, et al. 1990). Percent oil decreased on the average by 1.3% for the *pol* F_1 hybrid group relative to the *nap* F_1 hybrid group, giving a relative biological cost of 3% for the *pol* cytoplasm.

3.10 Conclusions

The commercial potential of developing F_1 hybrids in rapeseed-mustard appears very promising, as hybrid yield advantages equal to or even greater than those in corn, millet and rice have been demonstrated. The biological system, i.e. male sterility to produce F_1 hybrids, also exists. However, none of the CMS system available is perfect and all require improvement. The basic limitations to the presently available cytoplasmic sterility-fertility restoration systems include: unstable expression of male sterility (*pol* and *nap* systems), lack of fertility restorer genes (Rawat and Anand; and Ogura systems), female sterility in male sterile plants (*nigra* and *muralis* systems) or in restored F_1 hybrids (Ogura system).

The use of protoplast technology has been very fruitful, not only in refining existing CMS systems, but in the development of unique cytoplasmic combinations. It will have a still greater role in the rapid development of CMS lines and the introduction of fertility restorer genes. More efforts are needed to utilize self-incompatibility for producing hybrids. The price of commercial hybrid seed will reflect the expenditure for research and development as well as the cost of production, processing and marketing. The acceptable heterosis required to compensate for the increased seed cost will depend on the yield level of the traditional varieties. In countries where yield levels are very high, less yield heterosis (15-20%) will be enough to meet the additional expenditure. In countries with lower yield levels, a minimum of 25-35% yield heterosis will be of commercial value. Efficient and economic seed production will ultimately define the success of hybrids.

References

Amrithadevarathinam A, Arunachalam V, Murty BR (1976) A quantitative evaluation of intervarietal hybrids of *Brassica campestris* L. Theor Appl Genet 48: 1–8

Anand IJ (1987) Breeding hybrids in rapeseed and mustard. In: Proc 7th Int Rapeseed Conf, Poland, pp 79–85

Asthana AN, Singh CB (1973) Hybrid vigour in *Rai*. Indian J Genet Plant Breed 33: 57–63

Banga SS, Banga SK, Labana KS (1983) Nucleo-cytoplasmic interactions in *Brassica*. In: Proc 6th Int Rapeseed Conf, Paris, pp 602–606

Banga SS, Labana KS (1983a) Male sterility in Indian mustard (*Brassica juncea* (L.) Coss) II. Genetics and cytology of MS-1. In: Proc 6th Int Rapeseed Conf, Paris, pp 349–353

Banga SS, Labana KS (1983b) Production of F_1 hybrids using ethrel-induced male sterility in Indian mustard (*Brassica juncea* (L.) Coss). J Agric Sci Camb 101: 453–455

Banga SS, Labana KS (1984a) Heterosis in Indian mustard (*Brassica juncea* (L.) Coss). Z Pflanzenzuecht 92: 61–70

Banga SS, Labana KS (1984b) Ethrel induced male sterility in Indian mustard (*Brassica juncea* (L.) Coss). Z Pflanzenzuecht 92: 229–233

Banga SS, Labana KS, (1985) Male sterility in Indian mustard (*Brassica juncea* (L.) Coss). IV Genetics of MS-4. Can J Genet Cytol 27: 487–490

Banga SS, Labana KS, Banga SK, Singh B (1986) Experimental evaluation of male gametocides in Indian mustard (*Brassica juncea* (L.) Coss). SABRAO J 18: 31–35

Banga SS, Labana KS, Srivastava A (1988) Evaluation of '*Nap*' cytoplasmic male sterility in *Brassica napus* L. J Oilseeds Res 5: 13–16

Bannerot H, Boulidard L, Cauderon Y, Tempe J (1974) Transfer of cytoplasmic male sterility from *Raphanus sativus* to *Brassica oleracea*. Proc Eucarpia Meet Cruciferae Crop Sect 25: 52–54

Bannerot H, Boulidard L, Chupeau Y (1977) Unexpected difficulties met with the radish cytoplasm in *Brassica oleracea*. Cruciferae Newslett 2: 16

Barsby TL, Chuong PV, Yarrow SA, Sauching WU, Coumans M, Kemble RJ, Powell AD, Beversdorf WD, Pauls KP (1987) The combination of Polima CMS and cytoplasmic triazine resistance in *Brassica napus*. Theor Appl Genet 73: 809–814

Beversdorf WD, Erickson LR, Grant I (1985) Hybridization process utilizing a combination of cytoplasmic male sterility and herbicide tolerance. US Patent Trademark Off Ser 4517763

Bonnet A (1975) Introduction et utilisation d'une sterilité male cytoplasmique dans des variétiés précoces européennes de radis, *Raphanus sativus* L. Ann Amelior Plant, Paris, 25: 381–397

Chen BY, Heneen WK, Gertsson B, Hallden C (1990) Components of a cytoplasmic male sterility system in resynthesized and cultivated forms in oilseed rape (*Brassic napus* L.). Plant Breed 104: 20–25

Chopra VL, Jain SK, Swaminathan MS (1960) Studies on the chemical induction of pollen sterility in some crop plants. Indian J Genet Plant Breed 20: 188–199

Chowdhury JB, Das K (1966) Male sterility in *Brassica campestris* var. *yellow sarson*. Indian J Genet Plant Breed 26: 374–380

Chowdhury JB, Das K (1967a) Functional male sterility in *brown sarson*. Indian J Genet Plant Breed 27: 143–147

Chowdhury JB, Das K (1967b) Male sterility genes in *brown sarson*. Indian J Genet Plant Breed 27: 284–288

Chowdhury JB, Das K (1968) Cytomorphological studies on male sterility in *Brassica campestris* L. Cytologia 38: 195–199

Das K, Pandey BD (1961) Male sterility in *brown sarson*. Indian J Genet Plant Breed 21: 185–190

Davey V McM (1958) Root crops. In: Scott Plant Breed Stn Rep pp 17–18

Dhillon SS, Labana KS (1988) Outcrossing in Indian mustard *Brassica juncea* (L.) Czern and Coss. Ann Biol 4: 100–102

Dhillon SS, Labana KS, Banga SK (1990) Studies on heterosis and combining ability in Indian mustard (*Brassica juncea* (L.) Coss). J Res Punjab Agric Univ 27: 1–8

Dickson MH (1975) G1117A, G1102A and G1106A cytosterile broccoli inbreds. Hortic Science 10: 535

Dorsman C (1976) F_1 hybrids in some out-pollinating crops. In: Jannosy A Lupton FGH (eds). Heterosis in plant breeding. Proc 7th Cong Eucarpia, Conf, Budapest, pp 167–175

Eisikowitch D (1981) Some aspects of pollination of oilseed rape (*Brassica napus* L.). J Agric Sci Camb 96: 321–326

Erickson LR, Straus NA, Beversdorf WD (1983) Restriction patterns reveal origins of chloroplast genomes in *Brassica* amphiploids. Theor Appl Genet 65: 202–206

Erickson LR, Grant I, Beversdorf W (1986) Cytoplasmic male sterility in rapeseed (*Brassica napus* L.) I. Restriction patterns of chloroplast and mitochondrial DNA. Theor Appl Genet 72: 145–150

Fan Z, Stefansson BR (1986) Influence of temperature on sterility of two cytoplasmic male sterility systems in rape (*Brassica napus* L.). Can J Plant Sci 66: 221–227

Fan Z, Stefansson BR, Sernyk JL (1986) Maintainers and restorers of three male sterility inducing cytoplasms in rape (*Brassica napus* L.). Can J Plant Sci 66: 229–234

Fang GH, McVetty PBE (1987) Inheritance of male fertility restoration for the Polima CMS system in *Brassica napus* L. In: Proc 7th Int Rapeseed Conf, Poland, pp 73–78
Fang GH, McVetty PBE (1989) Inheritance of male fertility restoration and allelism of restorer genes for the Polima CMS system in oilseed rape. Genome 32: 1044–1047
Fu TD (1981) Production and research of rapeseed in the People's Republic of China. Cruciferae Newslett 6: 6–7
Fu TD, Yang G, Yang X (1990) Studies on "three-line" Polima cytoplasmic male sterility developed in *Brassica napus* L. Plant Breed 104: 115–120
Gallais A (1988) Heterosis: its genetic basis and its utilization in plant breeding. Euphytica 39: 95–104
Gaun Chung-yun, Wang Guo-Huai (1987) Studies on male sterility of rapeseed induced by chemical male gametocide. In: Proc 7th Int Rapeseed Conf, Poland, pp 243–251
Gemmell DJ, Bradshaw JE, Hodgkin T, Gowers S (1989) Self-incompatibility in *Brassica napus*: seed set on crossing 19 self-incompatible lines. Euphytica 42: 71–77
Gowers S (1973) Hybrid swedes. In: Scott Plant Breed Stn Ann Rep pp 12–13
Gowers S (1979) Self-incompatibility in *Brassica napus*. In: Proc Eucarpia Cruciferae Conf, Wageningen, pp 80–84
Gowers S (1980) The production of hybrid oilseed rape using self-incompatibility. Cruciferae Newslett 5: 15–16
Gowers S (1981) Self-pollination in swedes (*Brassica napus* spp. rapifera) and its implications for cultivar production. Euphytica 30: 813–817
Gowers S (1989) Self-incompatibility interactions in *Brassica napus*. Euphytica 42: 99–103
Grant I, Beversdorf WD (1985) Heterosis and combining ability estimates in spring-planted oilseed rape (*Brassica napus* L.). Can J Genet Cytol 27: 472–478
Gupta PR (1976) Heterosis in certain inter-varietal crosses in Indian mustard (*Brassica juncea* (L.) Czern and Coss). Indian J Agric Res 10: 125–128
Hayes JD (1968) The genetic basis of hybrid barley production and its application in western Europe. Euphytica 17: 87–102
Hayes JD, Foster CA (1976) Heterosis in self-pollinating crops, with particular reference to barley. In: Jannosy A., Lupton F.G.H. (eds.). Heterosis in plant breeding. Proc 7th Eucarpia Conf, Budapest, pp 239–256
Heyn FW (1973) Beitraege zum Auftreten unreduzierter Gameten und zur Genetik einiger Merkmale bei den Brassicae. Diss Georg-August Univ, Goettingen, pp 102
Heyn FW (1976) Transfer of restorer gene from *Raphanus* to cytoplasmic male sterile *Brassica napus*. Cruciferae Newslett 1: 15–16
Heyn FW (1978) Cytoplasmic genetic male sterility in *Brassica napus*. Cruciferae Newslett 3: 34–35
Heyn FW (1979) Introgression of restorer genes from *Raphanus sativus* into cytoplasmic male sterile *Brassica napus* and the genetics of fertility restoration. Proc 5th Int Rapeseed Conf, Malmö, Sweden, pp 82–83
Hinata K, Konno N (1976) Cytoplasmic male sterile strain of Yukina (*Brassica campestris* L.) produced by nuclear substitution. Jpn J Breed 26: 127–128
Hodyrev GA (1967) Method of pollination of Indian mustard under western Siberian conditions. Rec WK Oil crops 2: 17
Jarl CI, Grinsven Van MQJM, Mark Vanden F (1989) Correction of chlorophyll defective male sterile winter oilseed rape (*Brassica napus*) through organelle exchange: molecular analysis of the cytoplasm of parental lines and corrected progeny. Theor Appl Genet 77: 135–141
Jinks JL (1983) Biometrical genetics of heterosis. In: Frankel R (ed.). Heterosis—reappraisal of theory and practice. Springer Berlin Heidelberg New York, pp 1–46
Johnson AG (1971) Factors affecting the degree of self-incompatibility in inbred lines of Brussels sprouts. Euphytica 20: 561–573
Jones HA, Clarke AE (1943) Inheritance of male sterility in onion and the production of hybrid seed. Proc Am Soc Hortic Sci 43: 189–194
Koch H, Peters R (1953) Neue Gesichtspunkte der Rapszüchtung. Wiss Z. Martin Luther Univ Halle 2: 363–367

Labana KS, Badwal SS, Chaurassia BD (1975) Heterosis and combining ability analysis in *Brassica juncea* (L.) Czern and Coss. Crop Improv 2: 46–51

Labana KS, Banga SS (1984) Floral biology in Indian mustard (*Brassica juncea* (L.) Coss). Genet Agr 38: 131–138

Labana KS, Banga SK (1989) Transfer of Ogura cytoplasmic male sterility of *Brassica napus* into genetic background of *Brassica juncea*. Crop Improv 16: 82–83

Labana KS, Chaurassia BD, Singh B (1980) Genetic variation and association studies on some quantitative characters in the mutants of Indian mustard (*Brassica juncea* (L.). Czern and Coss) Indian J Agric Sci 50: 803–806

Labana KS, Jindal SK, Mehan DK (1978) Heterosis and combining ability in *yellow sarson* (*Brassica campestris* L. var. *yellow sarson*). Crop Improv 5: 50–55

Lee S, Zhang Y (1983) The utilization of genetic male sterility in Shanghai, China. In: Proc 6th Int Rapeseed Conf, Paris, pp 360–365

Lefort-Buson M, Dattee Y (1982) Genetic study of some agronomic characters in winter oilseed rape (*Brassica napus* L.) I Heterosis. Agronomie 2: 315–322

Lefort-Buson M, Dattee Y (1985a) Etude de l' heterosis chez le colza oléagineux d'hiver (*Brassica napus* L.) I Comparison de deux populations, l'une homozygote et l'autre héterozygote. Agronomie 5: 101–110

Lefort-Buson M, Dattee Y (1985b) Etude de l' heterosis chez le colza oleagineux d'hiver (*Brassica napus* L.). II Structure genétique d'une population de lignées. Agronomie 5: 201–208

Lefort-Buson M, Dattee Y, Guillot-Lemaine B (1987a) Heterosis and genetic distance in rapeseed (*Brassica napus* L.): Use of kinship coefficient. Genome 29: 11–18

Lefort-Buson M, Guillot-Lemoine B, Dattee Y (1987b) Heterosis and genetic distance in rapeseed (*Brassica napus* L.). Crosses between European and Asiatic selfed lines. Genome 29: 413–418

Liu H, Tingdeng Fu, Shaoniu Yang (1987) Discovery and studies on Polima CMS line. In: Proc 7th Int Rapeseed Conf, Poland, pp 69–72

Mackay GR (1976) Self-incompatibility in forage rape, *Brassica napus* L. var. *biennis*. Incom Newslett 7: 4–8

Mackay GR (1977a) A diallel cross method for the recognition of S-allele homozygotes in turnip, *Brassica campestris* L. ssp *rapifera*. Heredity 38: 201–208

Mackay GR (1977b) The introgression of S-alleles with forage rape, *Brassica napus* L. from turnip, *Brassica campestris* L. ssp *rapifera*. Euphytica 26: 511–519

Mathias R (1985) A new dominant gene for male sterility rapeseed, *Brassica napus* L. Z Pflanzenzuecht 94: 170–173

McVetty PBE, Edie SA, Scarth R (1990) Comparison of the effect of *nap* and *pol* cytoplasms on the performance of intercultivar summer oilseed rape hybrids. Can J Plant Sci 70: 117–126

Mesquida J, Renard M, Pellen-Delourme R, Pelletier G, Morice J (1987) Influence des secrétions nectarifers des lignées male stériles pour la production de semences hybrids F_1 de calza. Les Coloques INRA, 22-23 April 1987. Ste Sabine France, pp 269–280

Ogura H (1968) Studies on the new male sterility in Japanese radish with special reference to the utilization of this sterility towards the practical raising of hybrid seeds. Mem Fac Agric Kogoshima Univ 6: 39–78

Ohkawa Y (1984) Cytoplasmic male sterility in *Brassica campestris* ssp. *rapifera* L. Jpn J Breed 34: 285–294

Olsson G (1955) Investigations of the degree of cross pollination in white mustard and rape. Sver Utsaedesfoeren Tidskr 62: 311–322

Olsson G (1960a) Self-incompatibility and outcrossing in rape and white mustard. Hereditas 46: 241–252

Olsson G (1960b) Species crosses within the genus *Brassica*. II. Artificial *Brassica napus*. Hereditas 46: 351–386

Owen FY (1942) Male sterility in sugarbeet produced by complementary effect of cytoplasm and Mendelian inheritance. Am J Bot 29: 892

Patnaik MC, Murty BR (1978) Gene action and heterosis in *brown sarson*. Indian J Genet Plant Breed 38: 119–125

Pearson OH (1972) Cytoplasmically inherited male sterility characters and flavour components from the species cross *Brassica nigra* (L) Koch × *B. oleracea* L. J Am Soc Hortic Sci 97: 398–402
Pellan-Delourme R (1986) Etude de deux systémes de stérilite male cytoplasmique introduits chez le calza (*Brassica napus* L.) par croisements intergénériques avec *Raphanus* et *Diplotaxis*. Theses de doctorat, Univ de Rennes, Rennes, France
Pellan-Delourme R, Eber F, Renard M (1987) Male sterility restoration in *Brassica napus* with radish cytoplasmic male sterility. Proc 7th Int Rapeseed Conf, Poland, pp 226–233
Pellan-Delourme R, Renard M (1987) Identification of maintainer genes in *Brassica napus* L for the male sterility-inducing cytoplasm of *Diplotaxis muralis* L. Plant Breed 99: 89–97
Pelletier G, Primard C, Vedel F, Chetrit P, Remy R, Rosselle P, Renard M (1983) Intergeneric cytoplasmic hybridization in Cruciferae by protoplast fusion. Mol Gen Genet 191: 244–250
Pelletier G, Primard C, Vedel F, Chetrit P, Renard M, Pellan-Delourme R, Mesquida J (1987) Molecular phenotypic and genetic characterization of mitochondrial recombinants in rapeseed. Proc 7th Int Rapeseed Conf, Poland, pp 113–118
Penning de Vries FWT, Brunsting AHM, Van Lear HH (1974) Products, requirements and efficiency of biosynthesis: A quantitative approach. J Theor Biol 45: 339–377
Pinnisch R, McVetty PBE (1990) Seed production of hybrid summer rape in the field using the *pol* cytoplasmic male sterility system—A first attempt. Can J Plant Sci 70: 611–618
Polowick PL, Sawhney VK (1987) A scanning electron microscopic study on the influence of temperature on the expression of cytoplasmic male sterility in *Brassica napus*. Can J Bot 65: 807–814
Prakash S, Chopra VL (1988) Synthesis of alloplasmic *Brassica campestris* as a new source of cytoplasmic male sterility. Plant Breed 101: 253–255
Prakash S, Chopra VL (1990) Male sterility caused by cytoplasm of *Brassica oxyrrhina* in *B. campestris* and *B. juncea*. Theor Appl Genet 79: 285–287
Ram K, Krishna R, Chauhan YS, Katiyar RP (1976) Partial diallel analysis in the F_3 generation of Indian mustard. Indian J Agric Sci 46: 229–232
Rawat DS, Anand IJ (1979) Male sterility in Indian mustard. Indian J Genet Plant Breed 39: 412–415
Renard M, Mesquida J (1983) Pollinisation entomophile du colza male serile en production de semences hybrids F_1 dans différents régions de France. Proc 6th Int Rapeseed Conf, Paris, France, pp 552–557
Renard M, Mesquida J, Pellan-Delourme R, Pelletier G, Morice J (1987) Pollination des cybrids male steriles dans un système de culture mixte de colza. In: Variabilité genétique cytoplasmique et sterélité male cytoplasmique, INRA Paris, France, pp 281–292
Richards RA, Thurling N (1973) The genetics of self-incompatibility in *Brassica campestris* L. ssp. *oleifera* Metzg I. Characteristics of S-locus control of self-incompatibility. Genetics 44: 428–438
Rousselle P, Dosba F (1985) Restauration de la fertilite pour l'androsterilite genocytoplasmique chez le calza (*Brassica napus* L.) Utilization des *Raphano Brassica*. Agronomie 5: 431–437
Sachie K, Noborn K (1990) Inhibitors of cyanide-insensitive respiratory pathway induce male sterility in broccoli (*B. oleracea*). Jpn J Breed 40: 217–222
Schuster W (1969) Vergleich von zwei Zuchtverfahren in der Erhaltungszuechtung bei Winterraps. Z Pflanzenzuecht 62: 47–62
Schuster W, Michael J (1976) Untersuchungen ueber Inzuchtdepressionen und Heterosis effekte bei Raps (*Brassica napus oleifera*). Z Pflanzenzuecht 77: 56–66
Sernyk JL (1982) Heterosis and cytoplasmic male sterility in summer rape (*Brassica napus* L.). Ph D theses, Univ Manitoba, Canada
Sernyk JL (1983) Other known CMS sources (Polima and New *B. campestris*). In: Proc *Brassica* Workshop Hybrids Anal Chem Paris, France, p 11
Shelkudenko VG (1968) First results of evaluation of intervarietal hybrids of winter rape for heterosis. Sb Nauchno Issled Rab Maslichn Kult, pp 9–14 (In Russian)
Shelkudenko VG (1972) Early diagnosis of heterosis in rape hybrids. Sel Semenovod 37(1): 17 (in Russian)
Shiga T (1976) Studies on heterosis breeding using cytoplasmic male sterility in rape. Bull Nat Agric Sci 27: 1–101

Shiga T (1980) Male sterility and cytoplasmic differentiation In: Tsunoda S., Hinata S., Hinata K., Gómez-Campo C (eds.) *Brassica* crops and wild allies. Jpn Sci Soc, Tokyo, pp 205–221
Shiga T, Baba S (1971) Cytoplasmic male sterility in rape plant (*Brassica napus* L.) and its utilization in breeding. Jpn J Breed 23: 187–197
Shiga T, Ohkawa Y, Takayanagi K (1983) Cytoplasm types of European rapeseed (*Brassica napus* L.) cultivars and their ability to restore fertility in cytoplasmic male sterile lines. Bull Nat Inst Agric Sci Ser D 35: 103–124
Singh SP (1973) Heterosis and combining ability estimates in Indian mustard, *Brassica juncea* (L) Czern and Coss. Crop Sci 13: 497–499
Singh SP, Singh DP (1972) Inheritance of yield and other agronomic characters in Indian mustard (*Brassica juncea*). Can J Genet Cytol 14: 227–233
Takagi Y (1970) Monogenic recessive male sterility in oil rape (*Brassica napus* L.) induced by gamma-irradiation. Z Pflanzenzuechtg 64: 242–247
Thompson KF (1957) Self-incompatibility in narrow-stem kale, *Brassica oleracea* var. *acephala*. I. Demostration of a sporophytic system. J Genet 55: 45–60
Thompson KF (1972) Cytoplasmic male sterility in oilseed rape. Heredity 29: 253–257
Thompson KF (1975) Recessive self-incompatibility in *Brassica napus*. Proc 4th Int Rapeseed Conf, Giessen West Germany, pp 119–124
Thompson KF (1978) Application of recessive self-incompatibility to production of hybrid rapeseed. In: Proc 5th Int Rapeseed Conf, Malmö Sweden, pp 56–59
Thompson KF (1983) Breeding winter oilseed rape, *Brassica napus*. In: Coaker TG (ed) Advances in applied biology VII. Academic Press London, pp 1–104
Virmani SS, Edwards IB (1983) Current status and future prospects for breeding hybrid rice and wheat. Adv Agron 36: 146–214
Williams IH (1978) The pollination requirements of swede rape (*Brassica napus* L.) and of turnip rape (*B. campestris* L.). J Agric Sci Camb 91: 343–348
Yang, G, Fu TD (1990) The inheritance of polima cytoplasmic male sterility in *Brassica napus* L. Plant Breed 104: 121–124

CHAPTER 4

Physiological Constraints and Their Genetic Manipulation

N. Thurling

4.1 Introduction

Crop yield is the end product of a great many different processes operating within the plant for all or some part of the growing season. Its expression is controlled by a wide range of complex polygenic systems and is highly sensitive to environmental variation. Moreover, much of the variation in yield estimates from trials of cultivars conducted over different sites and seasons can often be attributed to genotype × environment interactions (Brennan and Byth 1979). Clearly, such interactions will greatly reduce the precision with which a breeder can identify high-yielding genotypes, especially in early segregating generations. Visual selection (McKenzie and Lambart 1961) and selection based on single-plant yields (Knott 1972) in early segregating generations of crosses in self-pollinating species have usually proved ineffective. Approaches to improving early generation selection for yield have centred on planting designs which reduce micro-environmental variation in the breeding block (Gardner 1961; Fasoulas 1973) or the use of alternative selection criteria (Bhatt 1988). The results of selection experiments involving special planting designs have been inconsistent, although some improvements have been proposed to improve their efficiency (Bos 1983). Alternative criteria of selection for yield must be simple to measure, significantly correlated with a plot yield and highly heritable. Most characters fulfilling these conditions would be morphological in type, but in certain cases are associated with physiological determinants of yield. The most widely used alternative criteria have been primary morphological components of seed yield such as the number of pods per plant, number of seeds per pod and weight per seed in the case of the oilseed brassicas. Although these characters would be appropriate in respect of the conditions listed above, they have often proved unreliable because of the existence of compensatory mechanisms resulting in reductions in one or more components in response to the selection for another specific component (Adams 1967). Various biometrical procedures have been proposed to overcome the problems associated with yield component compensation (Grafius and Thomas 1971), but their effectiveness in actual breeding programmes is yet to be properly evaluated. In certain instances, direct selection for one or other grain yield component has proved effective, as observed in the case of selection for seed weight in mustard (Gupta and Labana 1985).

Harvest index is another character which has been extensively studied, especially as an alternative criterion of selection for high yield. Traditionally, the harvest index is defined as the ratio of grain yield to total biomass yield (Donald 1962a), the latter usually referring to the total dry weight of aboveground plant parts. Although it is relatively simple to measure, the harvest index is a manifestation of complex photosynthetic, translocatory and storage processes associated with seed development. Its attraction to plant breeders stems largely from the fact that the improvement in

European wheat yields over the past 50 years has been accompanied by a marked increase in harvest index, but very little change in biomass production, even though there was no conscious selection for harvest index (Austin et al. 1980). Harvest index is often highly correlated with grain yield (Donald and Hamblin 1976) yet this relationship does not necessarily imply that it will be an effective criterion of selection for high yield in all situations.

Yield components and characters such as harvest index have been considered largely in respect of selection with genetically heterogenous populations. Various other morphological characters have also come under scrutiny in this regard because of their relationship to yield through some identifiable physiological process. Leaf type and leaf inclination variants in pea (Hedley and Ambrose 1981) and barley (Angus et al 1972) respectively were considered useful yield selection criteria because of their influence on crop canopy structure and the interception of photosynthetically active radiation. In both cases, yield improvement was associated primarily with the higher plant density at which these morphological variants could be grown. The value of a physiological trait as a selection criteria depends on the genetic characterization of the character, its agronomic benefit and finally the ease with which it can be measured in large-scale trials. Complex physiological yield determinants have not usually been regarded as practical selection criteria because of the difficulty of measuring the large number of plants which normally comprise a breeding population. Nevertheless, an understanding of the relationships between these physiological traits, simpler morphological characters and seed yield can be of particular significance to a breeding programme in providing a basis for defining crop yield ideotypes. The term ideotype was first proposed by Donald (1962b) to describe a conceptual model of any crop plant incorporating morphological components contributing to the higher yield of a monospecific crop in a specified environment. Such ideotypes provide a more rational basis for the development of a yield improvement programme, as the characteristics of the superior cultivar will be more precisely defined.

By providing information on the physiological diversity within a crop species, a physiologist should provide a sound basis for meaningful definitions of yield ideotypes. A major benefit of such collaboration is that highly heritable morphological characters related to more complex physiological processes will be utilized to improve the rate of progress in yield improvement. The most useful characters in this category would be those which can be rapidly measured with young seedlings, such as gibberellic acid insensitivity (O'Brien and Pugsley 1981).

The contributions of results from genetic and physiological studies with oilseed *Brassica* to their improvement will be the main theme of this chapter. However, before considering these in detail, some general aspects of the genetic diversity within the genus *Brassica* will be presented first.

4.2 The *Brassica* Gene Pool

Cultivar development in the species *Brassica napus*, *B. juncea* and *B. campestris* has been based largely on the exploitation of genetic variation within each species.

Although interspecific hybridization has been employed quite often to transfer genes for disease resistance from wild species to their domesticated relatives, this strategy has largely been ignored as an avenue of yield improvement. Genes from wild relatives can have an unexpectedly beneficial effect on yield, as was observed in studies with oats (Frey 1976). This study highlighted the potential of the largely untapped genetic resources of wild and weedy species in the agronomic improvement of their domesticated relatives. Such a situation is especially relevant to the oilseed *Brassica* species which, together with related species, comprise a vast reservoir of genetic variability. The cytogenetic organisation and utilization of this variability is described in more detail elsewhere in this Volume (Prakash and Chopra, Chap. 8). In the case of *B. napus*, an extremely wide range of parental combinations can be used because of the great diversity of domesticated types in *B. oleracea* and *B. campestris*. Olsson (1960), for example, used varieties of kale, Brussels sprouts, cabbage and kohlrabi (*B. oleracea*) in crosses with rape and turnip (*B. campestris*). The synthetic *B. napus* was usually completely fertile and morphologically indistinguishable from natural *B. napus*. Although the newly synthesized forms were agronomically inferior, selection amongst artificial biennial forms resulted in the isolation of fodder rape lines which outyielded the standard variety. This highlights the variability which can be generated in such crosses and the potential for its exploitation through selection. A commercial fodder rape cultivar has been developed in Japan from a cross between *B. campestris* ssp. *pekinensis* (Chinese cabbage) and *B. oleracea* var. *capitata* (Hosoda 1953). The same type of cross has also been utilized in the development of a new vegetable type (Hakuran) for the Japanese market (Nishi 1968).

The use of interspecific hybridization in the improvement of *B. napus* reflects a relatively narrow range of genetic diversity among cultivars. By contrast, the other source of rapeseed *B. campestris* ssp. *oleifera* is only one of the several distinct sub species grouped within *B. campestris*. These subspecies can be separated primarily on the basis of the plant part utilized by man into three groups—the leaf vegetable ssp. *pekinensis* (Chinese cabbage), ssp. *chinensis* (Chinese mustard) and ssp. *nipposinica* (curled mustard); the oilseeds ssp. *oleifera* (turnip rape), ssp. *trilocularis* (*sarson*) and ssp. *dichotoma* (*toria*) and a root vegetable ssp. *rapifera* (turnip). All these species are easily intercrossed to produce fertile F_1 hybrids and each, apart from some *trilocularis* cultivars, are self-incompatible. Because of the self-incompatibility, individual cultivars are highly outcrossing and genetically heterogenous. Sometimes desired improvement may require hybridization with another cultivar or even another subspecies. It is worthwhile combining the early vegetative vigour of ssp. *chinensis* with the early flowering and seed quality of ssp. *oleifera* as an avenue of yield improvement. The genetic base of less variable species like *B. napus* and *B. juncea* can be widened through hybridization of selected cultivars with related diploid or allotetraploid species or their resynthesis from putative diploid parents.

4.3 Morpho-Physiological Determinants of Yield

Definition of an ideotype must take into account features of a crop community which are essentially the same irrespective of species. A crop is a monospecific community comprising individuals with the same or similar genotypes. Thus, there is intense intra-genotypic competition for resources essential for growth so that a high-yielding crop is most likely to comprise individuals with a low competitive ability (Donald 1981). Except for fodder crops, only a portion of the crop has a significant economic value and it is the yield of this portion which must be maximized.

An ideotype can provide a valuable guide to breeders, particularly in selecting parents for a pedigree selection programme. The existence of an ideotype allows for a more rational screening of potential parents on the basis of characters used to define the ideotype. Following such screening, appropriate parental combinations may be identified which are complementary for the different characters required in the ideotype. Information from experiments with oilseed brassicas over the past 20 years provides a useful basis for defining simple yield ideotypes. Although these studies have been conducted in different regions, many aspects of relationships between growth, development and yield are similar irrespective of environment. There are, of course, certain plant characters which play a major role in adaptation to a local environment and ideotypes for different regions would vary in the case of these characters.

Seed production by plants is the culmination of two major, sequential processes.

1) Creation of a sink through production of flowers with the potential for further development after fertilization.

2) Supply of photosynthates from the source to a sink comprising all developing seeds on the plants.

Stems and roots are important sinks before flowering, after which reproductive organs rapidly become the primary sinks. Sink size may be expressed as the product of sink capacity and sink activity. Sink capacity is the maximum volume available for the accumulation of photosynthates and may be expressed in terms of the following for the oilseed brassicas:

$$\text{pods per unit area} \times \text{seeds per pod} \times \text{individual seed weight}$$

Sink activity is the capacity of the sink to establish a translocation gradient for the import of photosynthates. The source includes all plant parts with the ability to photosynthesize, but its size will be determined primarily by leaf area, or more specifically, by leaf area index. Leaf area index is the main determinant of the efficiency with which the radiant energy to be utilized in photosynthesis is intercepted by the crop canopy. However, it should be noted that certain plant organs, particularly fruits, can be a source of photosynthates at the same time as being powerful sinks. Although there is evidence to suggest that either the sink (Rawson and Evans 1970) or the source (Yoshida 1972) can be the main yield-limiting factor, there are probably situations where neither is limiting and the capacity for translocation between source and sink is of greatest significance (Evans 1975). Whichever is most important, our approaches to yield improvement through breeding and/or modifications to crop management will need to be based on some understanding of source-sink relationships. A number of physiological processes limit the crop productivity. These processes may be conveniently separated into five main groups.

1. Phenology.
2. Dry matter production and distribution.
3. Reproductive development.
4. Carbon/nitrogen metabolism.
5. Yield component relationships.

4.3.1 Phenology

The development of a rapeseed crop through the growing season may be characterized in terms of the timing of major phenological events such as seedling emergence, floral initiation, commencement of stem elongation, appearance of the first open flower, cessation of flowering and maturity. Intuitively, we would expect the timing of each developmental event to have some influence on seed yield. However, only time of flowering has been studied extensively in this regard. In areas with short growing seasons (Canada, southern Australia) where spring cultivars are normally grown, timing of flowering is crucial in enabling seed development to be completed before the onset of a severe environmental stress (drought in Australia, frost in Canada). Flowering time is, in fact, the major determinant of adaptation to those environments where the growing season is terminated by deteriorating climatic conditions.

The influence of flowering time on yield has been investigated in several studies. In western Australia, where there is a rapid decline in soil moisture during the spring, *B. napus* seed yields decrease markedly with delay in sowing between late autumn and mid-winter (Thurling and Vijendra Das 1979a). Delay in sowing meant that an increasingly greater part of the reproductive development phase was exposed to drought stress. At the same time, the duration of development before flowering decreased and the amount of dry matter accumulated before the reproductive phase declined. A similar reduction in seed yield was observed with successive delays in sowing between late autumn and spring in northern New South Wales (Hodgson 1978).

In Canada, timing of maturity is a major objective of cultivar improvement in *B. napus* (Campbell and Kondra 1977) because of the frost risk at the end of the growing season. Early maturing cultivars were obtained through a reduction in the period of pre-anthesis development, but without a reduction in seed yield (King and Kondra 1986). Early flowering plants of standard cultivars tend to be higher yielding then later flowering plants of the same cultivars (Campbell and Kondra 1978a). These plants were higher yielding as a result of greater biomass production and production of more lateral branches and pods. The duration of pre-anthesis development was therefore more important than the duration of post-anthesis development in determining high yield. Other studies with different cultivars (Degenhardt and Kondra 1981) have also shown that there is no direct relationship between seed yield and the duration of flowering or seed formation. In *B. juncea*, a 15-60 days delay of sowing reduced the seed yield by 12-66% (Saini 1984).

The selection of a cultivar for a short-season environment would involve tuning of developmental responses to temperature and day length. All *B. napus* spring cultivars are essentially facultative long-day plants, i.e. the time of flowering is shortened by an increase in day length. However, the photoperiod responses of *B. napus* spring cultivars are temperature-dependent (Thurling and Vijendra Das 1977; Mendham et al. 1990) and they are capable of flowering quite rapidly in short days (< 12 h) provided temperatures are low. This interaction between temperature and photoperiod in determining the time of floral initiation provides considerable scope for developing cultivars closely adapted to specific environments.

4.3.2 Dry Matter Production and Distribution

In simplest terms, rapeseed yield is determined by the dry matter produced by the crop during its life cycle and the proportion of that dry matter incorporated into seeds (harvest index). Total biomass yield reflects the duration of the period over which growth has occurred and the rate at which dry matter is accumulated by the crop. Various studies with *B. napus* spring (Thurling 1974a) and winter (Mendham and Scott 1975) cultivars under conditions in which water was not limited during seed development, showed that plant dry weight at anthesis was closely correlated with seed yield. Thus, in environments where early flowering is essential to the avoidance of stresses late in the season, a higher crop growth rate during vegetative development is an important pre-requisite for high seed yield. In an experiment conducted under suboptimal moisture conditions, although *B. campestris* flowered 24 days earlier than *B. napus*, the latter had a dry weight four times higher than *B. campestris* (Richards and Thurling 1978b). The mean seed yield of the *B. napus* cultivars was only 11% greater than that of the *B. campestris* cultivars, and both species had similar biomass yield at the final harvest. Most of the final biomass of *B. napus* was accumulated prior to anthesis and before the onset of severe drought stress, whereas the rapid growth of *B. campestris* after anthesis coincided with the drought stress (Richard and Thurling 1978b). The yield advantage of *B. napus* was thus explained in terms of the greater assimilate source accumulated before anthesis and available for mobilization to the developing seeds after anthesis. Further improvement in yield in this environment should be achieved through selection of genotypes having some optimal combination of flowering time and plant dry weight at flowering which is probably intermediate between those of the two rapeseed species.

Studies in *B. napus* have shown a strong positive association between seed yield and leaf area index (Allen and Morgan 1975) during the early stages of reproductive development. The cultivar Cresus had a greater leaf area index as well as the pods. Cresus also produced a higher number of larger seeds per pod. This was presumably due to the maintenance of a greater assimilate supply during a critical period of pod development and a more efficient transport of assimilates from the pod wall to developing seeds.

Studies with winter cultivars of *B. napus* (Mendham and Scott 1975) indicated that the plant must attain some critical size at the time of floral initiation for maximum yield in a given environment. Beginning of stem elongation has been described as the most critical phase of growth (Evans 1984). Decreasing the production of carbon assimilates at this stage led to production of fewer pods, and seeds which were lighter in weight. Similar results have also been obtained in *B. juncea* (Labana et al. 1987). The timing of flowering (or floral initiation) is crucial in the determination of seed yield, although it is also important with regard to the avoidance of environmental stresses in spring cultivars. Growth of plant canopy prior to some optimal flowering time must provide an assimilate source of sufficient magnitude to be fully exploited during reproductive development. Seed yield will then be dependent on the efficiency with which stored and current assimilates are utilized in seed development. The pattern of inflorescence development will be a major factor influencing this efficiency.

4.3.3 Reproductive Development

Reproductive development commences with the initiation of floral primordia at the apical meristem. The duration of the period between this event and the appearance of the first open flower can vary widely, depending on genotype and environment (Thurling and Vijendra Das 1979b). Although reproductive development prior to anthesis is probably important in determining sink size, this period is usually relatively short compared with the period of development between the appearance of the first open flower and maturity. It is during the later that the structure of the crop canopy changes most dramatically with the proliferation of lateral branches, the formation of a dense canopy of bright yellow flowers and the development of pods. As with other developmental phases, the characteristics of post-anthesis development can vary widely depending on genotype and environment. Such variation is normally associated with marked differences in seed yield.

The influence of genotype and environment on post-anthesis development was clearly demonstrated in comparisons between *B. napus* and *B. campestris* spring cultivars sown at different times (McGregor 1981). In all cases, there was a rapid development of successively lower lateral inflorescences immediately after the appearance of the first open flowers on the main stem inflorescence. However, the *B. campestris* cultivar produced about twice the number of inflorescences than the *B. napus* cultivar over a 2 week period in an early sowing. The superiority in inflorescence production of *B. campestris* was even greater in a late sowing. One other difference between these species was that all *B. napus* inflorescences tended to terminate about the same time, whereas later *B. campestris* inflorescences continued flowering after termination of flowering on early inflorescences. Monitoring of pod production and abortion in this study revealed that just over half the pods produced by *B. campestris* aborted in both early and late sowings. Percentage pod abortion in *B. napus* was 45 and 58% for early and late sowings respectively. Abortions were most prevalent in later-formed inflorescences of both species, although a greater proportion of inflorescences lost a high proportion of pods (> 80%) in *B. campestris*.

Another detailed analysis of flower and pod development in a *B. napus* spring cultivar (Tayo and Morgan 1975) revealed that the period of flowering for a single plant spanned an average of 26 days and that more than 75% of pods retained at maturity were produced by flowers opening within 14 days of anthesis. This wastage of reproductive units is a major factor limiting yield and an obvious approach to yield improvement would be to seek genotypes with developmental patterns minimizing excessive pod loss. A restriction in assimilate supply at the pod development stage is a primary cause of pod losses and reduced seed size (Tayo and Morgan 1979).

A substantial portion of incident radiation (60%) is reflected back by the mass of bright yellow flowers, which, together with shading of lower pods due to dense pod canopy, severely reduces the seed yield in *B. napus* (Mendham et al. 1984). The use of apetalous character will permit greater light penetration (Chay and Thurling 1989a).

In autumn sowings all spring cultivars produced high numbers of pods per unit area, but those retaining more seeds per pod had the much higher seed yields. A higher yielding cultivar such as Marnoo (Australian) appeared to have an inherent capacity to retain more seeds per pod than lower yielding Midas (Canadian) irrespective of the crop dry weight per pod at flowering. Thus, there is some scope for yield improvement through selection for high number of seeds per pod.

Substantial variation in seed weight per pod has been detected among selections from the *B. napus* spring cultivar Chena A differing in mean pod length (Chay and Thurling 1989b). A twofold increase in pod length was accompanied by an increase in seed weight per pod, but seed yield was not affected because of a compensatory decrease in pod number per plant. The magnitude of the increase in seed weight per pod was less than that in pod length, indicating a reduction in efficiency expressed as seed weight per unit length of pod, with increasing pod length. Monitoring of pod development after fertilization indicated that pods continue to elongate for about the next 12 days. This is in contrast to observations with *B. napus* spring cultivar Maris Haplona (Pechan and Morgan 1985), where pod elongation virtually ceased and the final number of seeds per pod was determined by about 8 days after fertilization. Surviving seeds continued to increase in size between 8 and 18 days after fertilization, as did the width of the pod. Growth of the pod wall appeared, in fact, to depend on the presence of seeds within the elongating region, so that there was a close relationship between final pod length and seed number per pod.

The previous discussion has indicated marked effects of genotype and environment on post-anthesis development with important consequences for final seed yield. Of particular significance has been the recognition of an optimal development pattern in which some specific balance of pod numbers and seed weight per pod should be identified in defining high yield ideotypes for different environments. The existence of an optimal phenotype was certainly indicated in the case of pod length, where lines producing pods of intermediate length appeared to have the highest yield potential. Definitions of such optimal phenotypes should take into account the wider phenomenon of source-sink relationships and particularly the aspects of carbon assimilate supply to the developing inflorescence.

4.3.4 Carbon/Nitrogen Metabolism

The rapeseed crop is somewhat unique in that the dense pod canopy is a major source of carbon assimilates for seed growth (Hozyo et al. 1972). However, both leaves (Freyman et al. 1973) and stems (Major and Charnetski 1976) also contribute assimilates to developing seeds. More detailed studies with a *B. napus* spring cultivar (Brar and Thies 1977) indicated that the primary source of assimilates for seed development changed as the plant developed. In this study, selected leaves were exposed to $^{14}CO_2$ when plants were 37, 52 or 101 days old and the proportion of labelled assimilates exported from a leaf was used as a measure of the contribution of this leaf to the development of other plant organs. Before flowering, fully expanded leaves contributed mainly to the development of the root system and younger leaves. Later, as plants were beginning to flower, these leaves primarily supported the structural development of stems bearing flowers. During the pod-filling stage, pod walls and stems bearing pods were photosynthetically active (28 and 41% respectively of leaves). Leaves, pod walls and stems contributed 37, 32 and 31% respectively of the ^{14}C assimilates incorporated into seeds at this time.

Similar studies conducted with a different *B. napus* spring cultivar (Major et al. 1978) also showed that pods were capable of assimilating $^{14}CO_2$ whilst still being strong sinks for assimilates from leaves and stems. In this case, different parts of field-grown plants were exposed to $^{14}CO_2$ when plants had completed flowering and seeds were enlarging in lower pods. ^{14}C assimilates produced in upper stems and leaves were largely exported elsewhere, but mainly to seeds and, to a lesser extent, pods. Significant amounts of ^{14}C assimilates were also produced in lower (older) pods, but these were retained in the pod wall or exported to seeds within the pod. A later study, also utilizing labelled carbon feeding, clearly indicated the increasing sink strength of seeds in a *B. campestris* spring cultivar over the period between commencement of flowering and the onset of pod ripening (Rood et al. 1984). Plants were exposed to $^{14}CO_2$ three times during this period and the radioactivity in different plant parts was measured at five different times after exposure. The slope of the regression of the percentage of total radioactivity in seeds on the percentage of total plant dry weight in seeds increased dramatically with successive treatments. Thus, as seeds matured, their capacity to attract assimilates from different sources greatly increased.

The changing importance of different photosynthesizing organs as sources of assimilates for developing seeds was also recognized in field studies with a *B. napus* winter cultivar in the United Kingdom (Chapman et al. 1984). Major sources of assimilates for developing seeds between early flowering and maturity were, in chronological order, upper leaves, stems and finally the pods themselves. It was also observed that the earliest and potentially largest pods on the main stem inflorescence and upper lateral inflorescences had seed weights lower than expected because of heavy shading by the pods above them. Improvement in yield might therefore be achieved through selection of an inflorescence structure permitting greater light penetration of the pod canopy. *B. napus* inflorescences can be markedly modified by various growth regulators such as paclo-butrazol (Scarisbrick et al. 1985) and ^{14}C-labelled assimilate distribution and utilization has been monitored in treated and untreated plants of a *B. napus* winter cultivar (Addo-Quaye et al. 1985). Increased amounts of $^{14}CO_2$ assimilation by leaves and reproductive parts of treated plants may have been achieved as a result of improved light interception by a more efficient crop canopy. Paclo-butrazol also changed the pattern of assimilate distribution during later reproductive development, more assimilates being translocated to reproductive parts of the main stem and uppermost lateral branches. This would explain the greater pod production at these sites and an overall yield increase of 16.9% compared with control plants (Addo-Quaye 1985).

Assimilate production by a rapeseed crop is greatly dependent on leaf area index which, in turn, is highly responsive to increased nitrogen supply to the plant (Scott et al. 1973). Increasing levels of nitrogen supply will therefore result in larger plants producing more pods and higher seed yield. The uptake of nitrogen by the rapeseed crop is large and can be in excess of 250 kg/ha (Racz et al. 1965). An increase in nitrogen uptake by the crop will usually be accompanied by a corresponding increase in biomass production, pod number and seed yield (Allen and Morgan 1972). However, average pod weight and average seed weight appear to be unaffected by increasing nitrogen supply even though the number of seeds per pod is higher at high nitrogen levels (Allen and Morgan 1972). This would suggest that nitrogen supply affects the amount of assimilates transported from leaves to juvenile pods, but is unlikely to affect assimilate production in the walls of larger pods during seed filling. Despite the responsiveness of rapeseed crops to applied nitrogen, there is a limit beyond which there is no further response to additional nitrogen or seed yield can even decline (Scott et al. 1973). This yield decrease is probably due, in part, to excessive pod production and shading within the canopy.

The high cost of nitrogenous fertilizers at the present time would make improved efficiency of nitrogen usage by the rapeseed crop an important breeding objective and an essential element of a yield ideotype. Improved efficiency could be approached either through an increase in yield of current levels of nitrogen supply or maintenance of current yields with a reduction in nitrogen supply. Such improvement would only be possible if inherent differences in nitrogen response exist between cultivars of the oilseed brassicas. Some evidence of cultivar variation in nitrogen response has been indicated in field studies with *B. napus* and *B. campestris* (Bhatty 1964; Grami and La Croix 1977). However, the clearest indication of genotypic variation in nitrogen response was provided by studies with a wide range of *B. napus* cultivars in western Australia (Yau and Thurling 1987a). Cultivar differences were detected for yield under a limiting nitrogen supply and yield response to an increase in levels of applied nitrogen. Differences in nitrogen uptake rate were observed only at intermediate nitrogen levels and in the efficiency of nitrogen utilization by the plant at low nitrogen levels. Although nitrogen uptake rate was correlated with the size of the root system, it was evident that cultivars also differed significantly in their capacity to absorb nitrate from the soil or solution culture.

4.3.5 Yield Component Relationships

Analyses of genetic and environmental variation in the yields of grain crops have often involved the partitioning of yield into a number of simple morphological components such as number of pods per plant (or per unit area), number of seeds per pod and weight per seed. These components develop sequentially in time and are, to some extent, interdependent (Adams 1967). Such a developmental pattern is frequently characterized by negative correlations among yield components which are of developmental rather than of genetic origin. The sequential development pattern of yield components means that the extent to which the first component in the sequence (e.g. pods per plant) utilizes the limited metabolic input will determine the expression of later-developing components (e.g. number of seeds per pod and seed size). Interdependence of these sequential components could well be magnified by an oscillatory input of metabolites which imposes limits of critical stages in the developmental sequence (Adams 1967). Mathematical analyses based on this concept have provided quantitative definitions of yield component systems enabling optimal yield component levels to be identified (Grafius and Thomas 1971).

Analyses of yield component relationships have revealed marked differences in patterns between *B. napus* and *B. campestris* grown at commercial density under stress-free conditions (Thurling 1974b). The number of pods per plant was most highly correlated with seed yield in *B. napus*, but number of seeds per pod was the main determinant of seed yield in *B. campestris*. In *B. napus*, number of pods per plant was negatively correlated with both seed number per pod and seed weight, there being no significant association between the latter. Pod number per plant was negatively correlated only with seed number per pod in *B. campestris*, but the latter was positively correlated with seed weight. The existence of negative correlations between yield components revealed a developmental interdependence between these in both *Brassica* species which would distort estimates of genotypic and environmental effects on each of them. A transformation procedure removing the influence of these developmental associations (Thomas et al. 1971) showed that variation in each yield component was due largely to genotype × environmental interactions rather than environmental factors as indicated in analyses of untransformed data. The pattern of yield component relationships in the oilseed brassicas was therefore characterized by significant genetic control only of characters early in the developmental sequence (e.g. pod number) and a strong environmental influence on character associations. This situation is considered to be particularly amenable to utilizing yield component data in predicting potential yields of different genotypes (Thomas et al. 1971). Yield prediction is more reliable because high-and low-yielding genotypes are, in most cases, clearly differentiated in terms of the expression of a specific yield component.

One major problem in understanding yield component relationships is the magnitude of differences in patterns between environments and populations of genotypes used in various studies. This is particularly evident in a study with three *B. napus* cultivars where yield component relationship characteristics varied widely among these cultivars (Campbell and Kondra 1978a). Pod number, for example, was correlated with seed yield in two of the three cultivars tested, but was not correlated with yield in the other cultivar. Both pod number and 1000-seed weight were significantly correlated with yield only in a cultivar in which seed size increased with more profuse branching. Branch number per unit area, and presumably pod number in *B. napus* were found to increase markedly with progressive increases in seeding rate, but this change was not associated with any clear yield increase (Degenhardt and Kondra 1981). Since there was virtually no change in seed size with variation in seeding rate, it would appear that any positive effect of increased branch number was nullified by a decrease in seed number per pod. The highest seed yields were usually obtained with lower seeding rates in an early sowing which were characterized by relatively few branches per unit area. A different result was obtained in experiments with winter *B. napus* in the United Kingdom (Mendham and Scott 1975), where both seed yield and pod number per unit area declined with successive delays in sowing over late summer and early autumn. Although there was a slight decrease in seed size over sowings, the rate of decline in yield was restricted by a compensatory increase in seed number per pod.

The influence of environment on yield component relationships was also evident in studies with *B. juncea*, which is widely grown in India as an oilseed crop (Singh and Chowdhury 1983). Differences between the lowest and highest yielding cultivars were significant for number of secondary branches in all four test environments, for seed size in three of the four environments and for number of seeds per pod in only one environment. The high-and low-yielding cultivars differed significantly only in secondary branch number in a rainfed, high-fertility environment, but in all three components in an irrigated low-fertility environment. These and the results of other studies referred to earlier all highlight the difficulty of defining an optimal combination of yield components for a specific environment. Nevertheless, the physiological implications of reduced pod number in respect of a more efficient canopy structure would require identification of genotypes capable of a substantial compensatory increase in seed weight per pod. Studies have shown that certain *B. napus* cultivars had the capacity to produce substantially higher number of seeds per pod than others with similar levels of pod production per unit area (Mendham et al. 1984).

4.4 Genetic Variation in Oilseed *Brassica* Yield Determinants and Its Utilization for Yield Improvement

Establishment and characterization of an appropriate genetic base is an essential starting point for any breeding programme. This should involve defining accessions in terms of their status, i.e. whether they are members of the primary gene pool (cultivars of the domesticated species), secondary gene pool (close relatives of the domesticated species) or tertiary gene pool (distant relatives of the domesticated species) (Harlan 1975). If yield is the primary breeding objective, then evaluation of accessions in terms of simple yield determinants such as flowering time and plant height would be undertaken at an early stage. This would provide the basis for selecting a much smaller group of accessions for more detailed evaluation of a wider range of yield determinants involving complex experiments in field and controlled environments providing estimates of genotypic and G × E interaction variance components as well as measures of relationships between characters. This will help in devising appropriate selection criteria.

Results from genetic studies of yield and various morphophysiological characters with relevance to yield improvement in the oilseed brassicas are reviewed in the following sections. The characters to be considered have been separated into three broad groups as follows:

1. Growth, development and yield.
2. Tolerance of environmental stresses.
3. Adaptation to new management strategies.

4.4.1 Growth, Development and Yield

Commencement of flowering is a major developmental event and the time between sowing and flowering is an important determinant of yield in short-season environments. Studies with a wide range of crop species have generally shown flowering time to be highly heritable, and it is often determined by relatively few major genes. High broad-sense and narrow-sense heritability estimates were obtained for flowering time in *B. napus* from diallel analyses of variation between spring cultivars and their F_1 hybrids sown at different times (Thurling and Vijendra Das 1979a). These studies indicated that gene interaction was not significant, but some dominance was present at the gene loci concerned. Genetic relationships differed considerably between sowings. In the case of the early winter sowing, parental array points were distributed evenly along the Vr, Wr regression line. By contrast, with the spring sowing, array points for all but one of the parents were clustered close to the origin of the Vr, Wr axis. The array point for a very late-flowering cultivar, Isuzu, was widely separated from the others, suggesting that it possessed the recessive allele of a major gene, determining vernalization requirement.

Examination of variation patterns in early segregating generations of crosses between three cultivars differing in vernalization response provided further evidence of major genes for vernalization requirement (Thurling and Vijendra Das 1979b). F_2 distributions for flowering time of crosses between Target (low vernalization requirement) on the one hand and Isuzu (high vernalization requirement) and Bronowski (moderate vernalization requirement) on the other hand were clearly bimodal when plants were grown under continuous light at 25 °C. The characteristics of these distributions indicated that the vernalization requirement of Bronowski was determined by duplicate recessive genes, whereas the high requirement of Isuzu was determined by two different genes which acted independently. Target, which was insensitive to vernalization under these conditions, possessed the dominant alleles of all four genes. Plants flowering earlier than Target were also detected in F_2 populations of crosses of Target with the other cultivars, this earliness being determined by genes from Isuzu and Bronowski which were not expressed in the parental background. Interactions of major genes with different genetic backgrounds can, therefore, generate unexpected flowering time phenotypes which should be utilized to good effect in yield improvement programmes.

Pre-anthesis development in *Brassica* species may be conveniently divided into several phases defined by the major events like floral initiation, commencement of stem elongation and appearance of first open flower. Controlled environment studies revealed substantial differences among cultivars in the duration of the stem elongation phase under a 15 °C, 12-h photoperiod regime (Thurling and Vijendra Das 1979a).

Development and dry matter production have been investigated in genetic analyses with crosses among three Canadian spring cultivars of *B. napus* (Campbell and Kondra 1978b). Analyses of F_1 and F_2 data revealed significant reciprocal differences, indicating the existance of nucleo-cytoplasmic interactions. Heritabilities for all characters were generally low, excepting for days to flower. Variation for duration of a specific developmental phase, the rates of flowering branch development, yield components and the production and distribution of dry matter, was largely of environmental origin. Thus, only flowering time would appear to have any potential as an alternative criterion of selection for yield in early generations.

Resynthesized *B. napus* genotypes have been used in extensive genetic studies (Kräling 1987). General combining ability estimates were negative for most growth and yield component characters in some of the resynthesized genotypes, but others had high positive values for seed yield per plant, dry matter at anthesis and pods per main stem. In top crosses of resynthesized genotypes to a commercial cultivar, hybrids were characterized by longer main stems and the production of more flowers and pods on the main stem than the commercial tester. This was associated with a marked reduction in lateral branch development. A greater potential for seed yield improvement was evident in crosses of the resynthesized × cultivar hybrids with a second cultivar to produce three-way hybrids in which the resynthesized component of the genome was only 25%. Therefore, resynthesized *B. napus* types, by broadening the genetic variability of breeding populations, could contribute to yield improvement in the long term.

Yield improvement can be achieved through manipulation of simple morphological characters. Modification of the pod canopy structure to permit greater infiltration of light is one aspect of yield improvement where variation in simple morphological characters can be utilized (Mendham et al. 1984). An apetalous mutant (Buzza 1983) might have some value in this respect. Comparisons between apetalous mutant and a commercial spring cultivar (Rao and Mendham 1985) showed that light penetration to the base of the inflorescence was about 30% greater in the apetalous mutant. The apetalous mutant plants produced a greater number of seeds per pod, but because of the greater number of unproductive pods it did not significantly outyield the commercial cultivar. Nevertheless, the incorporation of the apetalous character into commercial cultivars could provide a relatively simple route to yield improvement. Stable apetalous mutants have also been observed in *B. juncea* and *B. carinata* (Banga, unpubl.).

Another simple character likely to affect canopy structure is pod length. Long pod genotypes selected from the *B. napus* spring cultivar Chena A produced a greater weight of seeds per pod than standard pod length genotypes but did not significantly outyield the latter because of a compensatory reduction in pod number (Chay and Thurling 1989a). Genetic studies showed that pod length was determined by two complementary interacting genes (Chay and Thurling 1989b). While long pods were produced by dominant alleles, pods of intermediate length were produced when a dominant allele was present at one or other locus. Background genotype also influenced the expression of trait. In this case, selection for the desired character would be rapid and the breeding programme could be conducted in a controlled environment permitting rapid turnover of generations.

Yield improvement in *B. napus* and *B. juncea* would normally be caused on genetically heterogenous populations generated by hybridization between selected parents. In *B. campestris*, which is self-incompatible and freely outcrossing, cultivars are genetically heterogenous and yield improvement could be achieved through selection within a cultivar. The expected advances in yield from this type of selection were calculated for the Canadian spring cultivar Span of *B. campestris* ssp. *oleifera* (Thurling 1974c). Heritabilities obtained from analyses of variation between full and half-sib family means were relatively high (0.48-0.81) for a wide range of characters including yield, yield components, flowering time and leaf area at flowering. Using these heritabilities and genetic correlations between yield and another plant character, it was found that no correlated response in yield to selection for another character was as great as the direct response to yield selection. However, the expected genetic advance in yield from the simultaneous selection for a number of plant characters using a selection index technique was significantly greater than that expected from selection for yield alone. Expected yield responses to selection based on indices comprising measurements of yield components and vegetative characters at flowring were about 20% greater than that of the direct yield response. Disruptive selection could be particularly effective (Murty et al. 1972).

Although the genetic variability within most *B. campestris* ssp. *oleifera* cultivars is of sufficient magnitude to allow for significant improvements through selection, this variability may be considerably broadened by hybridization between different cultivars or by hybridization with cultivars from other subspecies.

4.4.2 Tolerance of Environmental Stresses

Crop yields in temperate regions are limited primarily by environmental stresses which can be either predictable or unpredictable. These stresses are normally most severe towards the end of the growing season, as in the case of drought and frost in southern Australia and western Canada respectively. These are predictable stresses which, to a large extent, may be circumvented by growing cultivars with a developmental pattern allowing for avoidance of these stresses. If the stress occurs at critical stages of plant development, such as at the commencement of flowering, they may cause substantial yield reductions. In this situation, the most effective approach is breeding for tolerance to the expected stresses.

Low Temperature. Suboptimal temperatures during spring in western Canada can delay seeding and thereby reduce potential yield in the short growing season. Studies of genotypic variation in low temperature germination and responses to selection for this character were undertaken with *B. napus* and *B. campestris* cultivars (Acharya et al. 1983). Heritabilities for germination and growth at 20 °C were close to zero in both species. However, heritabilities of the same characters in populations grown at 10 °C ranged between 0.49 and 0.91 in *B. campestris* and between 0.23 and 0.64 in *B. napus*. Significant responses to selection for germination at low temperature over two generations were obtained in six of the ten populations tested. The greatest responses in percentage emergence under simulated field temperatures were obtained with the cultivar Regent (46 to 89%) and the breeding line DI-63 (35 to 79%). In north India, frost is a major abiotic factor limiting the productivity of *B. juncea*.

Drought. Rapeseed and mustard crops can be exposed to drought stresses at any time during the reproductive phase. Imposition of artificial drought stresses at different stages after commencement of stem elongation indicated that both *B. napus* and *B. campestris* were most sensitive to drought stress at the commencement of flowering (Richards and Thurling 1978a). Genetic studies of drought tolerance in *B. napus* and *B. campestris* were therefore conducted with families either grown under irrigation from the commencement of flowering or drought-stressed from the same stage (Richards 1978a). Analyses were based on measurements of a wide range of growth and developmental characters in full and half-sib families derived from hybrid bulk composites of each species. These analyses were first used to assess environments for maximum selection response in seed yield under drought stress conditions, yield under irrigation, yield under drought and the ratio of these yield measurements being the criteria used. Selection for yield in a drought-stressed environment was found to be a more efficient strategy for yield improvement in a dryland situation than selection in an optimal environment or selection based on the drought response index.

Data obtained from the drought treatment were used to estimate heritabilities for a wide range of characters in both *B. napus* and *B. campestris* (Richards and Thurling 1979). The narrow-sense heritability of yield was much higher in *B. napus* (0.44) than in *B. campestris* (0.10), but only flowering time and 1000-seed weight had substantially higher heritabilities in the latter. Although no single yield-related character appeared to be a better criterion of selection for yield improvement in drought-prone environments, certain selection indices might be useful. One index incorporating measurements of yield, pod number and flowering time was more efficient than direct selection for yield in *B. napus*. The most efficient index in *B. campestris* was one combining measurements of yield, harvest index, 1000-seed weight and seed number per pod. Biochemical characters such as leaf proline accumulation and leaf chlorophyll stability were also associated with tolerance to simulated drought conditions in semi-controlled environments (Richards 1978b). Although these characters had moderately high broad sense heritabilities, they were not closely related to tolerance to natural drought conditions in the field and appeared to be of little use as alternative selection criteria.

4.4.3 Adaptation to New Management Strategies

Improvements in crop yield have been achieved through both the development of superior cultivars and the introduction of more efficient management practices. Ideally, these changes to a crop production system should be integrated in a way that allows for the selection of some optimal combination of plant type and management practice. Quite often, new management practices may be introduced without considering the possibility of a change in cultivar, with the result that the full benefits of a new management practice are not realized. It is feasible to breed crop cultivars that enable a new management practice to be introduced such as in the case of developing cultivars tolerant of a more effective herbicide. Three examples of this approach to improve the yields of oilseed brassicas are given in the following sections.

Herbicide Resistance. Improved rapeseed yields could be achieved in most growing areas by more effective control of broad leaf weeds using triazine herbicides. Current commercial cultivars are susceptible to these herbicides, but a weed biotype of *B. campestris* having a high level of triazine tolerance has been found in Canada (Maltais and Bouchard 1978). Reciprocal crosses between *B. campestris* biotypes resistant or susceptible to atrazine showed the resistance to be uniparentally inherited through the female parent and by cytoplasmic DNA (Souza Machado et al. 1978).

This cytoplasmic resistance to atrazine was subsequently incorporated into commercial cultivars used as male parents in each generation (Beversdorf et al. 1980). The presence of this cytoplasm in the backcross derivatives did not significantly affect flowering time, fertility or general plant vigour. Disruption of the photosynthetic Hill reaction in triazine-susceptible plants was not detected in the weedy *B. campestris* parent or the backcross derivatives when exposed to atrazine concentrations ranging between 10^{-7} and 10^{-4}M (Weiss and Beversdorf 1981).

One major problem encountered with triazine resistant oilseed brassicas is the reduced vigour and lower seed yield, presumably reflecting a less efficient photosystem II in plants lacking the ability to bind triazines (Gressel and BenSinai 1985).

Another approach to breeding for triazine resistance, or at least tolerance of a practical value in the crop situation, may be through utilization of nuclear polygenic variation detected in a study with *B. napus* (Karim and Bradshaw 1968). A later study involving a much wider range of *B. napus* cultivars and a highly variable hybrid bulk composite of *B. campestris* showed tolerance to simazine to have narrow-sense heritabilities of about 50% in both species (McGuire 1987). There was a marked response to mass selection within the *B. campestris* population, the level of tolerance to simazine more than doubling with two selection cycles. It has also been possible to incorporate genes for tolerance from *B. campestris* into *B. napus* through backcrossing, and to develop lines combining good agronomic performance with an adequate level of simazine tolerance for a commercial crop situation (McGuire and Thurling, unpubl.). In field trials involving both *B. napus* lines tolerant of simazine and a *B. napus* cultivar carrying cytoplasmic triazine resistance, the tolerant lines generally had seed yield double that of the latter at simazine levels which greatly supressed weed growth.

Shattering Tolerance. Direct harvesting of rapeseed crops is usually associated with significant loss of seeds (> 30%) because of pod shattering in the hot, dry environment at the time of harvest. Costs of production could be reduced by the introduction of shatter-tolerant cultivars for direct harvesting. Studies have been undertaken to screen for shatter tolerance in the oilseed brassicas using instruments which can measure the energy required to rupture a mature pod (Kadkol et al. 1984). Significant differences in measures of shatter tolerance were observed among cultivars of *B. napus*, *B. campestris* and *B. juncea*. However, there was also substantial within plant variation associated with pod size rather than with the position in the inflorescence from which the pod was obtained (Kadkol et al. 1985; 1989). This indicated that the sampling procedure was crucial to accurate assessments of genotype differences in shatter tolerance, and measurements of few pods from many plants adjusted for pod size was recommended. Extremely high levels of shatter tolerance were detected in *B. campestris* ssp. *trilocularis* cultivars which were characterized by the absence of an abscission (separation) layer in the region of attachment of the the pod walls to the replum (Kadkol et al. 1986a).

Shatter resistance has been found to be determined by recessive alleles at two or three gene loci which appeared to interact in a dominant epistatic manner (Kadkol et al. 1986b). Further biometrical analyses of variation among progenies derived by intercrossing of F_2 plants from the above cross revealed a high level of non-additive genetic variance for all measures of shatter tolerance (Kadkol et al. 1986c). Broad-sense heritabilities for all these characters were greater than 0.70 and, together with the presence of gene interaction, support the evidence for a small number of interacting genes revealed by Mendelian analysis. Control of shatter tolerance by a small number of major genes would suggest that the character could be transferred to the economically more important *B. napus* by recurrent backcrossing. This possibility was investigated with selfed progenies of first backcross derivates of crosses between *B. napus* cultivars and ssp. *trilocularis* lines. Although there was substantial variation in shatter tolerance in the BC_1, F_2 and BC_2, F_3 generations of different crosses, no plant was as tolerant as the ssp. *trilocularis* parent. Further breeding allowing for recombination between the appropriate genes should achieve shatter tolerance levels which are commercially adequate.

Nitrogen Uptake and Utilization. The feasibility of using measurements of N uptake and N utilization in breeding for improved yields under current N application rates was examined in an analysis of variation among full and half-sib families from a *B. napus* F_2 population (Yau and Thurling 1987b). Narrow-sense heritabilities for both N uptake rate and a N utilization index were very low and offered no additional advantages as selection criteria. However, selection based on an index combining information on flowering time, nitrogen uptake and utilization was expected to be more effective in improving yield than direct selection for yield. Such an index might have some value as a selection criterion in rapid generation advance breeding methods such as single-seed descent where comparisons are between large plots of relatively uniform families.

4.5 Conclusions

As indicated earlier, the main role of morpho-physiological characters in a yield-improvement programme would be in the selection of parents. This would be of primary importance in the self-fertile species *B. napus* and *B. juncea*, but should also be useful in hybridizing heterogenous *B. campestris* populations to broaden the variability in specific yield determinants. The effectiveness of these characters as selection criteria in early generations of pedigree selection in self-pollinators will depend on their heritabilities. Certain relatively simple characters having high heritabilities and some influence on yield have been identified. Foremost amongst these would be flowering time in the case of short-season environments. In such environments, early flowering is usually sought, and individual plants would be easily identified and tagged. Subsequent selection could be conducted within this group for other characters such as apetaly, pod length and perhaps seed size.

It is now possible to generate a large number of diverse pure lines in a single step through culture of pollen grains from an F_1 hybrid and subsequent colchicine treatment of the haploid plants. These lines could be compared on the basis of a previously defined ideotype,and lines selected at this stage would be advanced for further yield testing. Recurrent selection is an alternate strategy in which the selection of lines for inter-crossing in each selection cycle could be based on combinations complementary for various morphological and physiological yield determinants. As suggested earlier in this chapter, the transfer of germplasm between related *Brassica* species can be very rewarding.

A backcrossing technique was also found to be effective in transferring genes for fairly complex morphological characters between *B. napus* cultivars (Thurling 1982). In this case, yield of a spring cultivar, Target, was improved by introgression of genes for higher pod number per plant, greater dry matter accumulation during post-anthesis development and higher harvest index from the lower-yielding cultivar, Bronowski. The results of these studies with oats and *B. napus* highlight the potential of agronomically inferior exotic lines to donate useful genes for yield improvement. The selection of parents of hybrid cultivars could also be made on the basis of their complementary morphological and physiological characteristics. For example, in the study conducted by Sernyk and Stefansson (1983), the cultivars Regent and Gulliver had high means for harvest index and biomass production respectively. Since there was positive heterosis for both these yield determinants, the F_1 hybrid yield was substantially greater than that of either parent.

Response to selection for environmental stress tolerance in a pedigree scheme is unlikely to be very great in view of the low heritabilities observed for different stresses (e.g. Richards and Thurling 1979). In any event, accurate measurement of tolerance would presumably be difficult where large populations would need to be screened. An alternative approach, particularly in the case of the temperature stress tolerance, would be the use of pollen selection (see Zamir 1983 for review). It has been estimated that 60–70% of genes function similarly during the sporophytic and gametophytic phase of the life cycle. Thus, it might be possible to modify the sporophyte through selection based on gametophyte responses. The feasibility of this approach has been thoroughly assessed for low-temperature tolerance in tomato, and it has been shown that selection pressure applied to pollen during germination and tube growth in vivo will increase the tolerance of progenies obtained through fertilization by the more vigorous pollen (Zamir et al. 1982). One major advantage of this method is that very large numbers of genotypes can be simply screened within a glasshouse or growth room. The selection scheme would involve, for example, F_1 hybrid plants derived from a cross between a commercial cultivar and some stress-tolerant line being used to pollinate the commercial cultivar to initiate a backcrossing sequence. Following pollination, plants used as females would be transferred to a stress environment (e.g. low temperature) for a short period, where selection between pollen grains will occur. An investigation of this method with *B. napus* spring cultivars has revealed significant correlations between pollen tube growth at 10 °C and seedling growth under low temperature in field and controlled environments (Thurling, unpubl.).

References

Acharya SN, Dueck J, Downey RK (1983) Selection and heritability studies on canola/rapeseed for low temperature germination. Can J Plant Sci 63: 377–384

Adams MW (1967) Basis of yield component compensation in crop plants with special reference to the field bean, *Phaseolus vulgaris*. Crop Sci 7: 505–510

Addo-Quaye AA, Daniels RW, Scarisbrick DH (1985) The influence of paclobutrazol on the distribution and utilization of ^{14}C-labelled assimilate fixed at anthesis in oilseed rape (*Brassica napus* L.). J Agric Sci Camb 105: 365–373

Allen EJ, Morgan DG (1972) A quantitative analysis of the effects of nitrogen on growth development and yield of oilseed rape. J Agric Sci Camb 78: 315–324

Allen EJ, Morgan DG (1975) A quantitative comparison of growth and development and yield of different varieties of oilseed rape. J Agric Sci Camb 85: 159–174

Angus JF, Jones R, Wilson JH (1972) A comparison of barley cultivars with different leaf inclinations. Aust J Agric Res 23: 945–957

Austin RB, Bingham J, Blackwell RD, Evans LT, Ford MA, Morgan CL, Taylor M (1980) Genetic improvement in winter wheat yields since 1900 and associated physiological changes. J Agric Sci Camb 94: 675–689

Beversdorf WD, Weiss Leiman J, Erickson LR, Souzo Machado V (1980) Transfer of cytoplasmically inherited triazine resistance from birds' rape to cultivated oilseed rape (*Brassica campestris* and *B. napus*). Can J Genet Cytol 22: 167–172

Bhatt GM (1980) Early generation selection criteria for yield in wheat. J. Aust. Inst. Agric. Sci. 46: 14–22

Bhatty RS (1964) Influence of nitrogen fertilization on the yield, protein and oil content of two varieties of rape. Can J Plant Sci 14: 215–217

Bos I (1983) About the efficiency of grid selection. Euphytica 32: 885–893

Brar G, Thies W (1977) Contribution of leaves, stem, siliquae and seeds to dry matter accumulation in ripening seeds of rapeseed (*Brassica napus* L.). Z Pflanzenphysiol 82: 1–13

Brennan PS, Byth DE (1979) Genotype x environmental interaction for wheat. Aust J Agric Res 30: 221–232

Buzza GC (1983) The inheritance of an apetalous character in canola (*Brassica napus* L.). Cruciferae Newslett 8: 11–12

Campbell DC, Kondra ZP (1977) Growth pattern analysis of three rapeseed cultivars. Can J Plant Sci 57: 707–712

Campbell DC, Kondra ZP (1978a) Relationships among growth patterns, yield components and yield of rapeseed. Can J Plant Sci 58: 87–93

Campbell DC, Kondra ZP (1978b) A genetic study of growth characters and yield characters of oilseed rape. Euphytica 27: 177–183

Chapman JF, Daniels RW, Scarisbrick DH (1984) Field studies on ^{14}C assimilate fixation and movement in oilseed rape (*Brassica napus*). J Agric Sci Camb 102: 23–31

Chay PM, Thurling N (1989a) Variation in pod length in spring rape (*Brassica napus*) and its effect on seed yield and yield components. J Agric Sci Camb 113: 139–147

Chay PM, Thurling N (1989b) Identification of genes controlling pod length in spring rapeseed, *Brassica napus* L. and their utilization for yield improvement. Plant Breed 103: 54–62

Cooper JP (1974) The use of physiological criteria in grass breeding. Rep Welsh Plant Breed Stn Fer 1973, pp 95–102

Degenhardt DF, Kondra ZP (1981) The influence of seeding date and seeding rate on seed yield and yield components of five genotypes of *Brassica napus*. Can J Plant Sci 61: 175–183

Donald CM (1962a) In search of yield. J Aust Inst Agric Sci 28: 171–178

Donald CM (1962b) The breeding of crop ideotypes. Euphytica 17: 385–403

Donald CM (1981) Competitive plants, communal plants, and yield in wheat crops. In: Evans L.T., Peacock W.J. (eds.). Wheat science—today and tomorrow. Cambridge Univ Press. pp 223–248

Donald CM, Hamblin J (1976) The biological yield and harvest index of cereals as agronomic and plant breeding criteria. Adv Agron 28: 361–405

Evans LT (1975) The physiological basis of crop yield. In: Evans LT (ed) Crop Physiology Cambridge Univ Press, pp 327–355

Evans LT (1984) Pre-anthesis growth and its influence on seed yield in winter oilseed rape. Aspects Appl Biol 6: 81–90

Fasoulas A (1973) A new approach to breeding superior yielding varieties. Publ 3 Aristotlean Univ, Thessaloniki, p 42

Frey KJ (1976) Plant breeding in the seventies: useful genes from wild plant species. Egypt J Genet Cytol 5: 460–482

Freyman S, Charnetski WA, Crookston RK (1973) Role of leaves in the formation of seed in rape. Can J Plant Sci 53: 693–694

Gardner CO (1961) An evaluation of effects of mass selection and seed irradiation with thermal neutrons on yield of corn. Crop Sci 1: 241–245

Grafius JE, Thomas RL (1971) Stress: an analysis of its source and influence. Heredity 26: 433–442

Grami B, La Croix LJ (1977) Cultivar variation in total nitrogen uptake in rape. Can J Plant Sci 57: 619–624

Gressel J, BenSinai G (1985) Low interspecific competitive fitness in a triazine-resistant, nearly nucleus-isogenic line of *Brassica napus*. Plant Sci 38: 29–32

Gupta ML, Labana KS (1985) Effect of selection for seed size and its correlated response in Indian mustard. Crop Improv 12: 193–194

Harlan JR (1975) Crops and man. ASA, Madison, Wiscconsin

Hedley CL, Ambrose MJ (1981) Designing leafless plants for improving yield of the dried pea crop. Adv Agron 34: 225–277

Hodgson AS (1978) Rapeseed adaptations in northern New South Wales. II Predicting plant development of *Brassica campestris* L. and *Brassica napus* L. and its implications for planting time, designed to avoid water deficit and frost. Aust J Agric Res 29: 711–726

Hosoda T (1953) On new type of *Brassica napus* obtained from artificial amphiploids. Jpn J Breed 3: 44–50

Hozyo T, Kato S, Kabayashi H (1972) Photosynthetic activity of pods of rape plants (*Brassica napus* L.) and the contribution of pods to ripening of rapeseed. Proc Crop Sci Soc Jpn 41: 420–425

Kadkol GP, Beilharz VC, Halloran GM, Macmillan RH (1986a) Anatomical basis of shelter resistance in oilseed brassicas. Aust J Bot 34: 595–601

Kadkol GP, Halloran GM, Macmillan RH (1985) Evaluation of *Brassica* genotypes for resistance to shatter. II. Variation in siliqua strength culture and between accessions. Euphytica 34: 915–924

Kadkol GP, Halloran GM, Macmillan RH (1986b) Inheritance of siliqua strength in *Brassica campestris* L. I. Studies of F_2 and backcross populations. Can J Genet Cytol 28: 365–373

Kadkol GP, Halloran GM, Macmillan RH (1986c) Inheritance of siliqua strength in *Brassica campestris* L. II. Quantitative genetic analysis. Can J Genet Cytol 28: 563–567

Kadkol GP, Halloran GM, Macmillan RH (1989) Shatter resistance in crop plants. Criti Rev Plant Sci 8: 169

Kadkol GP, Macmillan RH, Burrow RP, Halloran GM (1984) Evaluation of *Brassica* genotypes for resistance to shatter. I Dev Lab test. Euphytica 33: 63–73

Karim A, Bradshaw AD (1968) Variation for triazine resistance in *Brassica napus*. Weed Res 8: 283–291

King JR, Kondra ZP (1986) Photoperiod response of spring oilseed rape (*Brassica napus* L.) and (*B. campestris* L.). Field Crop Res 13: 363–373

Knott DR (1972) Effects of selection for F_2 plant yield on subsequent generations in wheat. Can J Plant Sci 52: 721–726

Kräling K (1987) Utilization of genetic variability of resynthesized rapeseed. Plant Breed 99: 209–217

Labana KS, Banga SS, Gandhi K (1987) Source sink manipulation and its implications on biomass production in *Brassica*. In: Proc 7th Int Rapeseed Conf, Poland, pp 719–726

Major DJ, Bole JB, Charnetski WA (1978) $^{14}CO_2$ assimilation by stems, leaves and pods of rape plants. Can J Plant Sci 58: 783–787

Major DJ, Charnetski WA (1976) Distribution of ^{14}C-labelled assimilates in rape plants. Crop Sci 16: 530–532

Maltais B, Bouchard CJ (1978) A new birds rape (*Brassica rapa* L.) resistant to atrazine. Phytoprotection 59: 117–119

McGregor DI (1981) Pattern of flower and pod development in rapeseed. Can J Plant Sci 61: 275–282

McGuire GM (1987) Genetic variation in simazine tolerance of a *Brassica campestris* population. Proc. 7th Int. Rapeseed Congr, Poland, pp 469–470

McKenzie RIH, Lambert JW (1961) A comparison of F_3 lines and their related F_6 lines in two Barley crosses. Crop Sci 1: 246–249

Mendham NJ, Scott RK (1975) The limiting effect of plant size at inflorescence initiation on subsequent growth and yield of oilseed rape (*Brassica napus*). J Agric Sci Camb 84: 487–502

Mendham NJ, Russell J, Buzza GC (1984) The contribution of seed survival to yield in new Australian cultivars of oilseed rape (*Brassica napus* L.). J Agric Sci Camb 103: 303–316

Mendham NJ, Russell DJ, Jarosz NK (1990) Response to sowing time of three contrasting Australian cultivars of oilseed rape (*Brassica napus*). J Agric Sci Camb 114: 275–283

Mendham NJ, Shipway PA, Scott RK (1981) The effects of delayed sowing and weather on growth, development and yield of winter oilseed rape (*Brassica napus* L.). J Agric Sci Camb 96: 389–416

Murty BR, Arunachalam V, Doloi PC, Ram J (1972) Effects of disruptive selection for flowering time in *Brassica campestris* var. *brown sarson*. Heredity 28: 287–295

Nishi S (1968) Hakuran. An artificially synthesized heading *Brassica napus* L. JARQ 3: 18–21

O'Brien L, Pugsley AT (1981) F_3 yield response to F_2 selection for gibberellic acid insensitivity in eight wheat crosses. Crop Sci 21: 217–219

Olsson G (1960) Species crosses within the genus *Brassica*. II. Artificial *Brassica napus* L. Hereditas 46: 351–395

Pechan PA, Morgan DG (1985) Defoliation and its effects on pod and seed development in oilseed rape (*Brassica napus* L.). J Exp Bot 36: 458–468

Racz GJ, Webber MD, Soper RJ, Hedlin RA (1965) Phosphorus and nitrogen utilization by rape, flax and wheat. Agron J 57: 335–337

Rao MSS, Mendham NJ (1985) Proc. Fifth Aust. Rapeseed Agron and Breed Res Workshop Perth, W.A. 106–109

Rawson HH, Evans LT (1970) The pattern of grain growth within the ear of wheat. Aust J Biol Sci 23: 753–764

Richards RA (1978a) Genetic analysis of drought stress response in rapeseed (*Brassica campestris* and *B. napus*). I. Assessment of environments for maximum selection response for grain yield. Euphytica 27: 609–615

Richards RA (1978b) Variation between and within species of rapeseed (*Brassica campestris* and *B. napus*) in response to drought stress. III. Physiological and Physiochemical characters. Aust J Agric Res 29: 491–501

Richards RA, Thurling N (1978a) Variation between and within species of rapeseed *Brassica campestris* and *B. napus* in response to drought stress. I. Sensitivity at different stages of development. Aust J Agric Res 29: 469–477

Richards RA, Thurling N (1978b) Variation between and within species of rapeseed *Brassica campestris* and *Brassica napus* in response to drought stress. II. Growth and development under natural drought stress. Aust J Agric Res 29: 479–490

Richards RA, Thurling N (1979) Genetic analysis of drought stress response in rapeseed (*Brassica campestris* and *B. napus*). II. Yield improvement and application of selection indices. Euphytica 28: 169–177

Rood SB, Major DJ, Charnetski WA (1984) Seasonal changes in $^{14}CO_2$ assimilation and ^{14}C translocation in oilseed rape. Field Crop Res 8: 341–348

Saini JS (1984) Agronomic considerations for raising oilseeds production under irrigated conditions. Research and Development strategies for oilseed production in India. ICAR, New Delhi, pp 155–165

Scarisbrick DH, Addo-Quayo AA, Daniels RW, Mahamud S (1985) The effect of paclo-butrazol on plant height and seed yield of oilseed rape (*Brassica napus* L.). J Agric Sci Camb 105: 605–612

Scott R.K, Ogunremi EA, Ivins JD, Mendham NJ (1973) The effect of sowing date and season on growth and yield of oilseed rape (*Brassica napus*). J Agric Sci Camb 81: 277–285

Sernyk JL, Stefansson BR (1983) Heterosis in summer rape (*Brassica napus* L.). Can J Plant Sci 63: 407–413

Singh BP, Chowdhury RK (1983) Correlation and path coefficient analysis of seed yield and oil content in mustard (*Brassica juncea*). Can J Genet Cytol 25: 312–317

Souza-Machado Z, Bandeen JD, Stephenson GR, Lavigne LP (1978) Uniparental inheritance of chloroplast atrazine tolerance in *Brassica campestris*. Can J Plant Sci 58: 977–981

Stefansson BR, Kondra ZP (1975) Tower summer rape. Can J Plant Sci 55(1): 343–344

Tayo TO, Morgan DG (1975) Quantitative analysis of the growth, development and distribution of flowers and pods in oilseed rape (*Brassica napus* L.). J Agric Sci Camb 92: 363–373

Tayo TO, Morgan DG (1979) Factors influencing flower and pod development in oilseed rape. J Agric Sci Camb 103: 303–316

Thomas RL, Grafius JE, Hahn SK (1971) Stress: an analysis of its source and influence. Heredity 26: 177–188

Thurling N (1974a) Morphophysiological determinants of yield in rapeseed (*Brassica campestris* and *Brassica napus*): I. Growth and morphological characters. Aust J Agric Res 25: 697–710

Thurling N (1974b) Morphophysiological determinants of yield in rapeseed (*Brassica campestris* and *Brassica napus*): II. Yield components. Aust J Agric Res 25: 711–721

Thurling N (1974c) An evaluation of an index method of selection for high yield in turnip rape, *Brassica campestris* L. ssp. oleifera Metzg. Euphytica 23: 321–331

Thurling N (1982) The utilization of backcrossing in improving the seed yield of spring rape (*Brassica napus* L.). Z Pflanzenzuecht 88: 43–53

Thurling N, Vijendra Das LD (1977) Variation in the pre-anthesis development of spring rape (*Brassica napus* L.). Aust J Agric Res 28: 597–607

Thurling N, Vijendra Das LD (1979a) Genetic control of the pre-anthesis development of spring rape (*Brassica napus* L.). I. Diallel analysis of variation in the field. Aust J Agric Res 30: 251–259

Thurling N, Vijendra Das LD (1979b) Genetic control of the pre-anthesis development of spring rape (*Brassica napus* L.). II. Identification of individual genes controlling developmental patterns. Aust J Agric Res 30: 261–271

Thurling N, Vijendra Das LD (1980) The relationship between pre-anthesis development and seed yield of spring rape (*Brassica napus*). Aust J Agric Res 31: 25–36

Weiss J, Beversdorf WD (1981) Effects of atrazine concentrations on the hill reaction in isolated chloroplast of atrazine-resistant and susceptible oil seed rape (*Brassica napus* and *B. campestris*) genotypes. Can J Plant Sci 61: 723–726

Wilson D (1981) Breeding for morphological and physiological traits. In: Frey, KJ (ed.). Plant breeding. II. Iowa State Univ Ames, Ia, pp 233–290

Yau SK, Thurling N (1987a) Variation in nitrogen response among spring rape (*Brassica napus*). Cultivars and its relationships to nitrogen uptake and utilization. Field Crop Res 16: 139–155

Yau SK, Thurling N (1987b) Genetic variation in nitrogen uptake and utilization in spring rape (*Brassica napus* L.) and its exploitation through selection. Plant Breed 98: 330–338

Yoshida S (1972) Physiological aspects of green yield. Ann Rev Plant Physiol 23: 437–464

Zamir D (1983) Pollen gene expression and selection: application in plant breeding. In: Tanksley S.D., Otton TJ (eds.). Isozyme in plant genetics and breeding, part A. Elsevier, Amsterdam, pp 313–330

Zamir D, Tanksley SD, Jones RA (1982) Haploid selection for low temperature tolerance of tomato pollen. Genetics 101: 128–137

CHAPTER 5

Bioenergetic Limitations to Breeding Better Brassicas

C.R. Bhatia and *R. Mitra*

5.1 Introduction

Crop production is an energy conversion process in which the incident light energy is trapped and converted to chemical bond energy during photosynthesis. This energy is used for growth and maintenance of different plant tissues, part of which is allocated to protect the plant from insects and pathogens, and a part towards the reproductive effort (Bazzaz et al. 1987). Plants produce seeds for their self-perpetuation and hence store carbon, nitrogen and other nutrients for the initial growth of the future seedling. In *Brassica* oilseeds, the main reserves are in the form of lipids, proteins and carbohydrates. The high lipid content in the seeds is exploited for extracting oil. Thus, essentially, the breeding and agronomy research in the *Brassica* oilseeds aims at maximizing the trapping and conversion of solar energy—a free resource into nutritionally desirable lipids within the constraints of temperature, water, nutrients, biotic and abiotic stresses. The energy content of the oil (39.80 MJ/Kg) is much higher compared to that of the proteins (23.88 MJ/Kg) and carbohydrates (16.76 MJ/Kg). Because of the higher amount of oil, energy content of *Brassica* seeds (25.7 MJ/Kg) is higher than that of cereals (17 MJ/Kg) and pulses (18.3 MJ/Kg).

5.2 Conversion Efficiency of Substrate to Biomass

Conversion efficiency of substrate to biomass has been of interest to microbiologists and animal scientists for a long time. However, such estimates are relatively recent for higher plants (Rudolf 1971; Mooney 1972; Penning de Vries et al. 1974; McDermit and Loomis 1981). The quantitative relations between the substrate (glucose) and the end product were examined by Penning de Vries et al. (1974). Using the approaches of Penning de Vries et al. (1974), Mitra and Bhatia (1979) calculated the production values (PVs) for the synthesis of fatty acids, oil and seed biomass in different oilseed crops with specified oil content and fatty acid composition. A glossary of the bioenergetic terms used is presented in Table 1. Heat of combustion is a direct measure of the energy content. Production value (PV) and glucose equivalent (GE) represent carbon assimilates required to provide carbon skeletons and electrons for the end product. PV is based on biosynthetic pathways, while GE is based on proximate analysis or elemental composition.

Table 1. Terms used in bioenergetic computations. (Bhatia and Mitra 1988)

Name	Symbol	Definition	Unit
Heat of combustion	H°C	The enthalpy change for the reaction of combusting the chemical compound into constituent gaseous states, for example, amino acid at constant pressure at 298°K to yield CO_2, H_2O, N_2 (gas) and sulphur as H_2SO_4 (when applicable)	KJ/mol
Production value	PV	$\frac{\text{Weight of the end product}}{\text{Weight of substrate required for carbon skeletons and energy production (based on biochemical pathways)}}$	g/g
Inverse of production value	1/PV	Gram glucose required to make 1 g end product	g/g
Glucose equivalent	GE	Number of moles of glucose required to supply carbon skeletons and electrons to build 1 mole product (based on proximate analysis or elemental composition)	mol/mol
Glucose value	GV	$GV = \frac{(GE \times \text{mol. wt. of glucose})}{\text{Mol. wt. of compound}}$	g/g
Conversion of GV into PV	–	$PV = GV \times 0.88$	g/g

Oxygen requirement factor (ORF), carbon dioxide production factor (CPF), hydrogen requirement factor (HRF) and energy requirement factor (ERF), respectively, are the other measures used for oxygen consumed, CO_2 produced, $NADH_2$ and ATP requirement.

Based on Penning de Vries et al. (1974) and McDermitt and Loomis (1981).

5.3 Production Value of Individual Fatty Acid and Oil

Production values characterizing the conversion of glucose into commonly occurring fatty acids in plants estimated previously (Mitra and Bhatia 1979) are given in Table 2.

Since palmitic acid (16:0) is a common constituent of most vegetable oils, the glucose requirement of each fatty acid relative to it is given for comparison. The PV decreases with increase in chain length of the fatty acid. In general, the unsaturated fatty acids have low PVs in comparison to saturated fatty acids of similar chain length.

Based on the standard fatty acid composition, PV of mustard oil is estimated to be about 0.295. In terms of glucose ($1/PV \times 100$), it means that about 339 g of glucose would yield 100 g of oil. However, the PV will change with alterations in the fatty acid composition.

Table 2. Production cost and energy values of individual fatty acids excluding the cost of glucose uptake (Mitra and Bhatia 1979)

Name	No. of carbon atoms:double bond	PV	Percent glucose requirement relative to palmitic acid to synthesize 1 g product
Saturated			
Lauric acid	12:0	0.361	94
Myristic acid	14:0	0.349	97
Palmitic acid	16:0	0.340	100
Stearic acid	18:0	0.333	102
Arachidic acid	20:0	0.328	104
Behenic acid	22:0	0.323	105
Lignoceric acid	24:0	0.320	106
Unsaturated			
Palmitoleic acid	16:1	0.327	104
Ricinoleic acid	18:1	0.333	102
Oleic acid	18:1	0.323	105
Linoleic acid	18:2	0.311	109
Linolenic acid	18:3	0.300	113
Eicosenoic acid	20:1	0.318	107
Erucic acid	22:1	0.313	109
Nervonic acid	24:1	0.309	110

5.4 Seed Biomass Productivity

Like the PV for a chemical end product, the PV for seed biomass can be estimated based on its chemical composition. Sinclair and De Wit (1975) called it seed biomass productivity and defined it as gram of seed biomass produced per gram of photosynthate. Precise estimates of seed biomass productivity can be made for each genotype specifying its chemical composition. Based on proximate analysis seed biomass productivity for rape seed was estimated to be 0.399. In other words, about 251 g (1/0.399 × 100) of glucose would yield 100 g seed.

5.5 Bioenergetic Cost of Increasing Yield by Improving the Harvest Index

The energy content of the seed and crop residue is dependent upon their chemical composition. While the chemical composition and energy content of the seed are subject to considerable variation, the chemical composition and energy content of the crop residue is nearly constant. As shown by Sinha et al. (1982), the energy content of seed and crop residue of mustard on dry weight basis is about 25.7 and 13.9 MJ/Kg respectively. Thus, the seed energy content is about 85% higher than that of the crop residue. In cereals, this difference is only about 10%. Improvement in harvest index (Donald and Hamblin 1976) has contributed a great deal towards increasing the productivity of cereals (Bhatia et al. 1981). Harvest index (HI) based on biomass in *Brassica* oilseeds is around 20%; when calculated taking energy content into consideration it ranged between 39-45% in *B. juncea* and *B. campestris* (Sinha et al. 1982). This is nearly equal to that of high-yielding cereals, and explains why it has not been possible to significantly improve the harvest index in *Brassica* oilseeds by breeding.

Increase in HI from 25 to 40% in mustard would enhance energy harvest by about 10% (Table 3).

Table 3. Alternatives for increasing grain yield of mustard

	Biological yield (BY) Kg/ha	Grain yield (GY) Kg/ha	Harvest index (HI) %	Nitrogen requirement for seed biomass Kg/ha	Energy harvest MJ/ha	% increase in energy harvest
Baseline	8000	2000	25	73.6	136,000	
A	8000	3200	40	117.7	150,000	10.58
B	12800	3200	25	117.7	217,000	59.55
C	10667	3200	30	117.7	187,000	37.5
D	16000	3200	20	117.7	262,400	92.9

A BY remains constant, HI and GY increase
B HI remains constant, BY and GY increase
C BY, GY and HI increase
D BY and GY increase while HI decrease

Source: Breeding alternatives : Donald and Hamblin (1976)
Computational method : Bhatia et al. (1981)

Other options for increasing seed yield are also examined in Table 3. The assimilate requirement to achieve the seed yield of 3200 Kg/ha by the alternatives B, C and D are much greater. With higher seed yield, the nitrogen (N) requirement of the crop would also increase from 74 to 118 Kg/ha. *Brassica* crops are responsive to N fertilizers and if lodging can be prevented, mustard is reported to respond up to 160 Kg/ha (Singh and Chauhan 1982).

5.6 Increasing the Oil Content

As discussed earlier, the major constituents of the seed in *Brassica* crops are oil, protein and carbohydrates. An increase in oil content of these seeds must obviously be at the expense of either carbohydrates or protein, the former being preferable; but the energy requirement is greater when the oil content is increased at the expense of carbohydrates rather than of protein (Mitra and Bhatia 1979). An increase of 5% of oil in the seed entails an enhancement in photosynthate requirement by 4.6% (Table 4).

Table 4. Energy cost of increasing seed oil concentration in mustard at the cost of either carbohydrate or protein or both

Component	Amount g/100 g seed	PV	Equivalent glucose required (g)	% increase in photosynthate demand over standard cultivar
1. Standard cultivar				
Carbohydrate (CHO)	30	0.83	36.14	0
Protein	26	0.40	65.00	
Oil	40	0.295	135.59	
Minerals	4	—	—	
		Total:	236.73	
2. Cultivar with 5% more oil and 5% less carbohydrate				
CHO	25	0.83	30.12	
Protein	26	0.40	65.00	4.6
Oil	45	0.295	152.54	
Minerals	4	—	—	
		Total:	247.66	
3. Cultivar with 5% more oil and 5% less protein				
CHO	30	0.83	36.14	
Protein	21	0.40	52.5	1.8
Oil	45	0.295	152.54	
Minerals	4	—	—	
		Total:	241.18	
4. Cultivar with 5% more oil and 2.5% less protein and carbohydrate				
CHO	27.5	0.83	33.13	
Protein	23.5	0.40	58.75	3.2
Oil	45	0.295	152.54	
Minerals	4	—	—	
		Total:	244.42	

5.7 Changing the Composition of Fatty Acids

Based on the PVs of individual fatty acids given in Table 2, the cost of reducing erucic acid was estimated (Table 5). The calculations show that reduction of erucic acid content in rapeseed oil had no additional energetic cost (Table 5). Seed yields of low erucic acid cvs. Midas and Torch were equal to their respective normal erucic acid check cultivars.

Table 5. Seed composition, yield and production value of oil of low and normal erucic acid cultivars (Mitra and Bhatia 1979)

Cultivar/character	Rape		Turnip rape	
	Target	Midas	Echo	Torch
% oil	42.7	42.7	39.4	38.5
% protein	42.7	39.3	41.4	40.4
% erucic acid	43.2	0.4	27.7	0.1
% eicosenoic acid	13.8	2.9	12.9	0.9
% oleic acid	15.6	55.1	25.4	56.2
% linolenic acid	7.5	10.5	9.6	12.3
Yield (Kg/ha)	1610.0	1740.0	1420.0	1410.0
PV of oil	0.2929	0.2953	0.2941	0.2960
1/PV × 100 (g. glucose)	341	339	340	338

Fatty acid composition from Downey, Stringham, McGregor (1975) and yield data from Downey (1975).

5.8 Effect of Biotic and Abiotic Stresses

Brassica crops are affected by several biotic and abiotic stresses. Resistance or tolerance to various stress factors is an important component of the breeding programmes. The mechanism of resistance to biotic stresses could either be constitutive or induced after the attack. Different secondary plant products such as phenolics, glycosides and alkaloids have been implicated in plant-pest relationships. Such chemicals are synthesized and maintained in adequate amounts in plants so as to confer some protection from insects. Active resistance is due to activation of defence mechanisms in the host in response to infection. It is thus obvious that the host plant expends its resources to repel insects and restrict pathogens. The production cost of some such chemicals was estimated earlier (Mitra and Bhatia 1982).

Plants respond to abiotic stresses such as moisture and temperature by producing a wide range of biochemicals and by altering physiological reactions. Osmotic stress results in accumulation of small molecules such as glycine, betaine, proline, etc. The production of these osmoprotective molecules entails the withdrawal of carbon and nitrogen from the metabolic pool. In general, abiotic stresses are known to reduce yield as well as oil percentage. Under saline and sodic conditions oil content of mustard genotypes was reduced from 40.5 to 35.7% (Anonymous 1988).

5.9 Effect of Higher Temperature on Yield and Oil Percentage

Basically, *Brassica* oilseeds are crops of the temperate region. However, due to shortages in edible oil production, mustard cultivation is being extended to non-traditional areas. With increase in temperature, both growth and maintenance respiration are increased and consequently less of the assimilates are available for seed biomass. This is reflected in lower oil percentage as is evident from the data in Table 6 for standard cultivars grown at Junagadh and Anand in the western zone compared to Ludhiana, Sumerpur and Durgapura in the northwestern zone and at Pant Nagar and Kanpur in the central zone of India.

Table 6. Mean oil percentage in mustard cultivars grown in north-west, central and western zones in co-ordinated varietal trials (irrigated) (Anonymous 1988)

Cultivar	Northwest	Central	Western
Kranti	40.33	41.12	33.65
Krishna	39.94	42.30	33.10
Varuna	40.67	41.47	33.75

High temperature is also known to have negative effect on pod and seed growth, and also duration of pod filling in *B. napus* through direct effect on carbon balance of pod (Whitfield 1992). Kawanabe (1979) examined 35 oil-yielding plant species grown in temperate, subtropical and tropical regions in relation to the fatty acid composition of their oil. Tropical woody plants such as palms show a very high percentage of saturated fatty acids. Among the herbaceous species, crops grown in warm climates have a higher percentage of saturated fatty acids compared to the crops grown in cool regions. Kawanabe (1979) reports that seeds of *Brassica* species are an exception to this.

5.10 Other Negative Associations

Various negative correlations are observed in different *Brassica* crops between:

1. Number and mass of seeds.
2. Oil concentration and seed yield.
3. Protein and oil concentrations in seed.
4. Yield and crop duration.

It is difficult to visualize a genetic basis for such associations. In fact, they represent the competition for the same limited resources. If a large number of seeds are to be filled, the seed mass must be lower. When the seed mass is high, the available resources would not be sufficient for all the fertilized embryos to develop. We attribute these negative associations to limitation of both energy and nitrogen resources.

5.11 Conclusions

The above discussion indicates that most of the desired breeding goals have additional assimilate costs. However, the bioenergetic constraints pointed out should not be construed to imply that a simultaneous improvement of two or more parameters competing for energy resources in the plants is not possible. This would be true only when the available energy remains limited. Researches in breeding and agronomy are directed to enhancing the energy flow through cropping systems. Gains in overall energy resources have been made in the past, and will continue to be made in the future. Plant breeders can exercise the option within the limits of available energy to elevate productivity, to combat stresses, or to improve the end use quality (oil percentage or fatty acid composition). Yield levels under irrigation and good management will have to be enhanced first by selecting genotypes responsive to higher levels of nitrogenous fertilizers and capable of high yield at increased population densities. The genotypes should also have efficient translocation of assimilates from source to sink.

References

Anonymous (1988) Annual Progress Report of 33rd Annual Rabi Oilseeds Workshop, August 10-13, 1988. Dir Oilseeds Res Rajendranagar, Hyderabad-500 030

Bazzaz FH, Chairiell NR, Coley PD, Pitelka LF (1987) Allocating resources to reproduction and defence. Biosci 37: 58–67

Bhatia CR, Mitra R (1988) Bioenergetic considerations in the genetic improvement of crop plants. In: Biotechnol Trop Crop Improv Proc Int Biotechnol Workshop, 12-15 Jan 1987, ICRISAT Cent, Patancheru, Andhra Pradesh - 502 324, India

Bhatia CR, Mitra R, Rabson R (1981) Bioenergetic and energy constraints in increasing cereal productivity. Agric Syst 7: 105–111

Donald CM, Hamblin J (1976) The biological yield and harvest index of cereals as agronomic and plant breeding criteria. Adv Agron 28: 361–405

Downey RK (1975) Breeding rapeseed for the markets of today and tomarrow. In: Proc 8th Ann Meeting, Rapeseed Assoc Can Publ 38, pp 65–68

Downey RK, Stringham GR, McGregor DI (1975) Breeding rapeseed and mustard crops. Oilseed Pulse crops in Western Canada—a symposium, Winnipeg, Man, pp 157–183

Kawanabe S (1979) Fatty acid composition and iodine values of seeds of oil and fat crops in relation to climatic conditions. JPH J Trop Agric 23: 11–20

McDermitt DK, Loomis RS (1981) Elemental composition of biomass and its relation to energy content, growth efficiency, and growth yield. Ann Bot 48: 275–290

Mitra R, Bhatia CR (1979) Bioenergetic considerations in the improvement of oil content and quality in oilseed crops. Theor Appl Genet 54: 41–47

Mitra R, Bhatia CR (1982) Bioenergetic considerations in breeding for insect and pathogen resistance in plants. Euphytica 31: 429–437

Mitra R, Bhatia CR, Rabson R (1979) Bioenergetic cost of altering the amino acid composition of cereal grains. Cereal Chem 56: 249–252

Mooney HA (1972) The carbon balance of plants. Rev Ecol Sys 2: 315–346

Penning de Vries FWT, Brunsting AHM, Van Laar HH (1974) Products, requirements and efficiency of biosynthesis: a quantitative approach. J Theor Biol 45: 339–377

Rudolf N (1971) Szerves anyagok felhalomozodasanak nehany trovenyszerusege a kelaszos novenyekben. Agrartud 30: 229–234

Sinclair TR, De Wit CT (1975) Photosynthate and nitrogen requirements of seed production by various crops. Science 189: 565–567

Singh HG, Chauhan YS (1982) Factors limiting rapeseed-mustard production in north-western part of India. In: Bhargava SC, Jaiswal PL, and Gupta RS (eds.). Research and development strategies for oilseed production in India, Indian Counc Agric Res New Delhi, pp 77–84

Sinha SK, Bhargava SC, Goel A (1982) Energy as the basis of harvest index. J Agric Sci Camb 99: 237–238

Whitfield DM (1992) Effect of temperature and ageing on CO_2 exchange of pods of oilseed rape (*Brassica napus*). Field Crop Res 28: 271–280

CHAPTER 6

Oil and Meal Quality

K.L. Ahuja and *Shashi K. Banga*

6.1 Introduction

Rapeseed and mustard oil is primarily used for edible purposes, while the defatted meal is utilized as animal feed. Originally, rapeseed-mustard oil was considered inferior in quality to most other vegetable oils as it contained very high amounts of undesirable long chain fatty acids like eicosenoic acid (about 10%) and erucic acid (about 50%). The utilization of seed meal as a protein supplement was also limited due to the presence of glucosinolates. All these constituents were identified as the cause of some nutritional disorders and toxicity in animals and birds. Although comparable results are not available in humans, it was considered necessary by many countries to put a statutory limit on the intake of erucic acid and glucosinolates. The present status of rapeseed-mustard as the third most important source of vegetable oils is attributable to the success of plant breeders in developing low glucosinolates and zero erucic acid varieties. Quality-wise the improved rapeseed oil is now equivalent to peanut and olive oil. From the nutritional, cooking and stability points of view, possibilities exist for further alterations in the fatty acid composition in favour of increased levels of oleic and linoleic acids simultaneously with a reduction in linolenic acid content.

6.2 Oil Composition and Fatty Acid Biosynthesis

Fatty acids are long chain single carboxyl group containing organic acids, mostly of even carbon numbers, and are the building blocks of the majority of lipids. In cells or tissues they occur either saturated or unsaturated, containing 1, 2, 3 or more double bonds, mostly methylene interrupted, although conjugated fatty acids are also known. All saturated acids up to the chain length of ten carbon atoms are liquid and volatile, while higher saturated acids are solid and their melting point increases with increase in chain length. Unsaturated fatty acids are liquid at room temperature.

An important characteristic of rapeseed-mustard oil is the presence of erucic (*cis*-1, 3- decasenoic; 22:1 *n*-9) acid. This fatty acid is associated primarily with plants of the genus *Brassica* and some other plants such as lupins and meadowfoam, and constitutes about 50% of the total fatty acids. Genotypes, especially in *B. juncea*, are available where the erucic acid constitutes about 60-65% of the total fatty acids. These genotypes are important from the industrial point of view. The poly-unsaturated fatty acids, namely linoleic (*cis*, *cis*-9, 12-octadecadienoic; 18:2 *n*-6) and α-linolenic acid (*cis*, *cis*, *cis*-9, 12, 15-octadecatrienoic; 18:3 *n*-3) are also present in significant amount (20-25%) and confer liquidity on the oil. A higher amount of linolenic acid is as much of interest to nutritionists as is the reduced level of linolenic acid to millers and consumers. The nutritional role of α-linolenic acid is much debated, but recently a dietary need of 18:3 *n*-3 of 0.5% calories has been recommended (Holman 1981), as 18:3 is postulated to be a precursor of 20:5 n-3 in humans (Sanders and Younger 1981). Linoleic acid is also a surplus-cholesterol scavenger (Wolfram 1976). Of the saturated fatty acids, palmitic acid (16:0) and stearic acid (18:0) are present in very low quantities, totalling about 5%. They have been implicated in increasing thrombotic tendency in the blood platelets. The main biosynthetic pathway of the fatty acid biosynthesis is given in Fig. 1.

C 16 : 0 Palmitic acid	→	C 18 : 0 Stearic acid				
		↓				
		C 18 : 1 Oleic acid	→	C 20 : 1 Eicosenoic acid	→	C 22 : 1 Erucic acid
		↓				
		C 18 : 2 Linoleic acid				
		↓				
		C 18 : 3 Linolenic acid				

Fig. 1. Biosynthetic pathway to main fatty acids (Jönsson 1977)

As both the undesirable erucic acid and linolenic acid are the end products of a biosynthetic pathway, a reduction/elimination of these fatty acids is possible if the genetic block is achieved in the steps controlling the synthesis of erucic acid from oleic acid and linolenic acid from linoleic acid. The oleic acid has negative correlation with linoleic and linolenic acid on the one hand and erucic acid and eicosenoic acid on the other (Ahuja et al. 1984; Ahuja et al. 1989; Chen et al. 1977; Koyama et al. 1978). Because of the evident interdependence in the progenetic substrate, zero-erucic acid oil is marked by increased oleic acid (18:1), linoleic acid (18:2) and linolenic acid (18:3) contents.

6.3 Erucic Acid

6.3.1 Antinutritional Effects

Feeding experiments with animals have demonstrated that rapeseed oil (RSO) rich in erucic acid interferes in myocardial conductance and the peripheral vascular system, decreases survival time, shortens coagulation time and increases blood cholesterol (Renarid and McGregor 1976). Long-term feeding of a RSO diet altered mitochondrial morphology and caused cardiac degeneration or necrosis without lipidosis near the site of degeneration. The erucic acid accumulation in these lesions indicated that erucic acid is not metabolized efficiently. Monkeys receiving 20-25% mustard or rapeseed oil showed cardiac lipidosis with accumulation of erucic acid, and fibrosis (Gopalan et al. 1974; Ackman et al. 1977). Contrarily, some nutritionists believe that the imbalance between saturated and unsaturated fatty acids and not the erucic acid content is an important factor responsible for the deleterious effects caused by rapeseed oil (Hopkins et al. 1955; Walker et al. 1970). Human subjects from the countries usually not consuming high C 22:1 diet also have widespread but mild cardiac lipidosis due to high fat diet, irrespective of its nature (cf. Barlow and Duthie 1985). Their heart tissues also showed 1-5% C 22:1 fatty acid deposition. Hearts from inhabitants of India consuming C 22:1 fatty acid-rich mustard oil showed 1-10% C 22:1 fatty acid but had virtually no cardiac lipidosis (Shenolikar 1980). Even the subprimates fed on erucic fatty acid at 100 times more than that consumed by man (on a body weight basis) showed no long-term lesions in the myocardium (Barlow and Duthie 1985). Thus there is a need to study the effect of erucic acid-rich rapeseed oil on humans, as the experiments conducted with the animals may not be exactly relevant to humans.

6.3.2 Analytical Techniques

The fatty acid composition of the oil is normally determined by gas chromatography (Appelqvist 1968a,b) or paper chromatography (Thies 1971). The gas chromatographic method is useful for analyzing both large seed samples or even a single seed by using the half-seed technique. The results in both cases are comparable, reliable and reproducible. In the half-seed method, the analysis of the fatty acid content is made on the outer cotyledon of the freshly germinated seeds. The remainder of the embryo can give rise to a complete plant. The half-seed technique has proved an important milestone in the development of erucic acid-free varieties, as the fatty acid composition is determined by the genotype of the embryo in *Brassica* (Downey and Harvey 1963; Dorrell and Downey 1964). Analysis of the half seed ensures utilization of a known genotype in breeding programmes. For large seed lots, the turbidity method is a quick and rapid technique (McGregor 1977).

6.3.3 Genetic Regulation

Plants having a genetic block in the biosynthetic pathway towards eicosenoic acid and erucic acid were initially identified for the first time in Canada, both in summer rape (Stefansson et al. 1961) and summer turnip rape (Downey 1964). Later, the genotypes for zero-erucic acid were identified in *Brassica juncea* (Kirk and Oram 1981) in Australia. Genetic studies in rapeseed have demonstrated that the erucic acid content in rapeseed is controlled by multiple alleles. Homozygosity levels between 5-10, 10-35 and about 35% erucic acid have been reported (Jönsson 1977). These are controlled by alleles in one, one or two, and two loci respectively. Anand and Downey (1981) identified five genes in *B. napus* and designated them as e, Ea, Eb, Ec and Ed. They act in additive manner and result in erucic acid levels of < 1, 10, 15, 30 and 35% respectively. No direct evidence was obtained by these authors which could suggest that the designated genes are true alleles. Interspecific hybridization with zero erucic acid *B. campestris* suggested that the alleles Ea and Ed are located on C genome chromosomes. Occurrence of a single gene controlling high erucic acid content was later confirmed by Chen et al. (1988). Two genes showing no dominance and acting in an additive fashion have been implicated for erucic acid content in *B. juncea* (Kirk and Hurlstone 1983). Similar genetic control has been suggested in *B. carinata* (Fernandez-Escobar et al. 1988). Single non-dominant gene control in summer turnip rape was demonstrated (Dorrell and Downey 1964), although some other researchers suggest partial dominance or even overdominance (Møller et al. 1985). Despite a large number of attempts, the erucic acid heredity is still not very clear (Jönsson 1977).

6.4 Oleic, Linoleic and Linolenic Acids

Fatty acids, as previously indicated, are formed by a stepwise biosynthetic pathway in which oleic acid (18:1) either undergoes decreasing saturation to form linoleic acid and then linolenic acid or there is a further chain elongation to form eicosenoic acid and then erucic acid. Studies have revealed that oleic acid is the direct metabolic precursor of erucic acid (Bartkowiak-Broda and Krzymanski 1983) and is also utilized for linoleic acid and linolenic acid synthesis. On the other hand, the lack of a negative correlation between oleic acid and the polyunsaturated fatty acids means that the linoleic acid synthesis from oleic acid was independent of the erucic acid pathway in rapeseed. The pathway for the desaturation of oleic acid to linolenic acid could involve phospholipids in the endoplasmic reticulum and glycolipids that are exclusively localized in the plastids (Diepenbrock 1983). Hence, there are at least two possible mechanisms for linolenic acid synthesis during seed development. The metabolic end products of these mechanisms are triacylglycerol (TG), the major storage glycerolipid; and monogalactosyl diacylglycerol (MGDG), the major structural glycolipid in plastids. The presence of two different pathways for the conversion of oleic to linoleic and linoleic to linolenic acid suggests the possibility of genetic alternation of one without immediately affecting the other two fatty acids. In mutation experiments, Rakow (1973) obtained mutants with low and high linolenic acid contents respectively. The linoleic acid content was unchanged in both low and high linolenic acid mutants. A reduced level of linolenic acid is the ultimate goal for plant breeders as it easily becomes oxidized, resulting in taste impairment. Increased localization of the polyunsaturated fatty acids in the middle position of the triglycerides will also make them more stable (Axtell 1981). Hydrogenation of linolenic acid is uneconomical. An increased level of linoleic acid, a doubly unsaturated and essential fatty acid, is desirable from the nutritional point of view. Oleic acid, being thermostable, is desirable for cooking oil.

6.4.1 Analytical Procedures

The frequency of occurrence of the genotypes with decreased linolenic acid or increased linoleic acid is low. Therefore rapid and simple analytical methods are essential. Gas chromatographic determination of the individual fatty acids is time-consuming. A modification of a photometric method for the measurement of polyunsaturated fatty acids after alkaline isomerization is a simple and rapid method. Refinement of this method by the use of ethylene-glycol dimethyl ether instead of glycerine or glycerol not only shortens the time required for isomerization reaction, but also allows the use of low temperatures (Korver et al. 1967). With the photometric method, only the ratio between linolenic and linoleic acid is obtained. This is not disadvantageous for selection, as either lower linolenic or higher linoleic or both will result in the lower ratios and these are all desirable changes. A rapid and sensitive spot test for linolenic acid level in rapeseed is also usable. This involves the determination of linolenic acid in a half seed using 2-thiobarbituric acid with the oxidation products of linolenic acid (McGregor 1974). A test paper method for a specific quantitative estimation of linolenic acid (Thies and Nitsch 1974) is also available.

6.4.2 Genetic Regulation

Genetic studies have revealed that the concentration of linolenic acid in triacylglycerol (TG) was determined by nuclear and cytoplasmic gene interaction, whereas linolenic acid in monogalactosyl diacylglycerol was determined by cytoplasmic factors (Diepenbrock and Wilson 1987). Contrarily, Pleines and Friedt (1989) observed no reciprocal differences in F_2 thus ruling out the role of cytoplasmic factors in the inheritance of linolenic acid. The possible role of cytoplasmic and nuclear gene interaction for the inheritance of linolenic acid in TG significantly complicates the task to eliminate linolenic acid through plant breeding. Thus the characterization of linolenic acid in individual glycerolipids in rapeseed-mustard oil is very important for achieving significant breeding progress. The success in achieving mutants with very low linolenic acid level (Rakow 1973; Röbbelen and Nitsch 1975) has been significant.

Using induced mutants of *B. napus* cv. Oro, Roy and Tarr (1986) were able to develop near zero (1.60 to 1.87) linolenic acid lines of rapeseed (*B. napus*) along with moderately high (30%) linoleic acid (18:2) content from a complex cross. Similarly, some Indian mustard derivatives ($2n = 36$) of the cross *B. juncea* cv. RLM 198 × *B. napus* cv. Oro have linolenic acid up to 5% (Banga, unpubl.). Genotypes having very high oleic acid (73-79%) levels are available in *B. napus* (Pleines and Friedt 1988).

6.5 Glucosinolates

Defatted *Brassica* oilseed meal contains about 40% protein with a well-balanced amino acid composition (Miller et al. 1962). Its use in animal and human nutrition is, however, limited by its glucosinolate content. Although concentrated in the seed up to 5%, glucosinolates are also distributed throughout the vegetative tissue (Kajaer and Olesen-Larsen 1973). Of over 90 glucosinolates identified, the major seed glucosinolates are gluconapin (3-butenyl), glucobrassicanapin (4-pentenyl), progoitrin (2-hydroxy-3-butenyl), napoleiferin (2-hydroxy-4-pentenyl) and sinigrin (allyl). The biosynthetic studies of glucosinolates have revealed that all are derived from amino acids and most of them are the product of a common biosynthetic pathway. Amino acids pass through a sequence of consecutive oxidation, accompanied by decarboxylation *en route*, followed by sulphuration, glucosylation and finally sulphonation to yield glucosinolates (Kajaer and Olesen Larsen 1973). About 50% of the sulphur and more than 50% of the methionine of the cell is incorporated into glucosinolates in *Brassica* cultivars containing an average amount of these compounds (Josefsson 1970, 1972). The directed sulphate utilization is also confirmed, as the proportion of progoitrin to gluconapin is about 3:1 in the normal but 8:1 in low sulphur-grown rapeseed, showing that progoitrin synthesis was much more affected. Apparently the two groups of glucosinolates are synthesized from the same sulphur-containing precursors and the introduction of the hydroxy group into the precursor molecule of progoitrin is a relatively late step in the biosynthesis.

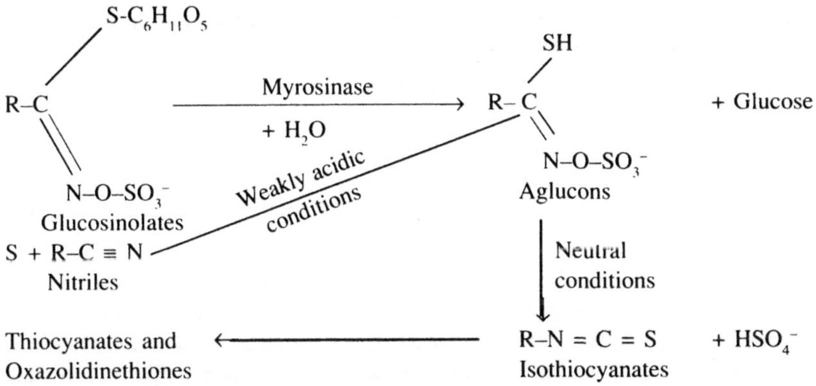

Fig. 2. The general structure of glucosinolates, and of the products formed by enzymatic hydrolysis

Following cellular disruption (Fig. 2) glucosinolates are broken down by the enzyme thioglucoside glucohydrolase (EC 3.2.3.1), commonly known as myrosinase, to yield sulphate, glucose and a range of other aglycone products, This enzyme is specifically present in idioblasts (cf. Röbbelen 1980) and is released when plant tissue is crushed. Sinigrin breaks down to allylisothiocyanate, which is the major flavour component in mustard. Progoitrin, although present in low quantities, has great significance because of the goitrogenic properties of its breakdown products. For further details of the chemistry and biosynthesis of glucosinolates, readers are referred to the comprehensive reviews on the subject by Kajaer (1976) and Underhill (1980).

6.5.1 Anti-Nutritional Effects

Contrary to the general belief, intact glucosinolates (without myrosinases) are capable of causing significant anti-nutritional and toxic effects (Bille et al. 1983). It has been experimentally demonstrated that the amount of glucosinolates in meals from low glucosinolate *B. napus* cvs. Tower and Bronowski which can be safely included in the diet of non-ruminants may be increased substantially over the recommended dose of high glucosinolate rapeseed meal (Clandinin and Robblee 1981). The acceptability of the feed concentrates to the diary cattle may be negatively influenced by increasing the proportion of high glucosinolate rapeseed meal (HGRSM) beyond 15%. The low glucosinolate rapeseed meal (LGRSM) was tolerable up to at least 20%. LGRSM type meal of *Brassica campestris* cv. Candle is acceptable even at higher levels (Thomke 1980). However, the inclusion of 34% of LGRSM lowered feed intake and milk yield. The use of rapeseed meal from *B. napus* cv. Tower (0.98 mg/g glucosinolates) as a partial replacement of soybean diet for pregnant and lactating swine caused no apparent reduction in their reproductive performance (Lewis et al. 1978). Glucosinolates originally present in the seed are concentrated in the meal, where they may be up to 8% of seed weight in *Brassica napus* and 3 to 4% in *B. campestris*. This difference is one of the main reasons why, in many experiments, the higher level of meal from *B. campestris* could be fed to animals without adverse effect, as compared to *B. napus* meal. There are also qualitative differences in the glucosinolates present in the meal of these two species.

Cleavage products like isothiocyanates, oxazolidinethiones and nitriles resulting from hydrolysis of glucosinolates are very toxic. They may affect the thyroid gland (Bell et al. 1972) and cause reduced appetite and productivity even when animals are fed with comparatively small amounts of rapemeal. Contrarily, Dietz et al. (1987) reported that rapeseed diet (*B. campestris* var. *toria*) in which the only glucosinolate was 3-butenyl glucosinolate or its hydrolytic products, caused no toxic effects in rats fed on this diet for 4 weeks. The enzyme myrosinase in the meal can be destroyed by heat treatment, but some microorganisms present in the intestinal system of animals produce enzymes that may cause hydrolysis of the glucosinolates (Oginski et al. 1965). Experimental evidence suggests that continued intake of progoitrin affected liver and kidney weights (Vermorel et al. 1986). Sinigrin,when taken orally, had a goiterogenic effect in rats (Nishie and Daxenbichler 1980); this effect may be mediated by prop-2-enyl isothiocyanates (Langer and Stolc 1965), which is then metabolised to thiocyanates.

In a study in which human volunteers consumed glucosinolate (especially progoitrin) rich Brussels sprouts daily for 28 days (McMillan et al. 1986), no effect was observed on serum TSH, T3 or T4, suggesting the possibility that in the absence of myrosinase (inactivated by cooking) the extent of the breakdown of glucosinolates might be limited. Similarly, neither glucotropaeolin (benzylglucosinolate) nor glucobrassicin showed any deleterious effect when fed at levels of 1 g and 0.5 g/kg respectively (Vermorel et al. 1986).

6.5.2 Variation of Glucosinolate Content in Crop Brassicas

Generally, each plant species contains more than one type of glucosinolate. Sinigrin is a major component of *B. oleracea*, while it is completely absent in *B. campestris* and *B. napus* (Josefsson 1972). On the other hand, progoitrin, yielding specifically potent goitrogen, is completely absent in *B. nigra* and *B. juncea* (Namai et al. 1972) but is present in *B. napus* (Thies 1977). Thus *B. nigra*, the B-genome donor to *B. juncea*, carries the main factors for sinigrin synthesis (Hemingway et al. 1961). Progoitrin is chiefly controlled by the genes located on the C-genome of *B. oleracea*. *B. campestris* has gluconapin as a dominant glucosinolate (Josefsson 1972). This is accompanied by somewhat higher values for glucobrassicin. Contradictory results are available in *B. juncea* through the analysis of samples from different origins (Hemingway et al. 1961; Vaughan et al. 1963). Samples from several countries contain only or mainly sinigrin, but those from India and Pakistan contained gluconapin as the major component in combination with sinigrin (Table 1). Butenyl producing glucosinolates are dominant in *B. juncea* samples from the Indian subcontinent, while European or Chinese *B. juncea* stocks are rich in allyl isothiocyanates (Downey 1990).

Table 1. Comparison of glucosinolates in different *Brassica* species (Anand 1974)

Brassicaceae	mg/g Oil-free meal				
	All	But	Pent	Ozt	Mtb
1. Indian					
B. campestris					
Yellow sarson	0.00	15.20	0.02	0.00	0.00
Brown sarson	0.00	10.90	0.20	0.00	0.00
Toria	0.00	13.50	0.10	0.00	0.30
B. juncea	1.50	12.40	0.10	0.00	0.00
2. Canadian					
B. campestris	0.00	2.50	2.00	3.80	0.10
B. napus	0.00	3.70	1.00	15.90	0.00
B. juncea	7.80	0.00	0.70	0.00	0.00
3. Swedish					
B. campestris	0.00	2.20	0.00	1.20	0.00
B. napus	0.00	1.30	0.00	3.70	0.00

All = Allylisothiocyanate, But = 3-butenyl isothiocyanate, Pent = 4-pentenyl isothiocyanate, Ozt = 5-vinyloxazolidine 2 thione, and Mtb = 4-methyl thiobutyl isothiocyanate.

6.5.3 Analytical Techniques

Currently popular techniques are based on:

1. Analysis of hydrolysis products.
2. Analysis of total glucosinolate content.
3. The gas chromatographic/high performance liquid chromatographic method for analysis of individual glucosinolates.

A number of methods based on estimation of split products like glucose, bisulphate ion, oxazolidinethione and isothiocyanates are available (McGhee et al. 1965; Tookey 1973). A very simple and commonly followed method is the estimation of glucose in the crushed seeds (about ten seeds) by using a glucose test paper impregnated with the enzyme system glucose oxidase, peroxidase and a chromogen (Lein 1970). Because of the simplicity of this test, 200-300 samples can be evaluated in a single day by one person. Some modifications in this technique to overcome the colour inhibitory effect of certain substances obtained after autolysis have been suggested (Van Etten et al. 1974; McGregor and Downey 1975). The modifications are, however, tedious, and not commonly followed. The thymol procedure originally suggested by Brzezinski and Mendelewski (1984) and later modified by Tholen et al. (1989) is useful for screening a large number of samples. The property of palladium to form coloured complexes with glucosinolates has also been exploited (Thies 1983). Gravimetric or titrimetric measurement of the sulphate released can be utilized for analysis of the total glucosinolates (McGhee et al. 1965). Separation of glucosinolates by high performance liquid chromatography (HPLC) has also been described (Spinks et al. 1984; Sang and Truscott 1984; Palmer et al. 1987). Another promising technique (Underhill and Kirkland 1971), especially suited to analyze trace amounts of glucosinolates, is gas chromatographic analysis of trimethylsilyl derivatives. This technique was later refined by Pearson (1974) and Thies (1976). Determination of total sulphur content through X-ray fluorescence spectroscopy (X-RF) is an attractive alternative due to the repeatability and reproducibility of the technique (Schung and Haneklaus 1988). The results obtained with this method correlated closely with the conventional reference methods (r^2 = 96-99%). Summarizing, it is practical to screen large germplasm or the early breeding materials by using simple techniques like the tes tape method, but the final product must be evaluated with more precise methods.

6.5.4 Genetic Regulation

The discovery of a Polish summer rape (*B. napus*) cultivar, Bronowski, having a very low level of glucosinolates (12 μmol/g defatted dry matter), gave a strong impetus to the attempts for the genetic elimination of potentially toxic glucosinolates from rapeseed meal (Josefsson and Appelqvist 1968). Further selection in this cultivar led to breeding lines containing only traces of glucosinolates. These lines have been used as gene sources for the production of low-glucosinolate materials in both summer and winter rape as well as in winter turnip rape. Later, genetic stocks of low glucosinolate summer turnip rape were also discovered (Downey et al. 1969; Lein 1970). A breeding line of *B. campestris* ssp. *chinensis* has also been used as a source of low glucosinolates in breeding summer turnip rape at Svalöf. In *B. juncea* and *B. carinata* no naturally occurring source for substantially low glucosinolate levels is available, though some variation has been reported in the literature. Recently, however, a *B. juncea* strain with glucosinolate level of 4 to 8 μmol/g of oil-free meal has been identified in Saskatoon (cf. Downey 1990).

Earlier genetic analysis from the segregating populations of various crosses involving Bronowski indicated the involvement of as many as 11 recessive alleles

(Kondra and Stefansson 1970) conditioning the low values of 3-butenyl-4-pentenyl- and 2-hydroxy-3-butenyl glucosinolates. Inheritance was determined by the maternal genotype rather than the embryo genotype. While higher concentrations of glucobrassicanapin were found to be overdominant, the high concentration of gluconapin was partially dominant to the lower concentration, and the high concentration of progoitrin was partially dominant to zero progoitrin. Josefsson (1971; 1973) compared the biosynthesis of glucosinolates in cv. Bronowski with cv. Regina containing an average glucosinolate level. The results indicated that a metabolic block in the biosynthesis of the main glucosinolates was located after the formation of 2-amino-6 (methylthio) caproic acid. This prevents the biosynthesis of 3-butenyl-glucosinolate in Bronowski which is situated before the synthesis of the intermediate 5-methyl thiopentanal oxime, as well as a block in the hydroxylation step.

6.6 Phenolic Constituents

With the development of double low (low erucic acid and glucosinolate content) varieties in *B. campestris* and *B. napus*, other anti-nutritional factors like phenolic components deserve attention. They contribute to the sharp and bitter flavour of the seed flour and may also produce undesirable dark-coloured complexes on oxidation (Fenton et al. 1980). Principal phenolic compounds have been identified as p-hydroxy benzoic trans-cinnamic and sinapic acids with small quantities of p-coumaric, ferulic, caffeic and chlorogenic acids (Kozolowska et al. 1983). Rapeseed contains approximately 1% sinapine, which is the ester of sinapic acid and choline (Josefsson and Uppström 1976). These components have been implicated in the "fishy" taint of egg yolk from some brown-egg-laying hens fed on rapeseed meal (Butler et al. 1982). This is caused by the interaction of choline, a genetic factor in the hens which adversely affects the synthesis of trimethyl amine oxidase, and a substrate in the rapeseed meal which drastically reduces the activity of hepatic trimethyl amine oxidase (Vogt 1980). A relatively narrow range of variation is available for the sinapine content in *B. rapa* and *B. napus*. This is 0.49-0.66 and 0.65-0.90% respectively on a fat-free dry weight basis (Kerber and Buchloh 1982). Mean values of 0.75 and 0.90% for *B. campestris* and *B. juncea* respectively have also been reported (Prasad and Rao 1979). *Eruca sativa*, on the other hand, is relatively sinapine-free (Kerber and Buchloh 1982).

No correlation has been observed between reduced glucosinolate content and lower level of sinapine in *B. napus* (Curtis et al. 1978). As a number of techniques for determination of sinapine content are available (Austin and Wolff 1968; Josefsson and Uppström 1976; Ismail and Eskin 1979; Fenwick 1979; Uppström and Johansson 1985), it is now possible to screen and breed varieties with reduced contents of sinapine (Mueller et al. 1978). Among various varieties of summer type *B. campestris*, Sv 83-36505 and Sv 83-36531 have a lower sinapine value (11 mg sinapine thiocyanate/g meal) than is observed in some commercial cultivars like Span (15 mg sinapine thiocyanate) (Uppström and Johansson 1985).

6.7 Phytic Acid

Relatively high amount of phytic acid in rapeseed-mustard meal (4%) is known to reduce the retention of zinc, phosphorus, calcium and magnesium when rapeseed-mustard meal was fed to growing broiler chicks (Nwokolo and Bragg 1977). This results from the strong chelating action of phytic acid, which can bind mono- and divalent metal ions to form complex phytate (Erdman 1979). The best way to solve this problem is to breed varieties low in phytic acid. Although the analytical procedure for phytic acid estimation is available (Uppström and Svensson 1980), phytic acid content is mostly influenced by environmental factors such as the availability of phosphorus in the soil. This will complicate the elimination of phytic acid by plant breeding.

6.8 Seed Colour

Black and dark brown seed coat colour is of normal occurrence in rapeseed-mustard. However, it has been possible to isolate spontaneous yellow-seeded mutants in Indian mustard and turnip rape. Although naturally occurring yellow-seeded variants are not available in *B. napus*, efforts to produce such lines are underway in different countries (Jönsson 1975; Chen et al. 1988; Chen and Heneen 1989). This is being done through the resynthesis of *B. napus* from the yellow seeded diploid parents, namely *B. oleracea* and *B. campestris*. Another approach is to make use of the C^Y genome of yellow-seeded *B. carinata*. However, the sterility of hybrids between these two species must be circumvented.

A change-over to yellow-seeded varieties entails many advantages. These include better meal quality with higher energy content due to low crude fibre. Discolouring of oil and lecithin is avoided, and light-coloured end products are obtained even without dehulling of seeds. Yellow-seeded lines normally have an extra 1-2% oil content. This is because the yellow seeds possess a thinner seed coat and higher oil content in the embryo. In addition, it is easier to determine the degree of ripeness in yellow seeds, as the occurrence of chlorophyll is not masked by the dark seed coat colour. Due to its relatively simple inheritance (Dhillon et al. 1986), breeding manipulation of this trait is simple once the gene sources are available.

6.9 Crude Fibre

The crude fibre content is mainly cellulose and hemicellulose. It amounts to ca. 15% in the defatted seed meal. These are mainly concentrated in the hull, which accounts for 12-20% of the seed weight, depending on the variety and seed shape. The small size of *B. napus* has been associated with a large proportion of seed coat rich in fibre (Appelqvist 1972). High crude fibre content results in the overall reduction of the feed value, and consequently a lower metabolizable energy (ME) value for animals fed on rapeseed meal than on soybean meal (Clandinin and Robblee 1981). ME value of rapeseed-mustard meal can be improved by developing cultivars with reduced hull content with an increased content of available carbohydrates in the cotyledons. The simplest and best technique for this is the development of yellow-seeded varieties, as yellow seed coat colour is associated with reduced crude fibre content. This will also minimize lignin content in meal, which adversely affects the protein and amino acid digestability in non-ruminant animals.

6.10 Breeding Strategies

In line with the international consensus on nutritional standards for rapeseed-mustard quality, the breeding mandate now is the development of 00 (zero erucic acid and low glucosinolate) or 000 (zero erucic acid, low glucosinolate and yellow seed colour) varieties. The upper limit for the glucosinolate content is 10 μmol/g defatted dry matter. As is evident (Table 2), the desirable genes were initially identified in the varieties/strains with low agronomic potential. For improvement in erucic acid content, the half-seed technique was routinely applied in backcross programmes to transfer genes preventing the synthesis of erucic acid into the agronomically superior recurrent parents (Jönsson 1973a,b; Morice 1974). This ensures the speeding up of the breeding process, as the selection for seeds free from erucic acid can be made on the seeds from F_1 material. Frequently, it is rewarding to delay the first backcross to the F_2 generation as a cycle of recombination provides a possibility of breaking certain undesirable linkages as well as for improving yield and other agronomic traits (Röbbelen 1975, Röbbelen and Leitzke 1975).

Table 2. Original gene sources for desirable quality traits

Character	Species	Genotype	Reference
Low erucic acid	*B. campestris*	Polish cultivar	Downey (1964)
		Sv. Torpe Sv. Bele	Jönsson (1977)
	B. napus	Liho	Stefansson et al. (1961)
	B. juncea	Zem 1, Zem 2	Kirk and Oram (1981)
Low linolenic acid	*B. napus*	Mutants of Oro	Rakow (1973) Roy and Tarr (1986)
High linoleic acid	*B. napus*	Breeding material	Jönsson and Uppström (1986)
Low glucosinolates	*B. napus*	Bronowski	Josefsson and Appelqvist (1968)
	B. campestris	Sv. Polish	Downey et al. (1969)
Low sinapine	*B. campestris*	Sv. 83-36505 Sv. 83-36531	Uppström and Johansson (1985)

The maternal influence on the heredity of the glucosinolates had been a complicating factor in breeding for low glucosinolate content. This meant that the selection for low glucosinolate was restricted only to Bronowski cytoplasm despite some adverse effects associated with it. Intervarietal hybridization and backcross methods have been used for the improvement of this trait (Jönsson et al. 1975). The use of doubled haploids produced through microspore culture from F_1 or F_2 plants can expedite the breeding process (Lichter et al. 1988). For the simultaneous improvement of both these characters, crosses are made between genotypes exhibiting low erucic acid content and low glucosinolate content. Segregants in F_2 or later generations combining double low (00) trait can be repeatedly backcrossed to a superior variety. Another method utilized earlier involved the parallel deployment of the genes for low glucosinolate content and low erucic acid content in two different superior backgrounds through the backcross technique. This was followed by hybridization between these two restructured populations and selection for 00 segregants. Now that the 00 types are available in excellent genetic backgrounds, except for *B. juncea* and *B. carinata*, the simultaneous transfer of both these characters is relatively simple. Triple low types can be produced by hybridizing double low types with yellow-seeded donors. Breeding experiments have indicated that even one or two backcrosses with productive varieties are enough for improving the agronomic performance of mutants with improved polyenoic fatty acid pattern (Brunklaus-Jung and Röbbelen 1987). Highly productive 000 types can be suitably restructured by introducing the genes for low linolenic acid through a simple backcross technique. Directional selection for high linoleic acid has proved very effective. (Laakso et al. 1983; Laakso 1985; Laakso et al. 1986). Reciprocal recurrent selection, especially in *B. campestris*, will prove very effective in the simultaneous improvement of all these traits.

6.11 Conclusions

The successful development of highly productive double low varieties in *B. campestris* and *B. napus* has proved that it is possible to breed for both oil and meal quality for nutritional purposes without any yield penalty. There is a need to extend these breeding gains to *B. juncea* and *B. carinata*. New possibilities exist for further increasing the amount of linoleic acid to 40-50%, while pegging down linolenic acid to less than 3%. A higher amount of oleic acid (up to 70%) is a desirable goal for a good cooking oil. There is also scope for increasing palmitic acid content to more than 10% for margarine production. Rapeseed meal can be further improved by reducing the tannins, phytic acid and crude fibre. Improvements in protein fractional components, especially lysine content, in the meal are also likely.

References

Ackman RG, Eaton CA, Sipos JC, Loew FM, Hancock D (1977) Comparison of fatty acid from high levels of erucic acid of RSO and partially hydrogenated fish oil in non-human primate species in a short term exploratory study. Nutr Dieta 25: 170–185

Ahuja KL, Labana KS, Raheja RK, Badwal SS (1984) Oil content and fatty acids variation in mutants of *Brassica juncea* L. J Oilseeds Res 1: 71–75

Ahuja KL, Batta SK, Raheja RK, Labana KS, Gupta ML (1989) Oil content and fatty acid composition of promising Indian *Brassica campestris* L. (*toria*) genotypes. Qualitas Pl Fd Hu Nutr 39: 155–160

Anand IJ (1974) Mustard oil glucosides of the Indian Brassiceae. Plant Biochem J 1:26

Anand IJ, Downey RK (1981) A study of erucic acid alleles in digenomic rapeseed (*Brassica napus* L.). Can J Plant Sci 61: 199–203

Appelqvist LÅ (1968a) Rapid methods of lipid extraction and fatty acid ester preparation for seed and leaf tissue with special remarks on preventing the accumulation of lipid contaminants. Ark Kemi 28: 551–570

Appelqvist LÅ (1968b) Lipids in Cruciferae I. Fatty acid composition in seeds of some Svalöf varieties and strains of rape, turnip rape, white mustard and false flax. Acta Agr Scand 18: 3–21

Appelqvist LÅ (1972) Chemical constituents of rapeseed. In: Appelqvist LÅ, Ohlsson R. (eds.). Rapeseed cultivation, composition, processing and utilization. Elsevier, Amsterdam, pp. 123–173

Austin FL, Wolff IA (1968) Sinapine and related esters in seed meal of *Crambe abyssinica.* J Agric Fd Chem 16: 132–135

Axtell JD (1981) Breeding for improved feeding quality. In: Frey K.J. (ed.). Plant breeding II Iowa State Univ Press, pp 365–414

Barlow SM, Duthie IF (1985) Long-chain monoenes in the diet. In: Padley R.B., Podmore J. (eds.). The role of fats in human nutrition. Ellis Harwood, pp 132–145

Bartkowiak-Broda S, Krzymanski J (1983) Inheritance of C-18 fatty acid composition in seed oil of zero erucic winter rape *Brassica napus.* In: Proc 6th Int Rapeseed Conf, Paris, 1983, pp 477–482

Bell JM, Benjamin BR, Giovannetti PM (1972) Histopathology of thyroids and livers of rats and mice fed on diets containing *Brassica* glucosinolates. Can J Anim Sci 52: 395–406

Bille N, Eggum BO, Jacobsen I, Olseno O, Soerensen N (1983) Antinutritional and toxic effects in rats of individual glucosinolates (± myrosinases) added to a standard diet. 1. Effects on protein utilization and organ weights. Tierphysiol Tierer naehr Futtermittelkd 49: 195–210

Brunklaus Jung E, Röbbelen G (1987) Genetical and physiological investigations on mutants for polyenoic fatty acids in Rapeseed (*Brassica napus* L.). III. Breeding behaviour and performance. Plant Breed 98: 9–16

Brzezinski W, Mendelewski P (1984) Determination of total glucosinolate content in rapeseed meal with thymol reagent. Z Pflanzenzeucht 93: 177–183

Butler EJ, Pearson AW, Fenwick GR (1982) Problems which limit the use of rapeseed meal as a protein source in poultry diets. J Sci Fd Agric 33: 866–875

Chen BY, Heneen WK (1989) Resynthesized oilseed rape (*Brassica napus* L.) and the development of yellow seeded type. In: Proc XII Eucarpia Congr, Göttingen, 15: 11–16

Chen BY, Heneen WK, Jönsson R (1988) Independent inheritance of erucic acid content and flower colour in the C-genome of *Brassica napus* L. Plant Breed 100: 147–149

Chen C, Huang YC, Lu Hung S (1977) Inheritance of the erucic acid in rapeseed. K'O Hsuch Fe Chan Yuen K'an 5: 1029–1034 (Chem Abstr 92: 137507)

Clandinin DR, Robblee AR (1981) Rapeseed meal in animal nutrition II Non-ruminant animals. J Am Oil Chem Soc 58: 682–685

Curtis F, Fenwick R, Heaney RK, Hobson-Frohock A, Land D (1978) Rapeseed meal and egg taint. In: Proc 5th Int Rapeseed Conf, Malmö, 1978, 300–302

Dhillon SS, Labana KS, Banga SK (1986) Genetics of seed coat colour in *Brassica juncea.* Ann Biol 2: 195

Diepenbrock W (1983) Genotypic differences between the contents of linolenic acid in galactolipids and triacyl glycerol from seeds of rape plants. In: Proc 6th Int Rapeseed Conf, Paris, pp 321–362

Diepenbrock W, Wilson RF (1987) Genetic regulation of linolenic acid concentration in rapeseed. Crop Sci 27: 75–77

Dietz HM, King RD, Shepherd DAL, Turvey A (1987) Antinutritional and toxics effects of glucosinolates and glucosinolate breakdown products in *B. campestris* var. *toria* in rats. Hum Nutr: Food Sci and Nut 41F: 213–223

Dorrell DC, Downey RK (1964) The inheritance of erucic acid content in rapeseed (*Brassica campestris*). Can J Plant Sci 44: 499–504

Downey RK (1964) A selection of *Brassica campestris* containing no erucic acid in its seed oil. Can J Plant Sci 44: 295

Downey RK (1990) Recent developments in oilseed brassicas. In: Proc the *Brassica* subnetwork II meeting Pantnagar, India, pp 4–8

Downey RK, Harvey BL (1963) Methods of breeding for oil quality in rape. Can J Plant Sci 43: 271–275

Downey RK Craig BM, Young CG (1969) Breeding rapeseed for oil and meal quality. J Am Oil Chem Soc 46: 121–123

Erdman JW (1979) Oilseed phytates: nutritional implications. J Am Oil Chem Soc 56: 736–740

Fenton TW, Leung J, Clandinin DR (1980) Phenolic components of rapeseed meal. J Food Sci 45: 1702–1705

Fenwick GR (1979) A micro method for the screening of individual seeds and cotyledons of *Brassica napus* and *Brassica campestris* (rapeseed) for low sinapine content. J Sci Fd Agric 30: 661–663

Fernandez-Escobar J, Dominguez J, Martin A, Fernandez-Maitinez JM (1988) Genetics of erucic acid content in interspecific hybrids of Ethiopian Mustard (*Brassica carinata* Braun) and Rapeseed (*B. napus* L.). Plant Breed 100: 310–315

Gopalan CD, Krishnamurthi D, Shenolikar IS, Krishnamachari KAVR (1974) Myocardial changes in monkeys fed on mustard oil. Nutr Metab 16: 352–365

Hemingway JS, Schofield HJ, Vaughan JG (1961) Volatile mustard oils of *Brassica juncea* seeds. Nature (London) 192: 993

Holman RT (1981) Essential fatty acids in nutrition and diseases. Chem Ind (London) 20: 704–709

Hopkins C.Y., Murray T.K., Campbell J.A. (1955). Optimum ratio of saturated to mono-unsaturated fatty acids in rat diets. Can J Biochem Physiol 33: 1047–1054

Ismail F, Eskin NAM (1979) A new quantitative procedure for determination of sinapine. J Agric Fd Chem 27: 917–918

Jönsson R (1973a) Breeding for improved composition of fatty acid in oil plants. I Decreased content of erucic acid insummer turnip rape by selection in Swedish market varieties. Sver Utsaedesfoeren Tidskr 83: 179–186

Jönsson R (1973b) Breeding for low erucic acid content in summer turnip rape (*Brassica campestris* L. var. *annua*). Z Pflanzenzuecht 69: 1–18

Jönsson R (1975) Yellow seeded rape and turnip rape II. Breeding for improved quality of oil and meal in yellow seeded materials. Sver Utsaedesfoeren Tidskr 85: 271–278

Jönsson R (1977) Erucic acid heredity in rapeseed (*Brassica napus* L.) and *Brassica campestris*. Hereditas 86: 159–170

Jönsson R, Josefsson E, Uppström B (1975) Breeding for low content of glucosinolates in rape and turnip rape. Sver Utsaedesfoeren Tidskr 85: 279–290

Jönsson R, Uppström B (1986) Quality breeding in rapeseed. In: Svalöf 1886-1986, Research and results in Plant Breeding, pp 173–184

Josefsson E (1970) Content of p-hydroxy benzyl glucosinolate in seed meals of *Sinapis alba* as affected by heredity, environment and seed part. J Sci Fd Agric 21: 94–97

Josefsson E (1971) Studies of the biochemical background to differences in glucosinolate content in *Brassica napus* L. II. Administration of some sulphur-35 and carbon-14 compounds and localization of metabolic blocks. Physiologia Pl 24: 161–175

Josefsson. E (1972) Conversion of indole-3-acetaldehyde oxime to 3-indolyl methyl glucosinolate in *Sinapis alba*. Physiol P1 27: 236–239

Josefsson E (1973) Studies of the biochemical background to differences in glucosinolate content in *Brassica napus* L. III. Further studies to localize metabolic blocks. Physiol P1 29: 28–32

Josefsson E, Appelqvist LÅ (1968) Glucosinolates in seed of rape and turnip rape as affected by variety and environment. J Sci Fd Agric 19: 564–570

Josefsson E, Uppström B (1976) Influence of sinapine and p-hydroxy benzyl glucosinolates on the nutritional value of rapeseed and white mustard meals. J Sci Fd Agric 27: 438–442

Kajaer A (1976) Glucosinolates in the cruciferae. In: Vaughan-Macleod A.S., Jones BMG (eds.). Biology and chemistry of Cruciferae. Academic Press London pp 207–219

Kajaer A, Olesen Larsen P (1973) Non-protein amino acids cyanogenic glucosides and glucosinolates. In: Geissman TA (ed) Biosynthesis. Specialist periodical report. Chem Soc London 2: 179–203

Kerber E, Buchloh G (1982) Sinapine in the tribe Brassicaceae. Angew Bot 56: 85–91 (Chem Abstr 97: 1241532)

Kirk JTO, Hurlstone CJ (1983) Variation and inheritance of erucic acid content in *Brassica juncea*. Z Pflanzenzuecht 90: 331–338

Kirk JTO, Oram RN (1981) Isolation of erucic acid free lines of *Brassica juncea*. J Aust Inst Agric Sci 47: 51–52

Kondra ZP, Stefansson BR (1970) Inheritance of the major glucosinolates of Rapeseed (*Brassica napus*) meal. Can J Plant Sci 50: 643–647

Korver O, Alderlieste ET, Boelhouwer C (1967) Analysis of linoleate and linolenate by isomerization with potassium tertiary butoxide in mixture of tertiary butanol and ethylene glycol dimethyl ether. J Am Oil Chem Soc 44: 484

Koyama Y, Fuju M, Kudo H (1978) Rapeseed and their extracted oils from different varieties. Yukagoku 27: 375–380 (Chem Abstr 89: 106119)

Kozolowska H, Rotkiewicz DA, Zodernowski R, Sosulski FW (1983) Phenolic acids in rapeseed and mustard. J Am Oil Chem Soc 60: 19–23

Laakso I (1985) Selection of high linoleic acid content in summer turnip rape (*Brassica campestris* L.). III Effects of selection on fatty acid composition. Acta Pharm Fenn 94: 51–57

Laakso I, Hiltunen R, Hovinen S (1983) Selection of high linoleic acid content in summer turnip rape (*Brassica campestris* L.). II. variation in linoleic acid content in successive generations. In: Proc 6th Int Rapeseed Conf, Paris, pp 607–612

Laakso I, Hovinen S, Hiltunen R (1986) Selection of high linoleic acid content in summer turnip rape (*Brassica campestris* L. ssp. *oleifera* var. *annua*). IV. Selection of improved oil yield. Acta Agric Scand 36: 347–351

Langer P, Stolc V (1965) Goitrogenicity of allylisothiocyanate—wide-spread natural mustard oil. Endrocrinol 76: 151–155

Lein KA (1970) Quantitative Bestimmungsmethoden für Samenglucosinolate in *Brassica* Arten und ihre Anwendung in der Züchtung Von Glucosinolataramen Raps. Z Pflanzenzuecht 63: 137–154

Lewis AJ, Aherne FX, Hardin RT (1978) Reproductive performance of Sows fed low glucosinolate (Tower) rapeseed meal. Can J Anim Sci 58: 203–208

Lichter R, DeGroot E, Fiebig D, Schweiger R, Gland A (1988) Glucosinolates determined by HPLC in the seeds of microspore-derived homozygous lines of rapeseed (*Brassica napus* L.). Plant Breed 100: 209–221

McGhee JE, Kirk LD, Mustakas GC (1965) Methods for determining thioglucosides in *Crambe abyssinica*. J Am Oil Chem Soc 42: 889–891

McGregor DI (1974) A rapid and sensitive spot test for linolenic acid levels in rapeseed. Can J Plant Sci 54: 211–213

McGregor DI (1977) A rapid and simple method for screening rapeseed and mustard seed for erucic acid content. Can J Plant Sci 57: 133–142

McGregor DI, Downey RK (1975) A rapid and simple assay for identifying low glucosinolate rapeseed. Can J Plant Sci 55: 191–196

McMillan M, Spinks EA, Fenwick GR (1986) Preliminary observations on the effect of dietary Brussels sprouts on thyroid function. Hum Toxicol 5: 15–19

Miller RW, Van Etten CH, McGrew CE, Wolff IA, Jones Q (1962) Amino acid composition of seed meals from 41 species of Cruciferae. J Agric Fd Chem 10: 426–430

Møller P, Rahman MH, Stølen O, Sørensen H (1985) Heredity of fatty acids and glucosinolates in oilseed rape. Possibilities for improvement of rape adapted for the growth conditions in the production and utilization of the cruciferous crop. Nijhoff/Junk, Dordrecht, pp 286–300

Morice J (1974) Selection d'une variete de colza sans acids erucique et sans glucosinolates. In: Proc. 4th Int. Rapeseed Conf, Giessen, West Germany, pp 31–47

Mueller MM, Ryl EB, Fenton T, Clandinin DR (1978) Cultivar and growing location difference on the sinapine content of rapeseed. Can Anim Sci 58: 579–583

Namai H, Kaji T, Hosoda T (1972) Interspecific and intervarietal variations in content of oxalolidinethione in seed meals of cruciferous crops. Jpn J Genet 47: 319–327

Nishie K, Daxenbichler ME (1980) Toxicology of glucosinolates, related compounds (nitriles, R-goitrin, isothiocyanates and vitamin U) found in Cruciferae. Food Cosmet Toxicol 18: 159–172

Nwokolo EN, Bragg DB (1977) Influence of phytic acid and crude fibre on the availability of minerals from four protein supplements in growing chicks. Can J Anim Sci 57: 475–477

Oginski EL, Stein AE, Greer MA (1965) Myrosinase activity in bacteria as demonstrated by the conversion of progoitrin to goitrins. In: Proc Soc Expt Biol Med 199: 360–364

Palmer MV, Yeung SP, Sang JP (1987) Glucosinolate content of seedlings, tissue cultures and regenerant plants of *Brassica juncea* (Indian mustard). J Agric Fd Chem 35: 262–265

Pearson S (1974) A method for determination of glucosinolates in rapeseed as TMS derivatives. In: Proc 4th Int Rapeseed Conf, Giessen, p. 381

Pleines S, Friedt W (1988) Breedng for improved 18-carbon fatty acid composition in rapeseed (*B. napus*). Fett Wess Technol 90: 167–171

Pleines S, Friedt W (1989) Genetic control of linolenic acid concentration in seed oil of rapeseed (*Brassica napus* L.). Theor Appl Genet 78: 793–797

Prasad A, Rao PV (1979) Tannin content in a few poultry feed ingredients. Indian J Anim Sci 49: 872–874

Rakow G (1973) Selection for content of linoleic and linolenic acids in rapeseed after mutagenic treatment. Z Pflanzenzuecht 69: 62–68

Renarid S, McGregor L (1976) Antithrombogenic effects of erucic acid poor rapeseed oils in the rats. Rev Fr Crops Cros 23: 393–396 (Chem Abstr 85: 1583695)

Röbbelen G (1975) Screening for oils and fats in plants. In: Frankel OH, Hawkes JG (eds.). Genetic resources for today and tomorrow. Univ Press, Cambridge 2: pp 231–247

Röbbelen G (1980) Breeding for low content of glucosinolates in rapeseed. In: Bunting ES (ed) Production and Utilization of protein in oilseed crops. Nijhoff-Junk. The Hague, pp 91–106

Röbbelen G, Leitzke B (1975) Stand und probleme der Zuechtung Erucasaürearmen Rapssorten in der Bundesrepublik Deutschland. In: Proc 4th Int Rapeseed Conf, Giessen, pp 63–71

Röbbelen G, Nitsch A (1975) Genetical and physiological investigations on mutants for polyenoic fatty acids in rapeseed (*B. napus* L.). I. Selection and description of new mutants. Z Pflanzenzuecht 75: 93–105

Roy NN, Tarr AW (1986) Development of Near-Zero linolenic acid (18:3) lines of Rapeseed (*Brassica napus* L.). Z Pflanzenzuecht 96: 218–223

Sanders TAB, Younger KM (1981) The effect of dietary supplements of w_3 polyunsaturated fatty acid composition of platelets and plasma choline phosphoglycerides. Br J Nutr 45: 613–616

Sang JP, Truscott RJW (1984) Liquid chromatographic determination of glucosinolates in rapeseed as desulfoglucosinolates. J Assoc Off Anal Chem 67: 829–833

Schung E, Haneklaus S (1988) Theoretical Principles for the indirect determination of the total glucosinolate content in Rapeseed and meal quantifying the sulphur concentration via x-ray fluorescence (X-RF Method). J Sci Fd Agric 45: 243–254

Shenolikar I (1980) Fatty acid profile of myocardial lipid in population consuming different dietary fats. Lipids 15: 980–982

Spinks EA, Sones K, Fenwick GR (1984) The quantitative analysis of glucosinolates in cruciferous vegetables, oilseeds and forage crops using high performance liquid chromotography. Fette Seifen Anstrichm 86: 228–231

Stefansson BR, Hougen FW, Downey RK (1961) Note on the isolation of rape plants with seed oil free from erucic acid. Can J Plant Sci 41: 218–219

Thies W (1971) Schnelle und einfache Analysen der Fettsäurerzusammensetzung in einzelnen Rape-Kotyledonen I. Gaschromatographische und papierchromatographische Methoden. Z Pflanzenzuecht 65: 181–202

Thies W (1976) Quantitative gas liquid chromatography of glucosinolates on the microliter scale. Fette Seifen Anstrichm 78: 231–234

Thies W (!977) Analysis of glucosinolates in seeds of rapeseed (*Brassica napus* L.). Concentration of glucosinolates by ion exchange. Z Pflanzenzuecht 79: 331–335

Thies W (1983) Complex formation between glucosinolates and tetra chloropalladate (II) and its utilization in plant breeding. Fette Seifen Anstrichm 84: 338–342

Thies W, Nitsch A (1974) Rapid analyses of the fatty acid composition in a single cotyledon of rapeseed. III. A spot test analysis for the "specific" determination of linolenic acid. Z Pflanzenzuecht 72: 74–83

Tholen JT, Shifeng S, Truscott RJW, Roger JW (1989) The thymol method for glucosinolate determination. J Sci Fd Agric 49: 157–165

Thomke S (1980) Review of rapeseed meal in animal nutrition: Ruminant animals. J Am Oil Chem Soc 58: 805–810

Tookey HL (1973) *Crambe* thioglucoside glucohydrolase (EC 3.2.3.1): separation of a protein required for epithiobutane formation. Can J Biochem 51: 1654–1660

Underhill EW (1980) Glucosinolates. In: Bell EA, Charlwood BV (eds.). Encyclopedia of plant physiology New series 8: 493–511 Springer-Verlag, Berlin, Heidelberg, New York

Underhill EW, Kirkland DF (1971) Gas chromatography of trimethylsilyl derivatives of glucosinolates. J Chromatogr 57: 47–54

Uppström B, Johansson M (1985) Determination of sinapine in rapeseed. Sver Utsaedesfoeren Tidskr 95: 123–128

Uppström B, Svensson R (1980) Determination of phytic acid in rapeseed meal. J Sci Fd Agric 31: 651–656

Van Etten CH, McGrew CE, Daxenbichler ME (1974) Glucosinolate determination in cruciferous seeds and meals by measurement of enzymatically released glucose. J Agric Fd Chem 22: 483

Vaughan JG, Hemingway JS, Schofield HJ (1963) Contributions to a study of variation in *Brassica juncea*. J Linn Soc (Bot) 58: 435–447

Vermorel M, Heaney RK, Fenwick GR (1986) Nutritive value of rapeseed meals: effects of individual glucosinolates. J Sci Fd Agric 37: 1197–1202

Vogt H (1980) Rapeseed meal in Poultry Rations In: Bunting ES (ed) Production and utilization of protein oilseed crops. Nijhoff/Junk. The Hague, 5:311–343

Walker BL, Lall SP, Bayley HS (1970) Nutritional aspects of rapeseed oil: Digestability, processing and influence of erucic acid on tissue lipids. Proc 3rd Int Rapeseed Conf, St-Adele, Canada 377-404

Wolfram G (1976) Essential fatty acids. Ernaehr Umsch 23: 267–270

CHAPTER 7

Embryo Rescue Techniques for Wide Hybridization

N. Inomata

7.1 Introduction

In crop improvement it is necessary to expand gene resources by selective introgression of alien genes in a good agronomic base. This is more important if the existing genetic variability in the breeding material is limited, as is the case in oleiferous *Brassica*. Wide hybridization can be used to transfer the desirable variability. Usually, it is difficult to produce such hybrids due to cross-incompatibility barriers. These may result from either failure to bypass the sexual constraints or hybrid breakdown. Cross-incompatibility is primarily caused by incompatible pollen-pistil interaction which precludes the formation of a viable zygote. This phenomenon, however, has not been investigated extensively (Shivanna 1982). Hybrid breakdown, hybrid inviability and hybrid sterility are some other recognizable post-fertilization barriers (Raghvan 1986). They may result from arrested embryo development, endosperm disintegration, abnormal development of ovular tissue or chromosome and genetic instability. Hybrid inviability may result from the action of specific genes or due to disharmony between the nucleus of one species and the cytoplasm of another, or adverse embryo-endosperm interaction. Recent developments in biotechnology and embryo rescue techniques may provide new genetic variability for the breeding of cruciferous crops (Chiang et al. 1978, 1980; McNaughton et al. 1978; Mohapatra et al. 1987). Various embryo rescue techniques are: ovary, ovule and embryo culture. In case of an incompatible cross, after a certain period of growth of the hybrid embryo the incompatible reaction becomes acute and it is imperative to culture the embryo or ovule before initiation of the incompatibility reaction (Takeshita et al. 1980). For ovule culture, younger ovules (10-21 days old) can be used, but for embryo culture, older ovules should be used, depending upon the size of the embryo. Embryos cultured during the globular stage may not survive in vivo but such embryos can survive during ovule and ovary culture. These techniques will help in broadening the genetic base of crop brassicas through the exploitation of secondary and tertiary gene pools. Various aspects and achievements of embryo rescue techniques are discussed in this chapter.

7.2 Ovary Culture

In many cases, the barriers to the development of the hybrid embryo occur very early during embryogenesis; embryos which abort at early stages are difficult to dissect out and culture. Under these circumstances, it is much easier to culture ovaries excised from the pollinated flowers. In the majority of cases, it is feasible to harvest the seeds from in vitro cultured ovaries in well-defined media supplemented with growth harmones and complex additives like coconut water, yeast extract, etc.

The conditions of the medium for ovary culture are standardized for *Brassica* (Inomata 1977; 1978; 1979; 1985a), and a large number of interspecific and intergeneric crosses have been successfully produced (Table 1). Various histological studies led to the belief that culture media containing casein hydrolysate, coconut water and yeast extract support the embryo and endosperm development (Inomata 1975, 1977). Available experimental data have indicated that the production of hybrid plants from cultured ovaries was better with a higher concentration of inorganic salts in the medium (e.g. Nitsch and Nitsch or Murashige and Skoog) than with a low concentration (e.g. White's, Nitsch and Heller). Supplementation of the media with 500 mg/l sucrose produced very good results. The best seed setting was observed on White's minerals, and on cultures incubated at 13–22 °C and 300-500 lμ continuous illumination (Inomata 1976). The culture of ovary explanted 2 to 9 days after pollination produced the best results. In this chapter, the example of ovary culture in *B. campestris* × *B. oleracea* is discussed in further detail as a case study. The composition of the culture media used was discussed earlier (Inomata 1985a). The results are presented in Table 2. The time of explant of the ovary was from 2 to 12 days after pollination (DAP). Production rate of the hybrids (number of hybrids obtained/number of capsules examined × 100) was best when ovaries were cultured 2 days after pollination.

The scope of the above experiment was increased to include three subspecies of *B. campestris* and three wild species of *B. oleracea*. The results are given in Table 3. About 50 ovaries were placed in the medium and many naked embryos protruding from undeveloped seed coat were obtained in each cross-combination. The production rate of the hybrids in this experiment was much higher than those observed in the experiment discussed earlier. The production of F_1 hybrids between cultivated *B. oleracea* and wild species is significant, as the present day cultivars originated from wild types (Snogerup 1980), and these may be used as excellent sources of resistance to various biotic and abiotic stresses. Similarly, hybrids have been produced between *B. napus* and wild species of *B. oleracea* through ovary culture (Table 4). Although hybrids could be obtained from all crosses, the production rate was low (1.2 to 13.8%, $\overline{X}$ = 5.3). This, however, compared favourably with hybrid production rate (0.86%) obtained after artificial pollination between *B. napus* and *B. oleracea* (Mizushima 1952). Even the cross between *B. napus* and autotetraploid *B. oleracea* (Chiang et al. 1977) had a production rate of only 6.7%. The morphological features of a number of F_1 hybrids produced by ovary culture, along with their respective parents, are shown in Fig. 1. The majority of the hybrids had the expected dihaploid chromosome number; some sesquidiploids were also identified.

7.3 Ovule Culture

Because of the presence of various pre-fertilization barriers in many cross-combinations, hybridization between many cultivated and wild species is not very successful. In this situation, in vitro fertilization followed by culturing the fertilized

Table 1. Recent results on the production of hybrids among the tribe Brassicaceae through ovary culture

Cross combination	No. of ovaries cultured (A)	No. of hybrids obtained (B)	Production rate (B/A× 100)	Reference
Brassica nigra × *Raphanus sativus*	954	4	9.4	Matsuzawa and Sarashima (1984)
Brassica campestris × *B. cretica*	89	135	151.7	Inomata (1985b)
Brassica campestris × *B. bourgeaui*	87	141	162.1	Inomata (1986)
Brassica napus × *B. juncea*	194	100[a]	51.5	Bajaj et al. (1986)
reciprocal cross	142	94[a]	66.2	"
Eruca sativa × *Brassica campestris*	218	2	0.9	Matsuzawa and Sarashima (1986)
Eruca sativa × *Brassica nigra*	243	0	–	"
Eruca sativa × *Brassica oleracea*	223	11	4.9	"
Brassica fruticulosa × *B. campestris*	115	133	115.6	Kumar et al. (1988)
Brassica campestris × *B. montana*	88	69	78.4	Inomata (1987)
Brassica juncea × *B. hirta*	370	9	2.4	Mohapatra and Bajaj (1987)
reciprocal cross	285	0[a]	–	"
Brassica juncea × *B. campestris*	80	11	13.8	Mohapatra and Bajaj (1988)
Moricandia arvensis × *Brassica oleracea*	118	320	271.2	Takahata (1988)
reciprocal cross	37	0	–	"
Diplotaxis siifolia × *B. juncea*	35	3	8.5	Batra et al. (1990)
Diplotaxis siifolia × *B. napus*	73	2	2.7	"
Brassica napus × *B. oleracea*	290	15	5.2	Present review
Brassica napus × *B. oleracea* (4*x*)	49	3	6.1	"
Brassica napus × *B. bourgeaui*	86	1	1.2	"
Brassica napus × *B. cretica*	78	8	10.3	"
Brassica napus × *B. montana*	24	1	4.2	"
Diplotaxis erucoides × *Brassica napus*	432	66	15.3	Delourme et al (1989)
Moricandia arvensis × *B. campestris*	80	36	45.0	Takahata and Takeda (1990)
Moricandia arvensis × *B. nigra*	32	8	25.0	"
Erucastrum gallicum × *B. juncea*	–	–	–	Batra et al (1989)
Erucastrum gallicum × *B. napus*	–	–	–	"

[a]Number of ovaries formed.

Table 2. Relation between the time of explanted ovary after pollination and the production of interspecific hybrids in *Brassica campestris* ssp. *chinensis* cv. Seppaku-taina × *B. oleracea* var. *alboglabra* cv. Senyo-shoyo through ovary culture

Day explanted after pollination	No. of ovaries explanted	No. of capsules examined (A)	No. of embryos further cultured			No. of seeds obtained	No. of seeds undeveloped	No. of undeveloped seeds further cultured	No. of hybrids obtained (B)	Production rate (B/A × 100)
			Late torpedo	Walking-stick	Full-grown embryo					
2	30	30	24	17	2	7	28	18	39	130.0
4	30	19	6	2	0	0	40	10	12	63.2
6	30	28	2	1	0	1	110	14	12	42.9
8	30	19	0	0	0	0	92	10	4	21.1
10	30	18	0	0	0	1	73	10	6	33.3
12	30	28	0	0	0	0	214	20	14	50.0
Total or mean	180	142	32	20	2	9	557	82	87	61.3

Table 3. Production of interspecific hybrids between *Brassica campestris* and wild species of *B. oleracea* through ovary culture. (After Inomata 1985b, 1986, 1987)

Cross combination	No. of ovaries explanted	No. of capsules examined (A)	No. of embryos further cultured			No. of. seeds obtained	No. of hybrids obtained (B)	Production rate (B/A × 100)
			Late torpedo	Walking-stick	Full-grown embryo			
subsp. *chinensis* × *B. bourgeaui* [a]	50	44	8	17	17	110	102	231.8
subsp. *pekinensis* × *B. bourgeaui* [b]	50	43	37	37	21	0	39	90.7
Total or mean	100	87	45	54	38	110	141	162.1
subsp. *chinensis* × *B. cretica* [c]	46	16	0	5	45	0	47	293.8
subsp. *dichotoma* × *B. cretica* [d]	50	44	24	12	20	0	34	77.3
subsp. *pekinensis* × *B. cretica* [e]	45	29	41	22	21	0	54	186.2
Total or mean	141	89	65	39	86	0	135	151.7
subsp. *chinensis* × *B. montana* [f]	50	44	17	32	18	14	54	122.7
subsp. *pekinensis* × *B. montana* [g]	50	44	6	14	15	1	15	34.1
Total or mean	100	88	23	46	33	15	69	78.4

[a] cv. Seppaku-taina × *B. bourgeaui* 120. [b] cv. Nozaki-hakusai No. 2 × *B. bourgeaui* 120. [c] cv. Seppaku-tania × subsp. *cretica* 35. [d] cv. Brown Sarson DS-2 × subsp. *cretica* 35. [e] cv. Nozaki-hakusai No. 2 × subsp. *cretica* 35. [f] cv. Seppaku-taina × *B. montana* 89. [g] cv. Nozaki-thakusai No. 2 × *B. montana* 89.

Table 4. Production of interspecific hybrids between *Brassica napus* and *B. oleracea*, and between *B. napus* and wild species of *B. oleracea* through ovary culture

Cross-combination	No. of ovaries explanted	No. of capsules examined (A)	No. of embryos further cultured			No. of. seeds obtained	No. of hybrids obtained (B)	Production rate (B/A × 100)
			Late torpedo	Walking-stick	Full-grown embryo			
subsp. *oleifera* × var. *acephala* [a]	100	93	0	0	3	2	5	5.4
subsp. *oleifera* × var. *alboglabra* [b]	50	50	0	0	0	2	1	2.0
subsp. *oleifera* × var. *alboglabra* [c]	100	97	0	0	0	4	4	4.1
subsp. *oleifera* × var. *capitata* [d]	50	50	0	0	0	5	5	10.0
subsp. *oleifera* × var. *capitata* (4*x*) [e]	50	49	0	0	0	3	3	6.1
subsp. *oleifera* × *B. bourgeaui* [f]	100	86	1	3	5	1	1	1.2
subsp. *oleifera* × *B. cretica* [g]	50	49	0	0	0	8	4	8.2
subsp. *oleifera* × *B. cretica* [h]	30	29	1	0	0	4	4	13.8
subsp. *oleifera* × *B. montana* [i]	49	24	2	0	0	0	1	4.2
Total or mean	579	527	3	3	8	29	28	5.3

[a] cv. Miho × cv. Portuguese kale. [b] cv. Aomori No. 1 × cv. Senyo-shoyo. [c] cv. Miho × cv. Senyo-shoyo. [d] cv. Aomori No. 1 × cv. Shutoku. [e] cv. Aomori No. 1 × autotetraploid cv. Shutoku. [f] cv. Miho × *B. bourgeaui* 120. [g] cv. Aomori No. 1 × subsp. *cretica* 35. [h] cv. Miho × subsp. *cretica* 35. [i] cv. Miho × *B. montana* 89.

Fig. 1 a–l. F_1 hybrids obtained through ovary culture and their parents. **a** *Brassica campestris* subsp. *chinensis* cv. Seppaku-taina. **b** *B. campestris* subsp. *dichotoma* cv. Brown Sarson DS-2. **c** *B. campestris* subsp. *pekinensis* cv. Nozaki-hakusai No. 2. **d** *B. oleracea* var. *alboglabra* cv. Senyo-shoyo. **e** *B. oleracea* var. *capitata* cv. Shutoku. **f** Autotetraploid *B. oleracea* var. *capitata* cv. Shutoku. **g** *B. bourgeaui* 120. **h** *B. cretica* subsp. *cretica* 35. **i** *B. montana* 89. **j** *B. napus* subsp. *oleifera* cv. Aomori No. 1. **k** A hybrid in Seppaku-taina × *B. bourgeaui* 120. **l** A hybrid in Nozaki-hakusai No. 2 × *B. bourgeaui* 120.

Fig. 1 m-w. **m** A hybrid in Seppaku-taina × subsp. *cretica* 35. **n** A hybrid in Brown Sarson DS-2 × subsp. *cretica* 35. **o** A hybrid in Nozaki-hakusai No. 2 × subsp. *cretica* 35. **p** A hybrid in Seppaku-taina × *B. montana* 89. **q** A hybrid in Nozaki-hakusai No. 2 × *B. montana* 89. **r** A hybrid in *B. napus* subsp. *oleifera* cv. Miho × Senyo-shoyo. **s** A hybrid in Aomori No. 1 × Shutoku. **t** A hybrid in Aomori No. 1 × autotetraploid Shutoku. **u** A hybrid in Miho × *B. bourgeaui* 120. **v** A hybrid in Miho × subsp. *cretica* 35. **w** A hybrid in Miho × *B. montana* 89.

ovules is a promising alternative to parasexual hybridization. Alternatively, the fertilized ovules can be dissected out from the ovary cultured for few days in a defined medium. This is especially suited for the case where very young ovules cannot be cultured due to ill-defined media requirements. Very few reports are available on ovule culture in *Brassica*. Interspecific hybrids between R-genome and C-genome species have been produced by ovule culture (Takeshita et al. 1980). Twenty-day old hybrid ovules were cultured on basic White's medium (1963), supplemented with coconut water (150 ml/l), napthalene acetic acid (2.5 mg/l) and kinetin (2.5 mg/l). Similarly, ovule culture was successfully used to produce interspecific hybrids, *B. juncea* × *B. hirta* (Mohapatra and Bajaj 1984), *B. juncea* × *B. napus* (Bajaj et al. 1986) and *Sinapis alba* × *B. napus* (Ripley and Arnison 1990). Supplementing the media with indoleacetic acid, kinetin and casein hydrolysate was helpful (Mohapatra and Bajaj 1984). It was also possible to produce *Eruca sativa* × *B. campestris* and *B. napus* × *Raphanobrassica* hybrids by dissecting out and culturing the enlarged ovules (Agnihotri et al. 1990a, b). The hybrid production rates obtained in these experiments were comparable to those from embryo culture. The ovule culture technique has also been used to study test tube fertilization in *Brassica* species.

7.4 Embryo Culture

Culture of hybrid embryos is now a routine method for interspecific and intergeneric hybridization. In the majority of instances, embryo abortion occurs during the mid to late stages of embryogenesis. It is generally believed that embryo abortion or death occurs as a result of no or abnormal endosperm development, which results in embryo starvation and abscision of the developing fruit. Laibach (1925) was the first to demonstrate the possibility of culturing embryos in a nutrient medium. He raised hybrid seedlings by culturing the immature embryos from *Linum parenne* × *L. autriacum* hybridization. Since then, embryo rescue has been widely practised to produce interspecific as well as intergeneric hybrids in about 70 crosses involving approximately 35 genera and 120 species (Collins and Grosser 1984).

In *Brassica*, Nishi et al. (1959) were the first to use embryo culture for producing hybrids between *B. campestris* and *B. oleracea*. The production rate of hybrids was higher than in vivo pollinations (Inomata 1985a). Nishiyama and Inomata (1966) made embryological studies of cross-incompatibility between $2x$ and $4x$ *Brassica*. The development process collapsed very early, leading to no seeds from the cross *B. campestris* ssp. *chinensis* ($2x = 20$) × *B. campestris* ssp. *pekinensis* ($4x = 40$). In the reciprocal cross the hybrid embryo developed to the globular stage and then failed to develop further due to endosperm degeneration. From the review of literature pertaining to embryo culture, it is clear that the time for embryo explant could be from 2 to 3 weeks after pollination. Table 5 shows recent results on the embryo-culture for production of hybrids among the Brassicaceae. The production of interspecific hybrids was comparatively easy in *B. juncea* × *B. napus*, its reciprocal cross and *B. oleracea* × *B. campestris*. Table 6 depicts the results of the production of intergeneric hybrids between *B. oleracea* × *Raphanus sativus* and *B. napus* × *R. sativus*. The production rate of the hybrids in the cross between diploid *B. oleracea* × *R. sativus* was 2 but

Table 5. Recent results on the production of hybrids from the tribe Brassicaceae through embryo culture

Cross-combination	No. of flowers pollinated (A)	No. of embryos cultured	No. of hybrids obtained (B)	Production rate (B/A × 100)	Reference
B. campestris × *B. oleracea*	30	–	1	3.3	Takeshita et al. (1980)
reciprocal cross	93	–	23	24.7	"
B. juncea × *B. napus*	–	126	92	–	Bajaj et al. (1986)
reciprocal cross	–	152	90	–	"
B. napus × *B. oleracea*	255	299	64	25.1	Quazi (1988)
reciprocal cross	73	35	8	11.0	"
B. napus × *B. oleracea* (4*x*)	164	33	15	9.1	"
reciprocal cross	160	54	13	8.1	"
B. oleracea × *Raphanus sativus*	100	8	2	2.0	Present data
B. oleracea (4*x*) × Raphanus sativus	135	49	21	15.6	"
R. sativus × *B. napus*	–	–	–	–	Paulmann (1987)
B. napus × *Raphanus sativus*	200	19	16	8.0	Present data
reciprocal cross	308	0	–	–	"

Table 6. Production of intergeneric hybrids between *Brassica oleracea* and *Raphanus sativus*, and between *B. napus* and *R. sativus* through embryo culture

Cross combination	No. of flowers pollinated (A)	No. of embryos cultured	No. of hybrids obtained (B)	Production rate (B/A × 100)
B. oleracea × *R. sativus* [a]	100	8	2	2.0
B. oleracea (4*x*) × *R. sativus* [b]	135	49	21	15.6
B. napus × *R. sativus* [c]	50	1	1	2.0
B. napus × *R. sativus* [d]	50	5	4	8.0
B. napus × *R. sativus* [e]	50	9	7	14.0
B. napus × *R. sativus* [f]	50	4	4	8.0

[a] cv. Shutoku × cv. Taibyo-sofutori. [b] Autotetraploid cv. Shutoku × cv. Taibyo-sofutori. [c] cv. Santana × Strain No. 1. [d] cv. Santana × Strain No. 2. [e] cv. Semu × Strain No. 1. [f] cv. Semu × Strain No. 2.

Fig. 2a-e. F_1 hybrids obtained through embryo culture and their parents. **a** *Raphanus sativus* cv. Taibyo-sofutori. **b** *Brassica napus* subsp. *oleifera* cv. Semu. **c** A hybrid in *B. oleracea* var. *capitata* cv. Shutoku × Taibyo-sofutori. **d** A hybrid in autotetraploid *B. oleracea* var. *capitata* cv. Shutoku × Taibyo-sofutori. **e** A hybrid in Semu × *R. sativus* strain No. 1

when autotetraploid *B. oleracea* was used, it improved to 15.6. The mean production rate for *B. napus* and *R. sativus* hybrids was 8 and it varied with the cultivar used. The F_1 hybrids and the corresponding parents are shown in Fig. 2. The morphological characters of the leaf were intermediate between the parents in the hybrids of *B. oleracea* × *R. sativus*, while maternal parent expression was more dominant in the *B. napus* × *R. sativus* cross.

7.5 Conclusions

Ovary, ovule and embryo culture as embryo rescue techniques are important for the improvement of curciferous crops by the utilization of alien genetic resources. Ovary culture is the best technique because of its simple manipulation. Although culture conditions for all three types of embryo rescue operations are very well defined in *Brassica*, there is a need to develop a completely synthetic culture medium without the use of such additives as casein hydrolysate and yeast extract etc., to ensure repeatability and quantity irrespective of the genotypes used in the hybrid combinations. Embryo implantation and ovule culture merit greater attention.

References

Agnihotri A, Gupta V, Lakshmikumaran MS, Shivanna KR, Prakash S, Jagannathan V (1990a) Production of *Eruca-Brassica* hybrid by embryo rescue. Plant Breed 104: 281–289

Agnihotri A, Shivanna KR, Raina SN, Lakshmikumaran MS, Prakash S, Jagannathan V (1990b) Production of *Brassica napus* × *Raphanobrassica* hybrids by embryo rescue: an attempt to introduce shattering resistance into *B. napus*. Plant Breed 105: 292–299

Bajaj YPS, Mahajan SK, Labana KS (1986) Interspecific hybridization of *Brassica napus* and *B. juncea* through ovary, ovule and embryo culture. Euphytica 35: 103–109

Batra V, Prakash S, Shivanna KR (1990) Intergeneric hybridization between *Diplotaxis siifolia*, a wild species and crop Brassicas. Theor Appl Genet 80: 537–541

Batra V, Shivanna KR, Prakash S (1989) Hybrids of wild *Erucastrum gallicum* and crop brassicas. Proc 6th Int Cong SABRAO, 1, pp 443–446

Chiang BY, Chiang MS, Grant WF, Crete R (1980) Transfer of resistance to race 2 of *Plasmodiophora brassicae* from *Brassica napus* to cabbage (*B. oleracea* spp. *capitata*). IV. A resistant 18-chromosome B_1 plant and its B_2 progenies. Euphytica 29: 47–55

Chiang BY, Grant WF, Chiang MS (1978) Transfer of resistance to race 2 of *Plasmodiophora brassicae* from *Brassica napus* to cabbage (*B. oleracea* var. *capitata*). II. Meiosis in the interspecific hybrids between *B. napus* and 2x and 4x cabbage. Euphytica 27: 81–93

Chiang MS, Chiang BY, Grant WF (1977) Transfer of resistance to race 2 of *Plasmodiophora brassicae* from *Brassica napus* to cabbage (*B. oleracea* var. *capitata*). I. Interspecific hybridization between *B. napus* and *B. oleracea* var. *capitata*. Euphytica 26: 319–336

Collins GB, Grosser JW (1984) Culture of embryos. In: Vasil I.K. (ed.). Cell culture and somatic cell genetics of plants. 1: 241–257

Delourme R, Eber F, Chevre AM (1989) Intergeneric hybridization of *Diplotaxis erucoides* with *Brassica napus*. I. Cytogenetic analysis of F_1 and BC_1 progeny. Euphytica 41: 123–128

Hosoda T (1950) On new types of *Brassica napus* obtained from artificial amphidiploids. I. A new type as a forage crop. Ikushu Kenkyu (Tokyo) 4: 91–95

Inomata N (1975) In vitro culture of ovaries of *Brassica* hybrids between 2x and 4x in *Raphanus sativus*. Jpn J Genet 50: 1–18

Inomata N (1976) Culture in vitro of excised ovaries in *Brassica campestris* L. I. Development of excised ovaries in culture media, temperature and light. Jpn J Breed. 26: 229–236

Inomata N (1977) Production of interspecific hybrids between *Brassica campestris* and *Brassica oleracea* by culture in vitro of excised ovaries. I. Effects of yeast extract and casein hydrolysate on the development of excised ovaries. Jpn J Breed 27: 295–304

Inomata N (1978) Production of interspecific hybrids between *Brassica campestris* and *Brassica oleracea* by culture in vitro of excised ovaries. II. Effects of coconut milk and casein hydrolysate on the development of excised ovaries. Jpn J Genet 53: 1–11

Inomata N (1979) Production of interspecific hybrids in *Brassica campestris* × *B. oleracea* by culture in vitro of excised ovaries. II. Development of excised ovaries on various media. Jpn J Breed 29: 115–120

Inomata N (1980) Hybrid progenies of the cross, *Brassica campestris* × *B. oleracea*. I. Cytological studies on F_1 hybrids. Jpn J Genet 55: 189–202

Inomata N (1983) Hybrid progenies of the cross, *Brassica campestris* × *B. oleracea*. II. Crossing ability of F_1 hybrids and their progenies. Jpn J Genet 58: 433–449

Inomata N (1985a) A revised medium for in vitro culture of *Brassica* ovaries. In: Chapman GP, Mantell SH, Daniels RW (eds) The experimental manipulation of ovule tissue. Longman, London, pp 164–176

Inomata N (1985b) Interspecific hybrids between *Brassica campestris* and *B. cretica* by ovary culture in vitro. Cruciferae Newslett 10: 92–93

Inomata N (1986) Interspecific hybrids between *Brassica campestris* and *B. bourgeaui* by ovary culture in vitro. Cruciferae Newslett 11: 14–15

Inomata N (1987) Interspecific hybrids between *Brassica campestris* and *B. montana* by ovary culture in vitro. Cruciferae Newslett 12: 8–9

Karpechenko GD (1924) Hybrids of *Raphanus sativus* L. × *Brassica oleracea* L. J Genet 14: 375–396

Kumar PBAN, Shivanna KR (1986) Interspecific hybridization between *Brassica fruticulosa* and *B. campestris*. Cruciferae Newslett 11: 18

Kumar PBAN, Shivanna KR, Prakash S (1988) Wide hybridization in *Brassica*-crossability barriers and studies on F_1 hybrids and synthetic amphidiploid of *B. fruticulosa* × *B. campestris*. *Sex Plant Repro.* 1: 234–239

Laibach F (1925) Das Taubwerden von Bastardsamen und die kuenstliche Auzucht frühausterbender Bastardembryonen. Z Bot 17: 417–459

Mastsuzawa Y, Sarashima M (1984) Intergeneric hybrids between *Raphanus sativus* and *Brassica nigra*. Cruciferae Newslett 9:29

Matsuzawa Y, Sarashima M (1986) Intergeneric hybridization of *Eruca, Brassica* and *Raphanus*. Cruciferae Newslett 11:17

McNaughton H, Ross CL (1978) Interspecific and intergeneric hybridization in the *Brassica* with special emphasis on the improvement of forage crops. Ann Rep Scott Plant Breed Sta 75–110

Mizushima U (1950) Karyogenetic studies of species and genus hybrids in the tribe Brassiceae of Cruciferae. Tohoku J Agric Res 1: 1–14

Mizushima U (1952) Karyo-genetical studies on Brassiceae. Gihodo, Tokyo, p 112

Mohapatra D, Bajaj YPS (1984) In vitro hybridization in an incompatible cross—*Brassica juncea* × *Brassica hirta*. Curr Sci 53: 489–490

Mohapatra D, Bajaj YPS (1987) Interspecific hybridization in *Brassica juncea* × *Brassica hirta* using embryo rescue. Euphytica 36: 321–326

Mohapatra D, Bajaj YPS (1988) Hybridization in *Brassica juncea* × *Brassica campestris* through ovary culture. Euphytica 37: 83–88

Nishi S, Kawata J, Toda M (1959) On the breeding of interspecific hybrids between two genomes "o" and "a" of *Brassica* through the application of embryo culture techniques. Jpn J Breed 8: 215–222

Nishiyama I, Inomata N (1966) Embryological studies on cross incompatibility between 2x and 4x in *Brassica*. Jpn J Genet 41: 27–42

Paulmann W (1987) Development of CMS using nucleo-cytoplasmic interactions between radish (*R. sativus*) and rape (*B. napus*). In: Proc. 7th Int Rapeseed Conf, Poznan-Poland, pp. 87–90

Quazi MH (1988) Interspecific hybrids between *Brassica napus* L. and *B. oleracea* L. developed by embryo culture. Theor Appl Genet 75: 309–318

Raghvan V (1986) Embryogenesis in angiosperms: a developmental and experimental study. Univ Press, Cambridge

Ripley VL, Arnison PG (1990) Hybridization of *Sinapis alba* L. and *Brassica napus* L. via embryo rescue. Plant Breed 104: 26–33

Shivanna KR (1982) Pollen-pistil interaction and control of fertilization, In: Johri B.M. (ed.). Experimental embryology of vascular plants. Springer, Berlin Heidelberg Tokyo New York, pp 131–174

Snorgerup S (1980) The wild forms of the *Brassica oleracea* group ($2n$ = 18) and their possible relations to the cultivated ones. In: Tsunoda, S, Hinata K, Gõmez-Campo C (eds) *Brassica* crops and wild allies. Jpn Sci Soc Tokyo, pp 121–132

Takada M, Maruyama Y, Kunieda H, Hibino Y, Ujihara K, Yai H, Etsugawa K, Doi H, Tsuda K (1987) Studies on the breeding of artificially synthesized *Brassica napus* Hakuran with head formation habit and the establishment of cropping system of the F_1 hybrids. Gifu Agric Res Cent 1: 1–185

Takahata T (1988) Production of intergeneric hybrids between *Moricandia arvensis* (L). DC and *Brassica oleracea* L. through ovary culture. Jpn J Breed 38 (suppl. 2): 338–339

Takahata T, Takeda T (1990) Intergeneric (intersubtribe) hybridization between *Moricandia arvensis* and *Brassica* A and B genome species by ovary culture. Theor Appl Genet 80: 38–42

Takeshita M, Kato M, Tokumasu S (1980) Application of ovule culture to the production of intergeneric or interspecific hybrids in *Brassica* and *Raphanus*. Jpn J Genet 55: 337–387

White PR (1963) The cultivation of animal and plant cells. Ronald Press, New York, pp 228

CHAPTER 8

Genome Manipulations

Shyam Prakash and *V.L. Chopra*

8.1 Introduction

Manipulations at the genome and chromosome level play an important role in plant improvement. Manipulation of entire chromosomal set, or genome, broadly falls into two categories: (1) doubling of the genome, i.e. autopolyploidy, and (2) combining two or more different sets, i.e. allopolyploidy. The artificial synthesis of alloploid brassicas has fascinated cytogeneticists, breeders and, of late, tissue culturists, their motive being the introduction of useful genes from diploids to alloploids. The earliest documented descriptions of interspecific hybrids are those of *B. napus* × *B. campestris* by Herbert in 1834 and *Raphanus sativus* × *B. oleracea* by Sageret in 1826. An entirely new species was synthesized from the cross *Raphanus sativus* × *B. oleracea* by Karpechenko (1928), and named *Raphanobrassica.* U (1935) produced the synthetic *B. napus* to substantiate Morinaga's views on genome relationships. In recent years, approaches like ovary culture, embryo rescue and somatic hybridization have helped the synthesis not only of the naturally occurring forms but also of new alloploid combinations of members from the secondary and tertiary gene pool, which was so far not possible. This has provided a spectrum of new genetic variation for exploitation in crop improvement (see Prakash and Chopra 1991). In this chapter, we present a comprehensive survey of different aspects of induced polyploids, both auto and allo, in crop brassicas.

8.2 The *Brassica* Genomes

Cytogenetical investigations on crop brassicas were initiated in the beginning of present century with the determination of chromosome numbers. Morinaga's extensive experiments (for references see Prakash and Hinata 1980), involving production of interspecific hybrids and study of their pairing behaviour during meiosis, led to the establishment of two groups of three species each : (1) monogenomic diploids comprising *B. nigra* ($2n$ = 16; BB), *B. oleracea* ($2n$ = 18; CC) and *B. campestris* ($2n$ = 20; AA); and (2) digenomic alloploids comprising *B. carinata* ($2n$ = 34; BBCC), *B. juncea* ($2n$ = 36; AABB) and *B. napus* ($2n$ = 38; AACC). The cytogenetic relationship among the *Brassica* species was presented by U (1935) in a scheme commonly referred to as U's triangle (Fig. 1). The diploid species form an aneuploid series (Manton 1932) and secondary polyploidy from an archetype with x = 6 has been implicated in their origin. This conclusion has been drawn on the basis of secondary associations of bivalents (Catcheside 1937; Haga 1938), chromosome pairing

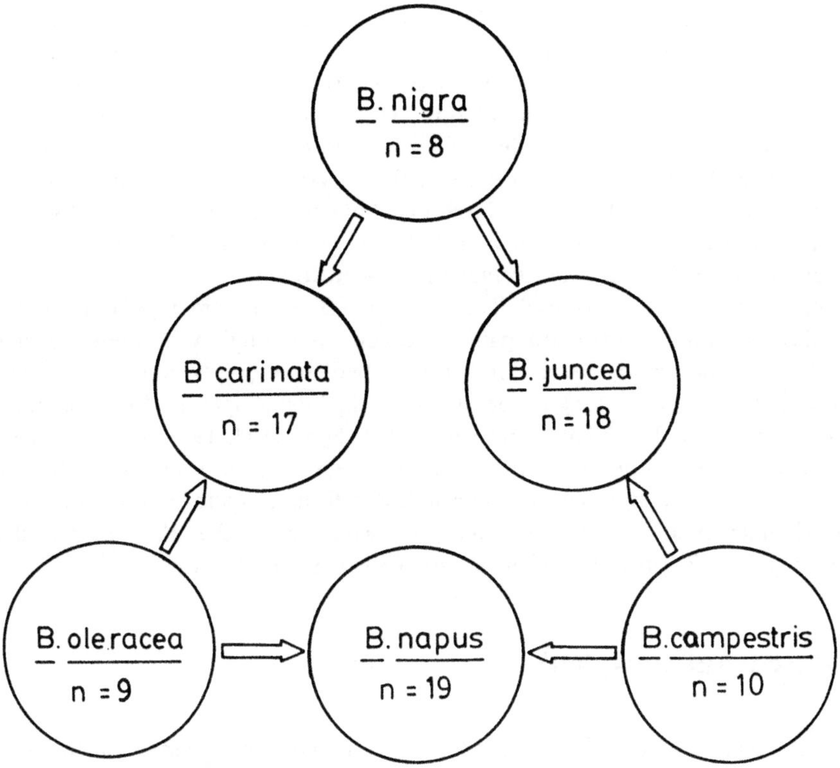

Fig. 1. Triangle of U (U 1935)

in haploids (Thompson 1956; Prakash 1973b; Armstrong and Keller 1981; 1982), pachytene chromosome morphology (Röbbelen 1960b; Venkateswarlu and Kamla 1971), and isozyme markers (Quiros et al. 1985). Pachytene chromosome analysis by Röbbelen (1960b) and Venkateswarlu and Kamla (1971) revealed that the diploids have six basic types of chromosomes. *B. campestris* genome is represented by AABCDDEFFF (tetrasomic for chromosomes A and D, and hexasomic for chromosome F); *B. oleracea* by ABBCCDEEF (tetrasomic for chromosomes B, C and E) and *B. nigra* by AABCDDEF (tetrasomic for chromosomes A and D). However, chromosomes of a type within or between the genomes have lost homoeology due to structural and/or genic alterations. This structural divergence is reflected in reduced pairing in the haploids and hybrids between monogenomics. Although all the cytogenetic-based reports suggest that the three basic species evolved from a common archetype, recent work on RFLPs (Song et al. 1988) and cpDNA restriction patterns (Yanagino et al. 1987) indicate a biphylatic origin: *B. campestris* and *B. oleracea* evolved from one archetype, and *B. nigra* from the other, probably *Sinapis arvensis*. The fact that up to nine bivalents are formed in *B. campestris* × *B. oleracea* hybrids, (Olsson 1960b) and up to eight bivalents in *S. arvensis* × *B. nigra* hybrid (Mizushima 1950a) substantiates this view. To assess chromosome homology between basic genomes, pairing has been studied in the relevant hybrids. Mizushima (1950a; 1980) has discussed these aspects extensively. In conjunction with the earlier works of Haga (1938) and Fukushima (1945), Mizushima (1968; 1980) considered three aspects:

(1) chromosome pairing in digenomic triploids such as AAB, BBC etc., (2) pairing in haploids of basic species, and (3) multivalent formation in induced allotetraploids. He suggested that A, B and C genomes can form 1, 2 and 2 autosyndetic bivalents respectively. He also proposed that in AC, AB and BC hybrids, 8, 3 and 4 bivalents respectively arise from allosyndetic pairing. He inferred that *B. campestris* and *B. oleracea* are closer to each other than either is to *B. nigra*—a fact later confirmed by RFLP analysis (Song et al. 1988), and studies of Attia and Röbbelen (1986) on chromosome pairing in AB, BC and AC hybrids. Attia and Röbbelen (1986) observed the chromosome associations were low in hybrids involving *B. nigra* (B) genome, while in an AC hybrid quadri- and pentavalents were formed.

Digenomic species *B. carinata, B. juncea* and *B. napus* are regular bivalent-forming allotetraploids. Restriction patterns of cp- and mtDNA of these species resolved their cytoplasmic donors and established *B. nigra, B. campestris* and *B. oleracea* as the respective female parents of the digenomic species (Erickson et al. 1983; Palmer et al. 1983; Ichikawa and Hirai 1983; Song et al. 1988; Palmer 1988). There are also indications of coinheritance of mt and cp genomes (Palmer 1988). The cp genomes have been conserved in *B. carinata* and *B. juncea* and slightly altered in *B. napus* (Palmer et al. 1983). DNA analysis has indicated that *B. juncea* is the oldest, followed by *B. carinata* and *B. napus* (Quiros et al. 1985).

8.3 Induced Autopolyploids

Morinaga and Kuriyama (1937) reported the first autoploid in *B. napus* which arose spontaneously. With the discovery of colchicine, induced tetraploidy became a routine procedure and tetraploids were obtained in all *Brassica* species. It was observed that, in general, tetraploids of monogenomic species were vigorous and put forth luxuriant growth and large-sized flowers. In contrast, tetraploids of digenomic species were weak. Considerable variations in chromosome pairing are observed in the C_1 generation in the tetraploids. Multivalents are more frequent in tetraploids of digenomics than in those of monogenomics. Tetraploid *B. nigra* showed, besides 16 bivalents, one or two ring or rod quadrivalents, and the mean quadrivalent frequency was 0.80% (Ramanujam and Deshmukh 1945; Olsson 1960a). In *B. oleracea*, the quadrivalent frequency was variable; 4.0 in var. *fruticulosa* (Howard 1939) and 1.50 in vars. *capitata, fimbriata* and *gongylodes* (McNaughton 1973b). In *B. campestris* fewer quadrivalents were found than expected, as Nishiyama (1949) observed only an occasional quadrivalent in ssp. *pekinensis, narinosa, nipposinica* and *rapifera*. Ramanujam and Deshmukh (1945) found, on an average, 1.94 quadrivalents per cell in oil forms of *B. campestris*. Meiosis is highly disorganized in tetraploids of *B. juncea* (Ramanujam and Deshmukh 1945), *B. carinata* and *B. napus* (Nishiyama 1949). Pollen fertility is conspicuously high in tetraploids of monogenomics (88-97%) and poor in digenomics (47-54%). Seed fertility is poor in monogenomics and very poor in digenomics. Olsson (1963) observed that in digenomic species tetraploidy raises chromosome numbers much above the optimum and is hence of no practical utility. Poor seed set is an impediment to utilization of autotetraploids particularly where seed is the economic product. Precisely for this reason, polyploids have been exploited only in vegetable and fodder forms. Several cultivars of importance have been bred and released. Turnip 'Svalöf Sirius' is a "mass-cross" between three induced tetraploid strains, and exhibit considerable increase in root and dry matter yields. It has also been reported that marrow-stem kale responds very favourably to autoploidy. Marrow-stem kale 'Svalöf Tema' is a highly productive tetraploid form released for fodder, it outyielded the diploid by 6% in dry matter and 5% in crude protein (Olsson and Ellerström 1980). A promising tetraploid fodder kale has also been obtained in Poland from a cross between tetraploids of cabbage and kale (Olsson and Ellerström 1980). Autotetraploid of oleiferous *B. campestris* var. *toria* in spite of sustained efforts to improve seed fertility, could not yield a commercial variety (Parthasarathy and Rajan 1953).

8.4 Synthesis of Natural Allopolyploids

8.4.1 Synthesis

Brassica carinata. Frandsen (1947) was the first to synthesize *B. carinata* by hybridizing tetraploid *B. oleracea* and *B. nigra.* Mizushima (1950b) crossed diploid *B. nigra* and *B. oleracea* var. *capitata.* Pearson (1972) used *B. oleracea* var. *italica broccoli* as the C genome donor. In all these reports, *B. nigra* was the female parent. Prakash et al. (1984) exploited different *B. oleracea* forms, such as vars. *capitata, botrytis, italica* and *alboglabra* and hybrids were obtained between diploids in reciporcal crosses.

Brassica juncea. Sinskaia (1927) made unsuccessful attempts to hybridize *B. campestris* and *B. nigra.* The first report on synthetic *B. juncea* was by Howard (1942), who obtained it from the cross of tetraploid *B. campestris* spp. *chinensis* and *B. nigra* doubled by colchicine treatment. The amphidiploid hybridized easily with natural *B. juncea.* A year later, Frandsen (1943), and Ramanujam and Srinivasachar (1943) reported synthesis of *B. juncea.* Frandsen (1943) employed reciprocal crosses between tetraploid *B. campestris* and *B. nigra,* while Ramanujam and Srinivasachar (1943) crossed diploid *B. campestris* spp. *oleifera* var. *brown sarson* with *B. nigra.* Olsson (1960a) concluded from extensive attempts at synthesis at the diploid and tetraploid levels that hybrids were obtained only when *B. campestris* was used as the female parent. He used three varieties of *B. campestris* ssp. *oleifera* and two of *B. nigra.* Srinivasachar (1965) crossed diploids of spp. *chinensis* and *rapifera* of *B. campestris* with *B. nigra* and obtained a synthetic *juncea* after chromosomal doubling. Prakash (1973a) employed a large number of varieties in *B. campestris* viz. ssp. *pekinensis*, *chinensis*, *japonica*, *narinosa* of leafy group; vars. *yellow sarson*, *brown sarson* and *toria* of ssp. *oleifera* and turnip (ssp. *rapifera*) as contributors of the A genome for the synthesis of *B. juncea. B. campestris* was always used as the female parent in crosses with *B. nigra.*

Brassica napus. To verify the hypothesis of Morinaga, U (1935) was the first to obtain a synthetic *B. napus* by hybridizing diploid *B. campestris* ssp. *oleifera* and ssp. *chinensis* with *B. oleracea* vars. *acephala* and *capitata.* In more recent times, attempts to synthesize *B. napus* have been spurred on by the economic importance of this species in agriculture. Crosses have been made at diploid and tetraploid levels, and by using both the species as the female parents. Frandsen (1947) crossed tetraploid *B. campestris* with *B. oleracea.* Reciprocal crosses between diploids of ssp. *rapifera* and *oleifera* of *B. campestris* and vars. *capitata* and *fimbriata* of *B. oleracea* were attempted by Rudorf (1950). Semi-synthetic forms of *B. napus* were produced by pollinating the primary AC hybrid with pollen of natural *B. napus* (Rudorf 1950). Another method of producing semi-synthetic *B. napus* was the production of *B. napocampestris* ($2n$ = 58, AAAACC) and *B. napoleracea* ($2n$ = 56; AACCCC) and pollinating the autoallohexaploids with *B. oleracea* and *B. campestris* respectively (Hoffman and Peters 1958; Grabiec 1967; Johnston 1974; McNaughton and Ross 1978). Japan has been the centre of vigorous activity for work on artificial *B. napus* (see Namai et al. 1980). The work was initiated by Mizushima (1946), and subsequent reports include those of Hosoda (1961), Nishi et al. (1962) Sarashima (1964, 1973), Namai (1971) and Inomata (1978). Sarashima (1964, 1973) in his comprehensive investigations, utilized, the vegetable morphotypes of *B. campestris* ssp. *pekinensis, chinensis* and *rapifera* and obtained hybrids in reciprocal directions at the $2x$ and $4x$ ploidy level. Introduction of *B. napus* as an oilseed crop in India was initiated in 1967 and synthetic forms were obtained from hybrids between *B. campestris* ssp. *oleifera* var. *brown sarson* × *B. oleracea* var. *botrytis* (Prakash 1980). Recently, Chen et al. (1988) have resynthesized *B. napus* for developing yellow-seeded forms. Light brown-seeded *B. oleracea* var. *alboglabra* was hybridized with yellow-seeded *B. campestris* for this purpose.

More recent developments in somatic cell manipulations have enabled production of *B. napus* through protoplast fusion (Schenck and Röbbelen 1982; Robertson et al. 1985; Jourdan et al. 1986; Taguchi and Kameya 1986; Terada et al. 1987; Sundberg et al. 1988; Rosen et al. 1988).

8.4.2 Crossability

The experience of many investigators has indicated that it is easier to produce *B. carinata* and *B. juncea* than to obtain *B. napus* from crosses of relevant parental species. As mentioned earlier, the natural crosses which led to the evolution of *B. carinata, B. juncea* and *B. napus* were *B. nigra* × *B. oleracea*, *B. campestris* × *B. nigra* and *B. oleracea* × *B. campestris.* This has been deduced from comparisons of cpDNA patterns. However, artificial synthesis has been accomplished from all possible combinations. Synthetic *B. carinata* with *B. nigra* and *B. oleracea* cytoplasms has been obtained with ease. For the synthesis of *B. juncea*, *B. campestris* has been successful only as a female, except for a report of Frandsen (1943), who obtained four plants from the cross $4x$ *B. nigra* × $4x$ *B. campestris.* All later attempts failed to hybridize *B. nigra* as female with *B. campestris*. Matsuzawa's (1984) work showed that embryo abortion in this cross is the cause of severe post-zygotic incompatibilities. Synthetic *B. napus* has been derived from the hybrids *B. campestris* × *B. oleracea* and *B. oleracea* × *B. campestris*, but mostly from the former.

Many barriers, mostly post-fertilization, limit hybridization between the basic species. Olsson (1960b) reported that genotype and the physiological state of the maternal parent play important roles in hybridization. Studies of Röbbelen (1960a) on pollen tube growth in interspecific crosses revealed that many crosses do not succeed because of inhibition of pollen-tube growth. Disturbed endosperm development and disharmonious interaction between the developing embryo and endosperm also cause abortion of hybrid embryos. Ways have been devised to overcome these barriers. Pre-fertilization barriers are overcome by grafting, mixed pollination, style excision, treatment of style with chemicals and embryo rescue (Table 1). It has also been

Table 1. Success rate of *B. campestris* × *B. oleracea* hybrids by application of different techniques

Methods and cross	Hybrids per 100 Pollinations	Reference
Graft		
B. campestris ×	7.865	Hosoda et al. (1963)
B. oleracea	7.071	Namai (1971)
Mixed pollination		
B. campestris ×	2.222	Feng (1955)
B. oleracea		Sarashima (1964)
Style excision		
B. campestris ×	5.615	Hosoda et al. (1963)
B. oleracea		
Embryo culture		
B. oleracea ×	6.815	Nishi et al. (1961)
B. campestris		
B. campestris ×	3.532	Nishi et al. (1970)
B. oleracea		Matsuzawa (1984)
Ovary culture		
B. campestris ×	8.333	Inomata (1978)
B. oleracea		
Supplemented with coconut milk and casein hydrolysate	23.214	Inomata (1978)

observed that hybridization between tetraploids yields more hybrids than crosses between diploids (Table 2). McNaughton (1963) has estimated the number of flowers required to be pollinated for obtaining one hybrid for the synthesis of *B. napus* (Table 3). His study indicates that 2*x* *B. oleracea* × 2*x* *B. campestris* cross is the most difficult and requires 12,238 pollinations to obtain a hybrid. In comparison, 192 pollinations will most likely yield a hybrid when crosses are attempted at the tetraploid level.

Post-fertilization barriers are overcome by in vitro techniques like ovary, ovule and embryo culture. These techniques have helped considerably in increasing the hybrid success frequency (Table 1) as examplified by the elegant experiments of Nishi et al. (1961) and Inomata (1978). Increased interest is also being shown in protoplast fusion to obtain somatic hybrids for combining cytoplasmic organelles.

Table 2. Frequency of successful hybrids at diploid and tetraploid level

Cross	No. of hybrids per 100 pollinations	Reference
For production of *B. juncea*		
2*x* *B. campestris* × 2*x* *B. nigra*	4.296	Ramanujam and Srinivasachar (1943)
	1.245	Olsson (1960a)
	0.787	Prakash (1973a)
2*x* *B. nigra* × 2*x* *B. campestris*	0.0	Olsson (1960a)
4*x* *B. campestris* × 4*x* *B. nigra*	2.257	Olsson (1960a)
		Frandsen (1943)
For production of *B. napus*		
2*x* *B. campestris* × 2*x* *B. oleracea*	0.545	U (1935)
	0.153	Olsson (1960b)
	0.299	Hosoda (1961)
	1.685	Namai (1971)
	0.291	Prakash and Raut (1983)
4*x* *B. campestris* × 4*x* *B. oleracea*	2.166	Frandsen (1947)
	0.578	Olsson (1960b)
2*x* *B. oleracea* × 2*x* *B. campestris*	0.659	Hoffman and Peters (1958)
	0.711	Nishi et al. (1961)
	0.0	Olsson (1960b)
	0.0	Hosoda et al. (1963)
	0.0	Namai (1971)
4*x* *B. oleracea* × 4*x* *B. campestris*	0.688	Olsson (1960b)
For production of *B. carinata*		
2*x* *B. oleracea* × 2*x* *B. nigra*	1.23	Prakash (unpubl.)
2*x* *B. nigra* × 2*x* *B. oleracea*	3.86	Matsuzawa (1984)
	2.29	Prakash (unpubl.)
4*x* *B. nigra* × 4*x* *B. oleracea*	4.25	Frandsen (1947)

Table 3. Number of pollinations required for the synthesis of *B. napus* (McNaughton 1963)

Cross-combination	Required number of pollinations per hybrid
2*x* *B. campestris* × *B. oleracea*	432
2*x* *B. oleracea* × *B. campestris*	15,238
4*x* *B. campestris* × 4*x* *B. oleracea*	211
4*x* *B. oleracea* × 4*x* *B. campestris*	192

8.4.3 Meiosis

In raw alloploids, meiosis is, in general, disturbed due to the presence of multivalent associations and univalents. With advancing generations, the cytological abnormalities decline and by generation A_7, the synthetics attain complete stabilization in meiotic behaviour. The literature is reviewed here according to species.

Brassica carinata. Frandsen (1947) observed near-regular meiosis in A_1 plants. Most PMCs had 17 bivalents and only occasional PMCs showed non-pairing of chromosomes in the form of two univalents. Mizushima (1950b) carried out a detailed study and reported the occurrence of 0-4 quadrivalents and 9-17 bivalents in generation A_1. In spite of these multivalent associations, anaphase I was fairly regular in terms of chromosome distribution. Mizushima and Katsuo (1953) further studied the progeny of these plants and observed that up to four quadrivalents persist till generation A_5. In generation A_6, numerical chromosome abnormalities were confined to an occasional trivalent and only a few univalents.

Brassica juncea. In *B. juncea*, variation in chromosome pairing in different generations of synthetics is small. An important feature is the very low frequency of multivalent associations in early generations. Frandsen (1943), and Ramanujam and Srinivasachar (1943) noticed mostly 18 bivalents in A_2 generation plants and regular bivalent formation afterwards. Olsson (1960a), and Prakash (1973a) found that multivalents were rare. In early generations, univalents can number up to eight per cell. Regularity of pairing improves quickly and the univalent number drops to four per cell by generation A_3. Complete meiotic stabilization, as evidenced from regular bivalent formation, is achieved by the A_6 generation.

Brassica napus. Most investigators reported the occurrence of the normal 19 bivalents in synthetic *B. napus*. Cells with 18II + 2I and 17II + 4I were observed in low frequency (U 1935; Frandsen 1947; Rudorf 1950; Olsson 1960b). A detailed analysis of meiotic pairing by Sarashima (1973), however, revealed a number of irregularities. Two quadrivalents, and univalents ranging from 2 to 12 were common from generation A_1 to A_2. An occasional trivalent was also noticed in the A_1 generation. The highest observed chromosome association was 2IV + 1III + 8II + 11 I. The number of quadrivalents and univalents diminished progressively with advancing generations. In generation A_4, the majority of the cells had regular chromosome pairing in the form of 19 bivalents. Prakash and Raut (1983) also noticed one to three quadrivalents in generations A_1 and A_2. Univalents up to eight persisted up to generation A_5. The synthetics achieved complete stabilization by generation A_5 or A_6.

Multivalents are believed to be formed due to intragenomic pairing (see Mizushima 1980). As the generations advance, structural changes in chromosomes accumulate so that pairing is restricted to homologues and only bivalents are formed. There is no conclusive evidence that the change from multivalent to bivalent formation is brought about by a genetic mechanism analogous to that present in wheat (Riley and Chapman 1958). However, such a system has been proposed for *B. juncea* (Prakash 1974), and naturally occurring autotetraploid of a wild species, *Brassica cossoneana* (Harberd 1976). The alloploids are as a rule inbreeders, although occasional outbreeding plants are observed in synthetic forms. The constituent parental genomes are, on the other hand, outbreeders. It is implied that for effective functioning of meiosis in the alloploid, a certain level of heterozygosity in the constituent genomes is required. However, chromosome doubling of F_1 interspecific hybrid and their inbreeding nature renders the alloploids completely homozygous. This causes a disturbance in balanced chromosome pairing, resulting in a decrease of chiasma frequency and the occurrence of univalents. As the generations advance, the constituent genomes adjust to a balanced interaction and cohesive functioning so that normalcy in pairing is attained and bivalents replace univalents. Consequently, the univalents disappear.

8.4.4 Fertility

Synthetics invariably show reduced pollen and seed fertility in early generations although some are fully fertile in A_1 itself, as has been recorded by Olsson (1960b) for *B. napus*. A few generations of selection for fertility introduce considerable improvement and are accompanied by improvement in chromosome pairing. Synthetic *B. carinata* was poorer in seed fertility as compared to *B. juncea* and *B. napus* in generation A_1. To attain high seed fertility, a longer time is necessary for *B. carinata* and *B. juncea* than for *B. napus*. Some *B. napus* forms were fully fertile in generation A_2 (Olsson 1960b). *B. juncea* synthetics attained around 96% fertility by generation A_4 or A_5; the majority of the plants had the same number of seeds per siliqua as the best natural cultivars (Olsson 1960a; Prakash 1973a). A detailed investigation on seed fertility in synthetic *B. napus* was undertaken by Sarashima (1973), who observed a gradual improvement: 24% in A_2, 36% in A_3, 59% in A_4 and 79% in the A_5 generation.

8.4.5 Agronomic Potential of Synthetic Alloploids

Brassica juncea. Work on synthetic *B. juncea* is confined to India because it is cultivated primarily in the Indian subcontinent and China (Prakash 1973a, 1980). Its cultivation is marginal in Europe.

A survey of synthetic *B. juncea* material reveals that those originating from leafy forms of *B. campestris* show vigorous growth. They have quick-growing large-sized leaves and the plants make a good source of fodder. In contrast, plants involving oil forms of *B. campestris* have small-sized leaves and a higher number of primary and secondary branches. Variation is also found for other yield-contributing parameters such as a large number of highly non-shattering pods and bold seed size (6-7 mg). The variation has been utilized for introgression of new sets of genes into existing *juncea* cultivars at different centres in India.

Brassica carinata. With the recent findings that *B. carinata* has particular promise as an oil crop under dryland conditions, work has been initiated on developing suitable genotypes through artificial synthesis in several countries including India (Prakash et al. 1984).

Synthetic forms exhibit several novel characteristics which can be exploited directly or by introgression into cultivated genotypes. These include early maturing (110 days against 195 days of natural *B. carinata*), increased number of primary and secondary branches, dwarf stature (85 cm against 230 cm of natural forms) and branches with high pod density. Some lines also possess reduced vegetative growth. A noteworthy feature revealed by our study on synthetics is that cytoplasm contributes considerably to several of the expressed morphological and physiological traits. It was observed that plants with *B. nigra* cytoplasm were tall (220-245 cm), had a stout stem and were late in flowering and maturity. Also, their large-sized leaves give the plants a bushy appearance. On the other hand, plants with *B. oleracea* cytoplasm were small in stature (150-179 cm), had weak stem, a reduced number of small-sized leaves and were early in flowering. Synthetics derived from *B. nigra* female were more productive (on an average 55 g seeds per plant) as compared to those with *B. oleracea* female (35 g seeds per plant).

Brassica napus. The enormous variation available in *B. campestris* and *B. oleracea* has been used for synthesizing three forms of *B. napus* (oil rape, fodder rape and swedes). Oil rape has received major attention in Europe and India, while fodder rape and swedes have been exploited in Japan and the European countries.

Synthetic alloploids of *B. napus* are inferior to natural cultivars in seed yield. They have, however, been used to advantage as germplasm in breeding programmes to evolve superior oil rapes. In an extensive programme of hybridization and repeated selection based on synthetic oil forms of *B. napus* (Olsson 1960b), several promising strains were evolved (Olsson 1986). These include Svalöf Panter, a selection from a cross between artificial rape 5 (4*x* Lembkes turnip rape × 4*x* green curled kale) and other artifical rapes. It was commercially released for its higer oil yield and comparatively rapid growth at low temperatures. Another selection, Svalöf Norde, from a cross between artificial rape (4*x* wild cabbage × 4*x* turnip rape) × Matador rape outyielded the Matador control both in seed and oil yield (Table 4). It is the most hardy Swedish rape cultivar, and possesses considerable resistance to *Peronospora* and *Verticillium*. Two other cultivars, Brink and Jupiter, both selected from the cross Sirus × Norde, are high seed yielders with very low erucic acid content (Table 5).

Table 4. Comparative performance of cultivar Matador and semisynthetic winter rape Norde (Olsson 1986)

Character	Metador	Norde
Seed yield, kg/ha	2940	3105
Seed yield, relative value	100	106
Oil content (%)	46.4	46.1
Oil yield, kg/ha	1164	1220
Oil yield, relative value	100	105
Winter hardiness (%)	74	81
Stem stiffness (%)	75	76
Resistance to *Peronospora* (%)	39	76
Resistance to *Verticillium* (%)	49	67

Table 5. Characteristics of synthetic cultivars Norde, Brink and Jupiter of *B. napus* (Olsson 1986)

Characters	Norde	Brink	Jupiter
Seed yield, kg/ha	2838	2699	3135
Oil content (%)	44.8	44.5	44.4
Oil yield, kg/ha	1086	1025	1186
Winter hardiness (%)	87	88	77
Erucic acid content (%)	52	1	1

Oil-yielding *B. napus* is a recent introduction in India. Exotic cultivars did not perform well because of their very late maturity and poor seed set due to photo- and thermosensitivity. Artificial synthesis of *B. napus* from indigenous and early-maturing constituent parents (*B. campestris* and *B. oleracea*) was initiated in 1967 to evolve early genotypes suitable for Indian conditions (Prakash 1980). Early and high-yielding strains were isolated in the A_{16} generation and from the segregating populations of the crosses between synthetics and exotic cultivars viz. Brutor, Tower and Maris Haplona (Prakash and Raut 1983). These selections are high-yielding, of medium maturity duration ranging from 151 to 158 days, and have a high degree of resistance to white rust (Table 6). However, all strains suffer from pod shattering (dehiscence of pods on maturity). The resultant seed loss limits their cultivation. This defect has now been rectified. Non-shattering of pods has been introgressed from *B. juncea* through non-homologous recombination between *B. nigra* and *B. oleracea* chromosomes (Prakash and Chopra 1988a, 1991). A recent report by Zaman (1989) indicates the possibility of introducing early maturing *B. napus* in Bangladesh also.

Fodder rape forms have been produced from crosses between leafy and rapiferous forms of *B. campestris* like ssp. *chinensis, narinosa, nipposinica, pekinensis* and *rapifera* (Olsson 1960b; Namai and Hosoda 1967, 1968; Ellerström and Sjödin 1973; McNaughton and Ross 1978). Hosoda (1950, 1953) bred a fodder rape, named CO, from a 56-chromosome AACCCC plant (*B. campestris* ssp. *pekinensis* × *B. oleracea* var. *capitata*) obtained by Mizushima (1946). Because of its very vigorous growth and strong winter hardiness, it was extensively cultivated in Japan as a forage crop

Table 6. Performance of synthetic strains of *B. napus* as against *B. juncea* cv. Varuna

Strains	Seed yield (kg/ha)	Maturity (days)	Oil content (%)
B. napus			
ISN-129	1508	157	42.6
ISN-706	1397	158	41.7
ISN-4	1466	158	42.9
B. juncea cv. Varuna	1384	150	41.6

Source: 1989 Report, All India Coordinated Workshop on *Brassica*

Table 7. Characteristics of Hakuran, *B. campestris* and *B oleracea* (Takasugi 1965)

Strains	Weight of head (kg)	No. of outer leaves	Height of head (cm)	Diameter of head (cm)	Sugar content (%)	Rate of heading
Hakuran SH-1 (Yoshin 1 × Matsushima)	1.46	26	27.0	21.0	7.4	95.4
B. oleracea var. *capitata* cv. Yoshin	1.40	14	12.0	27.0	7.0	100.0
B. campestris ssp. *pekinensis* cv. Matsushima	1.32	19	27.0	16.0	3.4	85.4

for spring harvest. Chinese cabbage is a major vegetable crop of Japan. However, its cultivation was threatened by its susceptibility to the soil-borne disease, soft rot, caused by *Erwinia aroideae*. Transfer of resistance to soft rot from cabbage led to the development of a new head-forming type of *B. napus* which was released under the name Hakuran in 1968 (Table 7). Hakuran is intermediate between chinese cabbage and cabbage in morphology. Its juicy, soft and low-fibre leaves taste like head lettuce.

Artificial swedes or rutabagas have been produced from the cross *B. campestris* ssp. *rapifera* × *B. oleracea* (Olsson et al. 1955), which excel over the existing cultivars. Hosoda et al. (1963), Namai and Hosoda (1968) and Kato et al. (1968) obtained rutabagas from the cross *B. campestris* ssp. *rapifera* × *B. oleracea* var. *gongylodes*. The resultant varieties, commonly known as SR types, are promising fodder forms in Japan. They are early, exhibit a high degree of winter hardiness and have good yields. McNaughton and Ross (1978) derived a rape which yielded more edible stem from the cross marrow stem kale: kohlrabi hybrid × turnip.

8.5 Synthesis of New Allopolyploids

Raphanobrassica. The intergeneric amphidiploid *Raphanobrassica* ($2n = 36$, RRCC) was synthesized and described for the first time by Karpechenko (1928). It had arisen from spontaneous chromosome doubling following the cross *Raphanus sativus* ($2n = 18$, RR) × *B. oleracea* ($2n = 18$, CC). *Raphanobrassica* does not occur in nature.

Several important investigations have been carried out. Particulars of synthesis, meiotic behaviour, fertility and agronomic potential of *Raphanobrassica* have been studied by many workers (Karpechenko 1928; Howard 1938; McNaughton 1973a; Iwasa and Ellerström 1981).

Raphanobrassica has been synthesized by hybridizing the diploids or autotetraploids, but *Raphanus* was always used as a female. Karpechenko (1924), Richharia (1937), Howard (1938) and Honma and Heecket (1962) used diploid forms for crossing. The amphidiploids arose following spontaneous chromosome doubling. McNaughton (1973a), on the other hand, mostly used autotetraploids. As mentioned earlier, *B. oleracea* is an extremely variable species and all its variations have been exploited in the synthesis of *Raphanobrassica*.

In spite of reported variable success of the cross *Raphanus* with *Brassica oleracea*, all investigators agree that it is a simple cross to attempt (0.2 hybrid seeds per pollination). The reciprocal hybrid *B. oleracea* × *Raphanus* is difficult to obtain, primarily due to failure of the *Raphanus* pollen tube to penetrate the *B. oleracea* style. Nevertheless, such hybrids have been obtained (Fukushima 1929; U et al. 1937; Makarova 1963; Moskov and Makarova 1969).

Raphanobrassicas are intermediate between the parents in most morphological characters but, on the whole, the *B. oleracea* parent dominates. The leaves are more like those of *B. oleracea* in that they are lyrate, glaucous and less hairy than those of *Raphanus*. Flower colour is generally pure white or veined with purple. The siliqua has a non-dehiscent, *Raphanus*-like, apical portion and a dehiscent *Brassica*-like basal portion. Chromosomes in the majority of the PMCs form 18 bivalents at M_I (Karpechenko 1928; Kondo 1942; McNaughton 1973a). However, occasional quadrivalent and trivalent was also observed (Howard 1938).

Raphanobrassicas, in general, are extremely low in seed fertility in the early generations. Karpechenko (1928), however, reported that the majority of his plants had nine to ten seeds per pod. Richharia (1937) and Howard (1938) reported considerably reduced seed fertility, ranging from 0.2 to 1.3 seeds per pod in generation A_2, four to five seeds per pod in A_3, increasing to eight seeds in A_5. McNaughton (1973a) found that a number of A_1 plants were completely sterile. They had empty and shrivelled anthers and produced no viable pollen. The plant produced very few seeds on self-pollination in A_1 (0.07 seeds/pollination) which was improved to 2.3 seeds/pollination after six generations of selection.

Several reasons have been ascribed to low seed fertility in spite of regular meiosis. Howard (1938) is of the view that non-homologous pairing in F_1 hybrids produced duplication-deficiency gametes which give rise to F_2 plants with fertility lower than true amphidiploids. McNaughton (1973a) discounted chromosomal abnormalities as the reason and suggested genetic imbalance as a more probable basis. This view is shared by Ellerström and Zagorcheva (1977), and Iwasa and Ellerström (1981).

During the 40 years after its first synthesis in 1928, no serious attempt was made to evaluate the agronomic potential of Raphanobrassicas. Around 1967, two independently conceived programmes were taken up at Svalöf and the Scottish Crop Research Institute to synthesize Raphanobrassicas and exploit them as possible forage crop and green manure. It was also envisaged to combine the disease resistance of fodder radish with the winter hardiness of kale. Moskov and Makarova (1969) were the first to report on the agronomic potential of Raphanobrassicas. At the Scottish Crop Research Institute, these were produced from autotetraploids of *B. oleracea* vars. *fruticulosa, fimbriata, acephala* and their hybrids (McNaughton and Ross 1978). Some of the strains exceeded forage rape in fresh weight and dry matter yield by 20%. It was also observed that RBs derived from marrow stem kale had better potential than others. They have resistance to *Plasmodiophora* (club root). *Peronospora* (downy mildew) and, in some instances, to *Erysiphe* (powdery mildew). RBs were synthesized at Svalöf with the aim of combining the rapid growth and high quality of *Raphanus* with the high productivity of marrow stem kale so that they could be used as green manure (Olsson and Ellerström 1980). Results from the preliminary trials showed that RBs produced almost the same yield at an early harvest as cv. Tema of marrow stem kale. However, they outyielded Tema by 17% in dry matter and had a much better quality due to a higher leaf to stem ratio. *Raphanobrassica* has also been suggested as a bridge species for transfer of genes for resistance to several diseases from *Raphanus* to *Brassica* (McNaughton 1982).

Prospects to exploit RBs as a new fodder crop appear to be good. However, the poor seed fertility and low dry matter content are constraints that need to be overcome. It is important to raise the seed fertility to at least ten seeds per siliqua from the present 2-3 seeds. Another difficulty is that one-fifth of the seed lodged in the apical *Raphanus*-like portion, which is hard and indehiscent, is difficult to extract.

Brassico-raphanus. Alloploid synthesized from the cross *B. campestris* × *R. sativus* ($2n$ = 38, AARR) is referred to as *Brassico-raphanus*. The term *Brassico-raphanus* was coined by Terasawa (1932) for the product from hybridizing *B. campestris* ssp. *chinensis* with *Raphanus*. This hybrid is difficult to obtain (0.1%) and has only seldom been reported. Tokumasu (1969) hybridized diploid *B. campestris* ssp. *japonica* with

Raphanus, while Ellerström and Sjödin (1973) crossed tetraploids of leafy oriental forms (such as *B. campestris* ssp. *chinensis* and *pervirides*) and tetraploid fodder radish. Plants were intermediate between the parents in general morphological characters and had white flowers like *Raphanus*. The meiosis was near normal with 19 bivalents at M1 and 19-19 distribution of chromosomes at anaphase 1 in generations A_1 and A_2. A few trivalents and one quadrivalent were occasionally noticed in the A_2 generation (Tokumasu 1976). However, the plants were highly sterile and set only a few seeds. In generation A_3, some plants with yellow flowers showed considerably improved pollen (71-84%) and seed fertility. Kato and Tokumasu (1976) investigated the origin of these yellow-flowered plants with increased fertility. They believed that the genes for flower colour are closely linked with those controlling embryo development. Homozygosity for the yellow-colour gene (y) and embryo-controlling gene(s) was attained due to segmental exchange of chromosomes between the homologous regions as manifested in quadrivalent formation in A_2 generation plants. The genetic reconstitution promotes development of embryos leading to higher fertility in yellow-flowered plants. In spite of the three times elevated level, seed fertility is still poor, and is a severe constraint to agronomic exploitation of *Brassico-raphanus*. It has been suggested that low seed fertility is caused by the failure of development of embryos after fertilization (Kato and Tokumasu 1989). This could be attributed to the fact that there exist several embryo growth-promoting genes, and in amphiploids, physiological imbalance caused by different genes derived from the different parents results in arresting the normal development of embryo and endosperm (Tokumasu and Kato 1988).

Brassica-napocampestris. In the beginning of the present century, extensive research was done to transfer desirable attributes from turnip (*B. campestris* ssp. *rapifera*) to swede rape (*B. napus*) by interspecific hybridization (Kajanus 1912; Sinskaia 1927). The 29-chromosome AAC hybrids were largely sterile. From such hybrids, Frandsen and Winge (1932) isolated one 58-chromosome plant that was partially fertile. It turned out to be an auto-allohexaploid of genetic constitution AAAACC. It was named *B. napocampestris*. Subsequently, synthesis of *B. napocampestris* has been attempted utilizing a wide range of available genetic variations in *B. campestris* with the aim of generating productive forms. The parents used for synthesis include ssp. *chinensis* (Yakuwa 1943), ssp. *oleifera* (Mizushima 1950b; Frandsen 1941; Jahr 1962; Röbbelen 1966) and ssp. *oleifera, rapifera, pekinensis* and *nipposinica* (Nwankiti 1970; McNaughton 1973b).

Synthesis can be accomplished by using *B. napus* as female (9.1 seeds per pollination) from the cross 2*x* *B. napus* × 2*x* *B. campestris* (McNaughton 1973b). The reciprocal cross gives hybrids only with difficulty, due to disturbed embryo development (Röbbelen 1966).

B. napocampestris plants are intermediate between the parents in morphological features. The leaves are thicker and the floral parts are larger.

Because of its auto-allohexaploid nature, multivalents are formed during meiosis. Up to nine IV and one VI were recorded by Mizushima (1950b) and one IV + two III by Frandsen (1941) and McNaughton (1973b). However, chromosome segregation at anaphase I was normal, as 90% cells contained 29 chromosomes at 2nd metaphase. Seed fertility varied from low to fairly good. Frequent aneuploids have also been observed in the progeny, as McNaughton (1973b) recorded plants with 55, 56 and 59 chromosomes in the A_3 generation.

Assessment of potential of *B. napocampestris* as an oilseed crop has been made at Svalöf with material originating from oilseed *B. napus*. The expectation was to combine rapid growth and winter hardiness of *B. campestris* with higher seed and oil yield of *B. napus*. However, *B. napocampestris* forms were found to be inferior to the diploid parent and of no practical value (Olsson 1986). Another objective of breeding is for green forage. A range of *B. napocampestris* forms has been produced from leafy *B. campestris* morphotypes (ssp. *pekinensis*, *chinensis*, *narinosa*, *nipposinica* etc.) at Svalöf (Nwankiti 1970) and at the Scottish Plant Breeding Station (McNaughton 1973b). It was observed that *napocampestris* gave a higher green-matter but a low dry-matter yield than fodder rape (Nwankiti 1975; McNaughton and Ross 1978). Therefore, efforts for their exploitation have been discontinued. In the mean time, it has been observed that primary triploid hybrid (29-chromosome, AAC) is vegetatively more vigorous than the fodder *B. napus* and *B. napocampestris*. It has been suggested that the triploid hybrids be used as fodder crop on a commercial scale. The triploid seed could be produced by exploiting self-incompatibility or male sterility systems in *B. napus*.

8.6 Chromosome Addition and Substitution Lines

Introgression of alien genetic variation is often hampered by the lack of pairing between two well-differentiated genomes. In such a situation individual chromosomes can be transferred intact from alien species by disomic addition or substitution lines. Apart from their use in the introgression of alien genetic variation, such stocks are also useful in the study of genome organization and evolution. Monosomic addition lines can be easily produced by first producing the amphiploid between recipient and donor species followed by repeated backcrossing with donor species and selecting plants with monosomic chromosome additions. Monosomic addition plants are then selfed to obtain disomic addition lines. Quiros et al. (1987) generated the *B. campestris-oleracea* chromosome addition lines by crossing and backcrossing natural amphiploid *B. napus* to the diploid parental species *B. campestris*. A total of eight disomic addition plants were identified and characterized. The disomic addition plants had an average pollen fertility of 91%. The addition lines were genetically characterized by genome-specific markers. The isozymes for 6PDG, LAP, PGI and PGM, and rDNA Eco I restriction fragments were found to possess the desired genome specificity. Duplicated loci for several of these markers were observed in *B. campestris* and *B. oleracea*, supporting the hypothesis that these diploid species are actually secondary polyploids. Stocks of *B. napus* with disomic chromosome additions from *B. nigra* have also been produced (Jahier et al. 1989).

The hybrid *B. napus* × *B. nigra* was backcrossed three times to *B. napus*. BC 3 plants with one or two additional chromosomes were either selfed or haplo-diploidized by anther culture followed by colchicine doubling. Thirty four disomic addition lines were produced. Their characterization is in progress. Studies are also underway for creating addition lines *B. napus-Diplotaxis erucoides* (Delourme 1989).

The addition lines are often unstable due to the low transmission frequency of the extra chromosome. The substitution lines, on the other hand, have the same chromosome number as the recipient parent and are expected to be meiotically more stable. The disomic chromosome substitution can be produced by crossing monosomics with disomic addition lines. The chromosome substitutions are also known to occur spontaneously in interspecific crosses. C-genome chromosome substitution lines in *B. juncea* have been reported to occur spontaneously in an interspecific cross between *B. juncea* and *B. napus* (Banga 1988). Cytological evidence suggests that at least three of the ten lines developed were substitutions for different chromosomes. The majority of these lines are meiotically stable and had normal vigour and fertility. This suggested that the genetic activity of some of the other substituted B chromosome is being compensated by a C chromosome. These stocks are important for their use in practical crop breeding and in certain basic genetic studies.

8.7 The Wild Allies

Genetic enriching of crop species from their wild relatives constitutes an important approach to broadening the base of genetic variability. An impressive spectrum of wild and weedy species closely related to crop brassicas exists, with extensive genetic diversity. These belong to several allied genera viz. *Diplotaxis, Erucastrum, Sinapis, Raphanus* etc. They are distributed from the Middle East to the northwest of India in the Saharo-Sindian region and occupy such diverse habitats as old coastal dunes, slopes of coastal volcanoes, stony pastures and arid to semi-arid regions. Although the chromosome numbers for quite a good number of them were determined as early as 1932 by Manton, they had received little attention in earlier years, and only recently have their collection, maintenance and exploitation been taken up. The credit for initiating investigations with wild species goes to Mizushima,who hybridized some of the wild species with crop species and elucidated the genomic homology with them (Mizushima 1968). Harberd (1972, 1976) carried out an extensive geno-systematical survey, classified them into cytodemes (47 such cytodemes are present) and referred to them collectively as *Brassica* coenospecies (Fig. 2). They were further studied by Takahata and Hinata (1983). There is wide variation in their somatic chromosome number which range from 14 to 44. The majority of them are diploids (39 cytodemes) and only eight cytodemes are polyploids.

This wild germplasm is the repository of many nuclear gene-controlled desirable attributes, such as resistance to diseases and pests, small thick dark green leaves with high photosynthetic efficiency, salt and drought tolerance and novel fatty acid composition. Crop brassicas host an array of pathogens and pests, and since wild cytodemes can exchange genetic material with them, the introduction of these genes appears to be a real possibility. Besides, their cytoplasm, on combining with the nuclear genome of crop brassicas could provide sources of cytoplasmic male sterility. So far, serious attempts have not been made to screen this germplasm for the different traits and only scanty information is available. At the same time,their exploitation is limited because of problems primarily in securing the hybrids and later of gene flow, as they belong to the secondary and tertiary gene pool. Hybridization barriers are

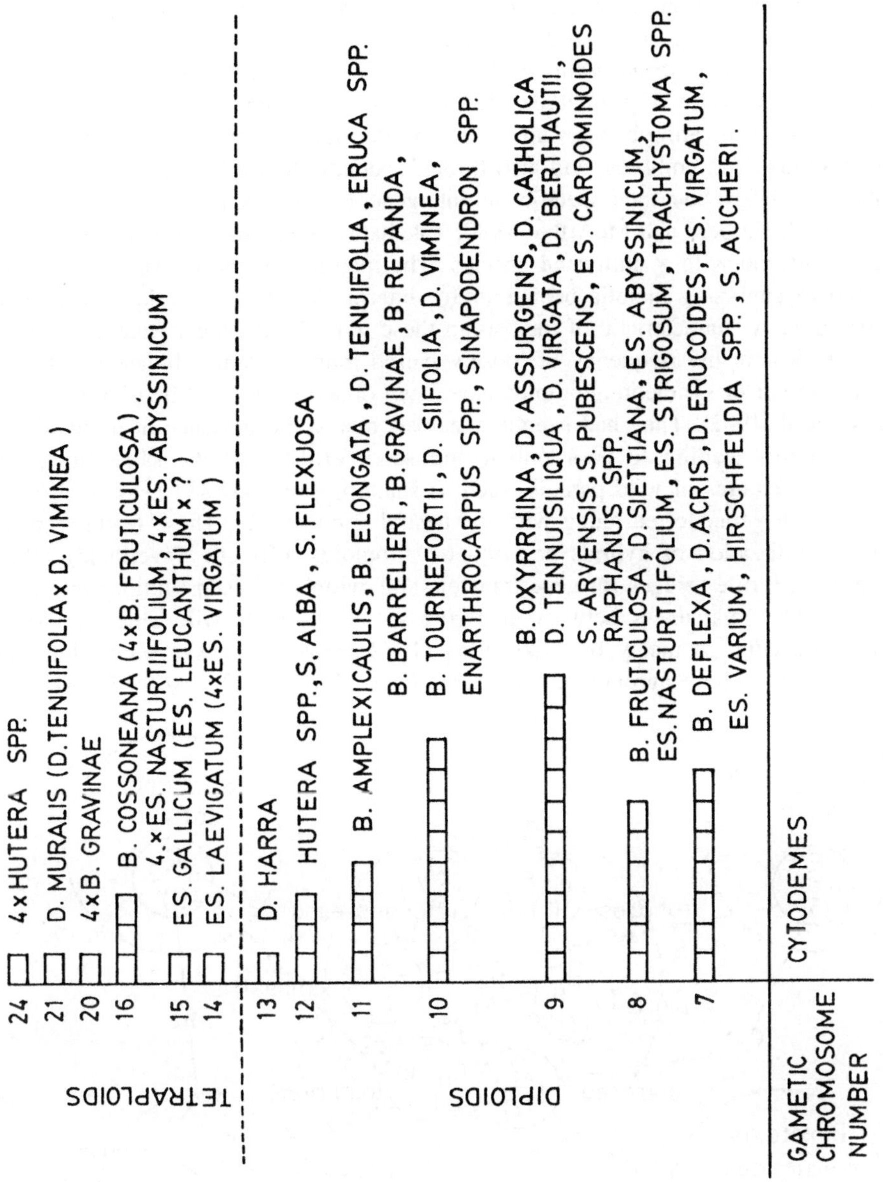

Fig. 2. Cytodemes in *Brassica* coenospecies (B = *Brassica*; D = *Diplotaxis*; ES = *Erucastrum*; S = *Sinapis*)

mostly post-fertilization and operate in only a few cases before fertilizaiton (Harberd 1976). Some of the species also display unilateral incompatibility, e.g. *B. oxyrrhina*, *B. tournefortii*, *Erucastrum gallicum*, *Enarthrocarpus* spp. are usually successful as female parents, while *B. fruticulosa*, *Sinapis pubescens* can be used only as male parents. Post-fertilization barriers are overcome through in vitro culture of pollinated ovaries, ovules and embryos, and also through somatic hybridization. As a result, a number of hybrids, both inter-specific and intergeneric, have been obtained (Mizushima 1950a, 1968; Harberd and McArthur 1980; Takahata and Hinata 1983). These include wild × wild and wild × cultivated species. The majority of the investigations were confined to analyzing meiotic behaviour for interpreting genomic relationships and homoeology. A characteristic of meiosis in these hybrids was the virtual absence of multivalents and low frequency of loosely paired monochiasmate bivalents. Only a limited number of synthetic alloploids have been obtained so far (Mizushima 1950b; Prakash et al. 1982). They had irregular meiosis due to the presence of multivalents and univalents leading to poor pollen and seed fertility. At the same time, the undesirable features of wild parents, such as shattering of pods and very small seed size, were also transferred along with the desirable characters. These features make the direct utilization of hybrids or synthetic alloploids difficult. Nevertheless, they can serve as fertile bridges for gene transfers and cytoplasmic substitutions into crop brassicas. The relatedness between *Brassica* species and the wild allies has been summarized (Fig. 3) nicely by Mizushima (1972). One of the synthetic alloploids which appears to have agronomic potential is *Eruca-Brassica* ($2n$ = 42, EEAA). This

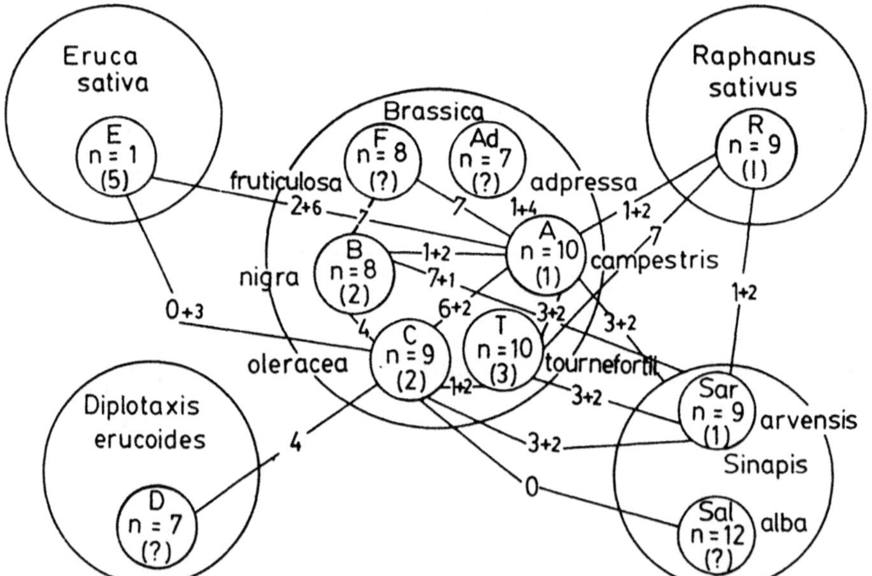

Fig. 3. Genome interrelationships in *Brassica* and allied genera. The *large-* and *medium-sized circles* denote the genera and the *small-sized ones* species. *A, B, C, . . .* are the genome symbols. The *figure in the species circle enclosed in parentheses* shows the maximum number of autosyndetic pairs that occur in the genome. The *figures on the connecting line between two genomes* show the maximum number of bivalents that appear in their hybrid. The *large-sized figure* shows the lower limit of the number of allosyndetic pairs between the genomes. The *sum of the large- and small-sized figures* shows the upper limit, if no autosyndesis occurs in the two genomes (after Mizushima 1972).

was obtained through embryo rescue and spontaneous chromosome doubling in the callus phase following hybridization between *Eruca sativa* ($2n$ = 22, EE) and *B. campestris* ($2n$ = 20, AA) (Agnihotri et al. 1990). The synthesis was attempted with the aim to combine the white rust resistance and adaptation to moisture stress of *Eruca* and the high productivity and oil content of *Brassica*. Alloploids showed regular meiosis (21 II at M_1) and attained high seed fertility (92%) in the A_4 generation and gave a seed yield of 18 q/ha.

To date, no gene of significance from wild germplasm has been introgressed into crop species. However, the cytoplasm of three wild species has been combined with crop nuclei to develop alloplasmic lines for inducing male sterility. A Japanese scientist, Ogura, in 1968 discovered sterility inducing cytoplasm in wild *Raphanus*. It was subsequently transferred to *B. oleracea* (Bannerot et al. 1974), *B. napus* (Rousselle and Renard 1978), *B. campestris* (Williams and Heyn 1981) and *B. juncea* (Labana and Banga 1989). Two more examples of cytoplasmic substitutions concerned *B. campestris* with *Diplotaxis muralis* cytoplasm (Hinata and Konno 1979) and *B. campestris* and *B. juncea* with *B. oxyrrhina* cytoplasm (Prakash and Chopra 1988b, 1990b). These reports strongly indicate that cytosteriles of variable origin can be obtained by exploiting the vast reservoir of *Brassica* coenospecies.

8.8 Conclusions

Cytogenetical investigations on *Brassica* were initiated in Japan and Russia in the second and third decade of the present century. This type of work later extended to India, Sweden and Canada. As a result, spectacular achievements have been recorded.

Past results conclusively demonstrate that synthetic alloploids are of great practical value. They provide an extended range of genetic variability combining the variations available in diploid constituent parents. At the same time, interactions between the parental genomes also give rise to novel combinations. It is well established that alloploid species originated long before the variations in diploid species evolved; thus they did not partake in the formation of alloploids. As two of the basic species, viz. *B. oleracea* and *B. campestris*, are highly polymorphic, usage of these morphotypes will result in new variations. Production of somatic hybrids is another way to generate further variability. Apart from circumventing pre- and post-fertilization barriers, this will result in novel cytoplasmic combinations and mitochondrial recombinations. In the majority of the investigations, synthesis has been attempted by hybridizing the diploids followed by chromosome doubling of the sterile hybrid. This results in complete homozygous types. It would be desirable to attempt hybridization between tetraploids, which would have two advantages (1) the frequency of hybrids is greater at the tetraploid level as compared to diploids, and (2) hybrids between tetraploids are expected to be genetically variable due to the heterozygosity of the tetraploid parents. Heterozygosity in synthetics can also be maximized by the fusion of unreduced gametes. Our observations on synthetic *B. carinata* with reciprocal cytoplasms strongly suggest cytoplasmic effects on morphological and physiological characteristics. It would be interesting to synthesize alloploids with reciprocal cytoplasms.

Genetic variability in alloploid species can also be enlarged by their reconstruction following interspecific hybridization between digenomic species (see Prakash and Chopra 1990a). The advantages of this method are: (1) improved seed set due to buffering provided by two sets of one genome, and (2) a high amount of chromosome pairing leading to genetic exchange between genomes that carry homologous chromosome segments. These approaches will result into reconstituted alloploids with genetic information from the third non-constituent parent (Table 8).

Table 8. Reconstituted alloploids following hybridization between digenomic brassicas

Hybrid combination	Somatic chromosome number and genomic constitution	Chromosome pairing range at MI	Reconstituted species
B. carinata × *B. juncea*	35, ABBC	8II + 19I - 16II + 3I	*B. carinata* BBCC[a] *B. juncea* AABB[c]
B. napus × *B. carinata*	36, ABCC	9II + 18I – 15II + 6I	*B. napus* AACC[b] *B. carinata* BBCC[a]
B. juncea × *B. napus*	37, AABC	10II + 17I - 14II + 9I	*B. juncea* AABB[c] *B. napus* AACC[b]

[a] Superscript denotes the introgression of genes from the indicated genome.

References

Agnihotri A, Gupta V, Lakshmikumaran MS, Shivanna KR, Prakash S, Jagannathan V (1990) Production of *Eruca-Brassica* hybrid by embryo rescue. Plant Breed 104: 281–289

Armstrong KC, Keller WA (1981) Chromosome pairing in haploids of *Brassica campestris*. Theor Appl Genet 59: 49–52

Armstrong KC, Keller WA (1982) Chromosome pairing in haploids of *Brassica oleracea*. Can J Genet Cytol 28: 323–329

Attia T, Röbbelen G (1986) Cytogenetic relationship within cultivated *Brassica* analysed in amphihaploids from the three diploid ancestors. Can J Genet Cytol 28: 323–329

Banga SS (1988) C-genome chromosome substitution lines in *Brassica juncea* (L) Coss. Genet 77: 81–84.

Bannerot H, Boulidard L, Cauderon Y, Tempe J (1974) Transfer of cytoplasmic male sterility from *Raphanus sativus* to *Brassica oleracea*. Proc. Eucarpia Meet, Dundee Scotland, pp 52–54

Catcheside DG (1937) Secondary pairing in *Brassica oleracea*. Cytologia Fuji Jub Vol 366–378

Chen BY, Haneen WK, Jonsson R (1988) Resynthesis of *Brassica napus* L. through interspecific hybridization between *B. alboglabra* Bailey and *B. campestris* L. with special emphasis on seed colour. Plant Breed 101: 52–59

Delourme R, Eber F, Chevre AM (1989) Intergeneric hybridization of *Diplotaxis erucoides* with *B. napus*. I. Cytogenetic analysis of F_1 and BC_1 progeny. Euphytica 41: 113–122

Ellerström S, Sjödin J (1973) Species crosses in the family Brassiceae aiming at creation of new fodder crops. In: New ways in Fodder Crop Breeding. Proc Meet Fodder Crop Sec, Eucarpia, pp 26–28

Ellerström S, Zagorcheva L (1977) Sterility and apomictic embryosac formation in *Raphanobrassica*. Hereditas 87: 107–109

Erickson LR, Straus NA, Beversdorf WD (1983) Restriction patterns reveal origins of chloroplast genomes in *Brassica* amphidiploids. Theor Appl Genet 65: 201–206

Feng W (1955) An interspecific cross of *Brassica-B. pekinensis Rupr* × *B. oleracea* var *fimbriata* Mill. Acta Bot Sinica 4: 63–70

Frandsen HN, Winge Ö (1932) *Brassica napocampestris*, a new constant amphidiploid species hybrid. Hereditas 16: 212–218

Frandsen KJ (1941) Beitrage Zur Cyto-genetik der *Brassica napus* L. der *Brassica campestris* L. und deren Basterden, sowie der amphidiploiden *Brassica napocampestris*. K Vet Landbohoejsk Arsskr: 59–90

Frandsen KJ (1943) The experimental formation of *Brassica juncea* Czern et Coss. Dan Bot Ark 11: 1–17

Frandsen KJ (1947) The experimental formation of *Brassica napus* L. var. *oleifera* DC. and *Brassica carinata* Braun. Dan Bot Ark 12: 1–16

Fukushima E (1929) Preliminary report on *Brassica-Raphanus* hybrids. Proc Imp Acad Tokyo 5: 48–50

Fukushima E (1945) Cytogenetic studies on *Brassica* and *Raphanus* I. Studies on inter-generic F_1 hybrids between *Brassica* and *Raphanus*. J Dep Agric Kyushu Imp Univ 7: 281–400

Grabiec B (1967) Investigation on interspecific hybrids of the genus *Brassica* with the aim of utilizing them in winter rape breeding. III. Investigations on synthetic and semisynthetic forms of rape. Hodowla Resl Aklim Nasienn 11: 303–314

Haga T (1938) Relationship of genome to secondary pairing in *Brassica*. Jpn J Genet 13: 277–284

Harberd DJ (1972) A contribution to the cytotaxonomy of *Brassica* (Cruciferae) and its allies. Bot J Linn Soc 65: 1–23

Harberd DJ (1976) Cytotaxonomic studies of *Brassica* and related genera. In: Vaughan JG, MacLeod AJ, Jones BM (eds). The biology and chemistry of the Cruciferae, pp 47–68

Harberd DJ, McArthur ED (1980) Meiotic analysis of some species and genus hybrids in the Brassiceae. In: Tsunoda S, Hinata K, Gõmez-Campo C (eds.) *Brassica* crops and wild allies. Jpn Sci Soc, Tokyo, pp 65–87

Hinata K, Konno N (1979) Studies on a male sterile strain having the *Brassica campestris* nucleus and the *Diplotaxis muralis* cytoplasm. I. On the breeding procedure and some characteristics of the male sterile strain. Jpn J Breed 29: 305–311

Hoffman W, Peters P (1958) Versuche zur Herstellung Synthetischer und Semisynthetischer Rapsformen. Züchter 28: 40–51

Honma S, Heeckt O (1962) Investigations on F_1 and F_2 hybrids between *Brassica oleracea* var *acephala* and *Raphanus sativus*. Euphytica 11: 177–180

Hosoda T (1950) On new types of *Brassica napus* obtained from artifical amphidiploids. I. A new type as a forage crop. Ikushu Kenkyu 4: 91–95

Hosoda T (1953) On the breeding of *Brassica napus* obtained from artificially induced amphidiploids. II. Fertility of artificially induced *napus* plants. Jpn J Breed 3: 44–50

Hosoda T (1961) Studies on the breeding of new types of *napus* crops by means of artificial synthesis in genomes of genus *Brassica*. Mem Tokyo Univ Agric Educ 7: 1–94

Hosoda T, Namai H, Goto J (1963) On the breeding of *Brassica napus* obtained from artificially induced amphidiploids. III. On the breeding of synthetic rutabaga (*Brassica napus* var. *rapifera*). Jpn J Breed 13: 99–106

Howard HW (1938) The fertility of amphidiploids from the cross *Raphanus sativus* × *Brassica oleracea*. J Genet 36: 239–273

Howard HW (1939) The cytology of autotetraploid kale *Brassica oleracea*. Cytologia 10: 77–87

Howard HW (1942) The effect of polyploidy and hybridity on seed size in crosses between *Brassica chinensis*, *B carinata*, amphidiploid *B. chinensis-carinata* and autotetraploid *B. chinensis*. J Genet 43: 105–119

Ichikawa H, Hirai A (1983) Search for the female parent in the genesis of *Brassica napus* by chloroplast DNA restriction patterns. Jpn J Genet 58: 419–424

Inomata N (1978) Production of interspecific hybrids in *Brassica campestris* × *B. oleracea* by culture in vitro of excised ovaries. I. Development of excised ovaries in the crosses of various cultivars. Jpn J Genet 53: 166–173

Iwasa S, Ellerström S (1981) Meiosis disturbances, aneuploidy and seed fertility in *Raphanobrassica*. Hereditas 95: 1–9

Jahier J, Chevre AM, Tunguy AM, Eber F (1989) Extraction of disomic addition lines-*Brassica napus-B nigra*. Genome 32: 408–413

Jahr W (1962) Befruchtungsbiologie und Allopolyploidie bei der Artkreuzung von Sommerraps und China Kohl (*Brassica napus* f typica Poskichal × *B pekinensis* Rupr var *cylindrica* Tsen et Lee). Züchter 32: 216–225

Johnston TD (1974) Transfer of disease resistance from *Brassica campestris* L to rape (*B napus* L). Euphytica 23: 681–683

Jourden PS, Mitschler MA, Earle ED (1986) Production and characterization of somatic hybrids between *Brassica oleracea* and *B campestris*. Cruciferae Newslett 11: 84–85

Kajanus B (1912) Genetische Studien an *Brassica*. Z Induk Abstamm Vererbungsl 6: 137–139

Karpechenko GD (1924) Hybrids of *Raphanus sativus* L × *Brassica oleracea* L. J Genet 14: 375–396

Karpechenko GD (1928) Polyploid hybrids of *Raphanus sativus* × *Brassica oleracea* L. Z Induk Abstamm Vererbungsl 48: 1–85

Kato K, Namai H, Hosoda T (1968) Studies on the practicality of artificial rutabaga SR lines obtained from interspecific crosses between Shogoin-kabu (*B. campestris* ssp. *rapifera*) and kohlrabi (*B. campestris* var. *gongylodes*). II. Feeding value of SR lines. J Jpn Grassl Sci 14: 177–181

Kato K, Tokumasu S (1976) The mechanisms of increased seed fertility accompanied with the change of flower colour in *Brassicoraphanus*. Euphytica 25: 761–767

Kato K, Tokumasu S (1989) *Brassicoraphanus* (*Brassica japonica* Sieb × *Raphanus sativus* L.). Mem Coll Agr Ehime Univ 33: 157–170

Kondo N (1942) A new *Raphanobrassica* from the cross 4x *Raphanus sativus* L. × 4x *Brassica oleracea* L. Jpn J Genet 18: 123–130

Labana KS, Banga SK (1989) Transfer of 'Ogura' cytoplasmic male sterility of *Brassica napus* into genetic background of *Brassica juncea*. Crop Improv 16: 82–83

Makarova GA (1963) Segregation in the progeny of intergeneric hybrid *Raphanus napus* to *Brassica oleracea* L. Agrobiologia 849–858

Manton I (1932) Introduction to the general cytology of the Cruciferae. Ann Bot 46: 509–556

Matsuzawa Y (1984) Studies on interspecific hybridization in *Brassica*. III. Cross compatibility between *B. nigra* and *B. campestris* and *B oleracea*. Jpn J Breed 34: 69–78

McNaughton IH (1963) The scope and problems involved in synthesizing new amphidiploid and autotetraploid fodder brassicas in group *B. napus* L., *B. campestris* L., *B. oleracea* L. Ann Rep Scott Plant Breed Stn 1963, pp 1–21

McNaughton IH (1973a) Synthesis and sterility of *Raphanobrassica*. Euphytica 22: 70–88

McNaughton IH (1973b) *Brassica napocampestris* L. ($2n = 58$) Synthesis, cytology, fertility and general considerations. Euphytica 22: 301–309

McNaughton IH (1982) *Raphanobrassica* in retrospect and prospect. Cruciferae Newslett 7: 34–40

McNaughton IH, Ross CL (1978) Interspecific and inter-generic hybridization in the Brassiceae with special emphasis on the improvement of forage crops. Ann Rep Scott Plant Breed Stn 1978, pp 75–110

Mizushima U (1946) Some amphidiploids in the Cruciferae. Ikushu to Nogei 1: 31–32, pp 67–68

Mizushima U (1950a) Karyogenetic studies of species and genus hybrids in the tribe Brassicaceae of the *Cruciferae*. Tohoku J Agr Res 1: 1–14

Mizushima U (1950b) On several artificial allopolyploids obtained in the tribe Brassicaceae of the *Cruciferae*. Tohoku J Agr Res 1: 15–27

Mizushima U (1968) Phylogenetic studies on some wild *Brassica* species. Tohoku J Agr Res 19: 83–99

Mizushima U (1972) Evolution of species in Brassiceae and their breeding. Kagaku to Seibutsu 10: 78–85

Mizushima U (1980) Genome analysis in *Brassica* and allied genera. In: Tsunada S, Hinata K, Gõmez-Caomp C (eds.) *Brassica* crops and wild allies, Jpn Sci Soc Tokyo, pp 89–108

Mizushima U, Katsuo K (1953) On the fertility of artificial amphidiploid between *B. nigra* Koch and *B. oleracea* L. Tohoku J Agric Res 4: 1–14

Morinaga T, Kuriyama H (1937) On the autopolyploids of the rape. Cytologia Fujü Jub Vol 967–969

Moskov BS, Makarova GA (1969) High yielding intergeneric hybrids in the Cruciferae. Bull Appl Bot Genet Pl Breed Ser 2, 8: 92–102

Namai H (1971) Studies on the breedng of oil rape (*Brassica napus* var. *oleifera*) by means of interspecific crosses between *B. campestris* ssp. *oleifera* and *B. oleracea*. I. Interspecific crosses with the application of grafting method or the treatment of sugar solution. Jpn J Breed 21: 40–48

Namai H, Hosoda T (1967) On the breeding of *Brassica napus* obtained from artificially induced amphidiploids III. On the breeding of synthetic rutabaga (*Brassica napus* var. *rapifera*). Jpn J Breed 117: 194–204

Namai H, Hosoda T (1968) Studies on the practicality of artificial rutabaga SR lines obtained from interspecific crosses between Shogoin-kabu (*Brassica campestris* ssp. *rapifera*) and Kohlrabi (*B. oleracea* var. *gongylodes*). I. Productivity of SR lines. J Japan Grassl Sci 14: 171–176

Namai H, Sarashima M, Hosoda T (1980) Interspecific and intergeneric hybridization breeding in Japan. In: Tsunoda S, Hinata K, Gõmez-Campo C (eds). *Brassica* crops and wild allies. Jpn Sci Soc, Tokyo, pp 191–204

Nishi S, Kawata J, Toda M (1961) Studies on the embryo culture in vegetable crops. I. Embryo culture of immature Crucifer embryos. Bull Nat Inst Agric Sci 9: 58–128

Nishi S, Kawata J, Toda M (1962) Studies on the embryo culture in vegetable crops. II. Breeding of interspecific hybrids between cabbage varieties and Chinese cabbage varieties through the application of embryo culture techniques. Bull Hortic Res St Japan Ser A 1: 111–156

Nishi S, Toda M, Toyoda TO (1970) Studies on the embryo culture in vegetable crops. III. On the conditions effecting the embryo culture of interspecific hybrids between cabbage and Chinese cabbage. Bull Hortic Res St, Japan Ser A 9: 75–100

Nishiyama I (1949) Studies on artificial polyploid plants. XII. On the production of polyploids in *Cruciferae*. Jpn J Genet Suppl Vol 2: 19–23

Nwankiti O (1970) Cytogenetic and breeding studies with *Brassica*. I. Cytogenetic experiments with *Brassica napocampestris*. Hereditas 66: 109–126

Nwankiti O (1975) Cytogenetic and breeding studies with *Brassica*. III. Aspects on breeding with *Brassica napocampestris*. L. Z Pflanzenzuecht 75: 286–296

Olsson G (1960a) Species crosses within the genus *Brassica*. I. Artificial *Brassica juncea* L. Hereditas 46: 171–222

Olsson G (1960b) Species crosses within the genus *Brassica*. II. Artificial *Brassica napus* L. Hereditas 46: 351–396

Olsson G (1963) Induced polyploids in *Brassica*. In: Akerberg E, Hagberg A (ed) Recent Researches in Plant Breeding, Svalof 1944-1961. John Willey, New York, pp 179–192

Olsson G (1986) Allopolyploids in *Brassica*. In: Olsson G (ed) Svalöf 1886-1986. Research and Results in Plant Breeding. Stockholm, pp 114–119

Olsson G, Ellerström S (1980) Polyploidy breeding in Europe. In: Tsunoda S, Hinata K, Gomez-Campo C (eds). *Brassica* crops and wild allies. Japan Sci Soc, Tokyo, pp 167–190

Olsson G, Josefsson A, Hagberg A and Ellerström S (1955) Synthesis of the ssp. *rapifera* of *Brassica napus*. Hereditas 41: 241–249

Palmer JD (1988) Intraspecific variation and multicircularity in *Brassica* mitochondrial DNAs. Genetics 118: 341–351

Palmer JD, Shields CR, Cohen DB, Orton TJ (!983) Chloroplast DNA evolution and the origin of amphidiploid *Brassica* species. Theor Appl Genet 65: 181–189

Parthasarthy N, Rajan SS (1953) Studies on the fertility of autotetraploids of *Brassica campestris* var. *toria*. Euphytica 2: 25–36

Pearson OH (1972) Cytoplasmically inherited male sterility characters and flavour components from the species cross *Brassica nigra* (L) Koch × *B. oleracea* L. J Am Soc Hort Sci 97: 397–402

Prakash S (1973a) Artificial synthesis of *Brassica juncea* Coss. Genetica 44: 249–263

Prakash S (1973b) Haploidy in *Brassica nigra* Koch. Euphytica 22: 613–614

Prakash S (1974) Probable basis of diploidization of *Brassica juncea* Coss. Can J Genet Cytol 16: 232–234

Prakash S (1980) Cruciferous oilseed in India. In: Tsunoda S, Hinata K, Gõmez-Campo C (eds.) *Brassica* crops and wild allies. Jpn Sci Soc, Tokyo, pp 151–163

Prakash S, Chopra VL (1988a) Introgression of resistance to shattering in *B. napus* from *Brassica juncea* through non-homologous recombination. Plant Breed 101: 167–168

Prakash S, Chopra VL (1988b) Synthesis of alloplasmic-*Brassica campestris* as a new source of cytoplasmic male sterility. Plant Breed 101: 253–255

Prakash S, Chopra VL (1990a) Reconstruction of alloploid *Brassicas* through non-homologous recombination: Introgression of resistance to pod shatter in *Brassica napus*. Genet Res Camb 56: 1–2

Prakash S, Chopra VL (1990b) Male sterility caused by cytoplasm of *Brassica oxyrrhina* in *B. campestris* and *B. juncea*. Theor Appl Genet 79: 285–287

Prakash S, Chopra VL (1991) Cytogenetics of Crop Brassicas and their allies. In: Tsuchiya T, Gupta PK (eds). Chromosome Engineering in plants: Genetics, Breeding and Evolution. Elsevier Sci Publ, Netherlands 2: 161–189

Prakash S, Gupta S, Raut RN and Anita Kalra (1984) Synthetic *Brassica carinata*. Cruciferae Newslett 9: 36

Prakash S, Hinata K (1980) Taxonomy, Cytogenetics and Origin of crop Brassicas. Opera Bot 55: 1–57

Prakash S, Raut RN (1983) Artificial synthesis of *Brassica napus* and its prospects as an oilseed crop in India. Indian J Genet Plant Breed 43: 283–291

Prakash S, Tsunoda S, Raut RN, Gupta S (1982) Interspecific hybridization involving wild and cultivated genomes in the genus *Brassica*. Cruciferae Newslett 7: 28–29

Quiros CF, Kianian SF, Ochoa O, Douches D (1985) Genome evolution in *Brassica*: Use of molecular markers and cytogenetic stocks. Cruciferae Newslett 10: 21–23

Quiros CF, Ochoa O, Kianian SF, Douches D (1987) Analysis of the *Brassica oleracea* genome by the generation of *B. campestris-oleracea* chromosome addition lines: Characterization by isozymes and rDNA genes. Theor Appl Genet 74: 258–266

Ramanujam S, Deshmukh MJ (1945) Colchine-induced polyploidy in crop plants. III. Oleiferous Brassiceae. Indian J Genet 5: 63–81

Ramanujam S, Srinivasachar D (1943) Cytogenetical investigations in the genus *Brassica* and the artificial synthesis of *Brassica juncea*. Indian J Genet 3: 73–88

Richharia RH (1937) Cytological investigations of *Raphanus sativus*, *Brassica oleracea* and their F_1 and F_2 hybrids. J Genet 34: 19–44

Riley R, Chapman V (1958) Genetic control of cytologically diploid behaviour of hexaploid wheat. Nature, London 182: 713–715

Röbbelen G (1960a) Ueber die Kreuzungsunvertraeglichkeit verschiedener *Brassica*-Arten als Folge cines gehemmten Pollenschlavchwachstums. Züchter 30: 300–312

Röbbelen G (1960b) Beitraege zur Analyse des *Brassica*-Genomes. Chromosoma 11: 205–228

Röbbelen G (1966) Beobachtungen bei interspezifischen *Brassica*-Kreuzungen insbesondere über die Entstehung matromorpher F_1-Pflanzen. Angew Bot 39: 205–221

Röbertson D, Earle ED, Mutschler MA (1985) Synthesis of atrazine-resistant *Brassica napus* through protoplast fusion. Cruciferae Newslett 10: 88–89

Rosen B, Hallden C, Haneen WK (1988) Diploid *Brassica napus* somatic hybrids: Characterization of nuclear and organellar DNA. Theor Appl Genet 76: 197–203

Rousselle P, Renard M (1978) Study of a cytoplasmic male sterility in rapeseed. Cruciferae Newslett 3: 40–41

Rudorf W (1950) Ueber die Erzeugung und die Eigenschaften synthetischer Rapsformen. Z Pflanzenzuecht 29: 35–54

Sarashima M (1964) Studies on the breeding of artificially synthesized rape (*Brassica napus*). I. F_1 hybrids between *B. campestris* group and *B. oleracea* group and the derived F_2 plants. Jpn. J Breed 14: 226–236

Sarashima M (1973) Studies on the breeding of artificially synthesized forage rape (*Brassica napus* ssp. *oleifera*) by means of interspecific crosses between *B campestris* and *B. oleracea*. Sppl Bull Coll Agric Utsunomiya Univ 1–117

Schenck HR, Röbbelen G (1982) Somatic hybrids by fusion of protoplasts from *Brassica oleracea* and *B. campestris*. Z Pflanzenzuecht 89: 278–288

Sinskaia EN (1927) Geno-systematical investigations of cultivated *Brassica*. Bull Appl Bot Genet Pla Breed 17: 3–166

Song KM, Osborn TC, Williams PH (1988) *Brassica* taxonomy based on nuclear restriction fragment length polymorphisms (RFLPs). 1. Genome evolution of diploid and amphidiploid species. Theor Appl Genet 75: 784–794

Srinivasachar D (1965) Comparative studies on the amphidiploids of *Brassica campestris* × *B. nigra*, *B. chinensis* × *B. nigra* and *B. rapa* × *B. nigra*. Indian J Genet 5: 63–81

Sundberg E, Landgren M, Glimelius K (1988) Fertility and chromosome stability in *Brassica napus* resynthesized by protoplast fusion. Theor Appl Genet 75: 96–104

Taguchi T, Kameya T (1986) Production of somatic hybrid plants between cabbage and chinese cabbage through protoplast fusion. Jpn J Breed 36: 185–189

Takahata Y, Hinata K (1983) Studies on cytodemes in the subtribe *Brassicinae*. Tohoku J Agric Res 33: 111–124

Takasugi K (1965) Breeding of synthesized species Hakuran. I. Breeding by segregation. Rep Gifu Agric Exp St 32–34

Terada R, Yamashita Y, Nishibayashi S, Shimamoto K (1987) Somatic hybrids between *Brassica oleracea* and *B. campestris*: selection by the use of iodoacetamide inactivation and regeneration ability. Theor Appl Genet 73: 379–384

Terasawa Y (1932) Tetraploide Bastarde von *Brassica chinensis* L. × *Raphanus sativus* L. Jpn J Genet 7: 183–185

Thompson KF (1956) Production of haploid plants of marrow stem kale. Nature, (London) 78: 748

Tokumasu S (1969) Intergeneric hybrids between *Brassica japonica* and *Raphanus sativus*. Mem Coll Agr Ehime Univ 14: 285–298

Tokumasu S (1976) The increase of seed fertility of *Brassicoraphanus* through cytological irregularity. Euphytica 25: 463–470

Tokumasu S, Kato M (1988) Chromosomal and genetic structure of *Brassicoraphanus* related to seed fertility and presentation of an instance of improvement of its fertility. Euphytica 39: 145–151

UN (1935) Genome analysis in *Brassica* with special reference to the experimental formation of *B. napus* and its peculiar mode of fertilization. Jpn. J Bot 7: 389–452

UN, Mizushima V, Saito K (1937) On diploid and triploid *Brassico-Raphanus* hybrids. Cytologia 8: 319–326

Venkateswarlu J, Kamla T (1971) Pachytene chromosome complements and genome analysis in *Brassica*. J Indian Bot Soc 50A: 442–449

Williams PH, Heyn FW (1981) The origin and development of cytoplasmic male sterile Chinese cabbage. In: Talekar S, Griggs TD (eds) Proc Int Symp Chinese Cabbage. Asian Vegetable Res Dev Cent, Tainam, Taiwan, China, pp 293–300

Yakuwa K (1943) On allopolyploids obtained from 4x *Brassica chinensis* L × *Brassica napus* L. Jpn J Genet 19: 229–234

Yanagino T, Takahata Y, Hinata K (1987) Chloroplast DNA variation among diploid species in *Brassica* and allied genera. Jpn J Genet 62: 119–125

Zaman MW (1989) Introgression in *Brassica napus* for adaptation to the growing conditions in Bangladesh. Theor Appl Genet 77: 721–728

CHAPTER 9

Manipulation of Cytoplasmic Genomes

Stephan A. Yarrow

9.1 Introduction

The presence of cytoplasmic male sterility (CMS) in oilseed crops and its importance in developing F_1 hybrids has led to a plethora of research activity in cytoplasmic genome manipulations. This has involved both traditional breeding methodologies and, relatively more recent, protoplast technology. The backcross substitution of the nucleus of one species into the cytoplasmic background of the alien species has been utilized extensively in *Brassica* to produce alloplasmic lines. Many of these alloplasmic lines are CMS. Alternatively, the protoplast fusion provides a possibility of producing any desired nuclear/cytoplasmic combination and, more significantly, any chloroplast/mitochondrial combination within the cytoplasm. The resultant cytoplasmic hybrids are referred to as cybrids. The ability to substitute specific chloroplast populations or plastomes independent of the mitochondrial genome (chondriome) or vice versa is, with a few exceptions, a technique exclusive to the protoplast fusion. This is in comparison to the backcross substitution, where only whole cytoplasm can be transferred.

9.2 Important Cytoplasmic Traits in *Brassica* Crops

Cytoplasmic male sterility (CMS), in *Brassica* at least, is encoded by mitochondrial DNA (mtDNA) in the cytoplasm (Pelletier et al. 1983; Barsby et al. 1987a). A number of different CMS systems are now available in oilseed *Brassica* crops. These were created by alloplasmic substitutions of the cytoplasm from other *Brassica* species and related wild allies (see Banga, Chap. 3, this Vol.). From a biotechnological point of view, much work has been done on the Ogura (*ogu*) CMS system discovered originally by Ogura (1968) in Japanese radish (*Raphanus sativus*). This was later transferred to *Brassica* (Bannerot et al. 1974). Major limitation of this system is that the chloroplasts in *ogu* cytoplasm, in combination with *Brassica* nucleus, are suboptimal at low temperature (Bannerot et al. 1977). Similarly, the efforts at the production of restorer lines have also met with limited success. Attempts have been made to correct both the restorer and low-temperature problems in *Brassica* by developing fresh CMS lines with *Raphanus sativus* cytoplasm. Similarly, the newly developed CMS lines in *B. campestris* and *B. juncea*, based on *B. oxyrrhina* cytoplasm, also suffer from low-temperature leaf yellowing (Prakash and Chopra 1990). These situations illustrate the significance of chloroplast complement in *Brassica* cytoplasm. The defective plastomes in both these systems have now been replaced by normal chloroplasts through protoplast fusion.

Another example is the modified cytoplasm isolated from wild *B. campestris* by Matais and Bouchard (1978) that accords resistance to triazine herbicide atrazine. This cytoplasmic atrazine resistance (*atr*^{R}) is caused by a single point mutation in the psbA gene of the plastome, which confers a decreased affinity of triazine herbicides to the binding site in the thylakoid membrane (Reith and Straus 1987). *Atr*^{R} has been transferred to *B. napus*, to develop the canola variety Triton (Beversdorf et al. 1980). Further still, the *atr*^{R} trait could prove very useful in combination with CMS, as proposed by Beversdorf et al. (1985), whereby lines developed with both the traits are used as females for commercial F_1 production (see later).

9.3 "Traditional" Cytoplasmic Manipulations

9.3.1 Alloplasmic Substitutions

Alloplasmic substitutions by traditional crossing methods have allowed the transfer of different cytoplasmic borne traits (e.g. CMS and *atr*^{R}) from wild-growing *Brassica* species or closely related species (e.g. *R. sativus* and *Diplotaxis muralis*) to cultivated forms of *Brassica*. However, the process involves extensive backcrossing to convert the initial interspecific or intergeneric hybrids back to a commercially acceptable form. Additional backcrossing is then necessary to further transfer the cytoplasmic trait to the variety or line of choice. In the case of winter-planted forms of rapeseed, the process is further time-consuming due to vernalization requirements at each crossing step. As described by Barsby et al. (1987b), the total time required to introduce CMS from one winter line to another is approximately 3 years (Fig. 1). Another drawback to the conventional approach is that, generally, alloplasmic transfers can only be performed between sexually compatible species, inhibiting the possibility of introducing cytoplasmic traits, such as novel CMS systems, into rapeseed from sexually incompatible but related species. The technique of embryo rescue, however, can overcome these incompatibilities in some cases. For example, Ayotte (1986) transferred *atr*^{R} from *B. napus* to *B. oleracea* via embryo rescue, overcoming the incompatibilities between these two species. Similarly, it has been possible to transfer *B. campestris* and *B. juncea* genomes into *B. oxyrrhina* cytoplasm via embryo rescue (Prakash and Chopra 1990). Whether alloplasmic substitutions are performed by conventional means or with the assistance of embryo rescue, only the whole cytoplasms (essentially the chondriome and plastome combined) can be transferred, through maternal inheritance.

9.3.2 Pollen Transmission

According to conventional wisdom, mitochondria are inherited exclusively from the maternal parent in higher plants. *Brassica* species were no exception until Erickson and Kemble (1990) demonstrated paternal inheritance of mitochondria in *B. napus* following sexual crosses between a *atr*^{R} canola line (as females) and a *pol* CMS line.

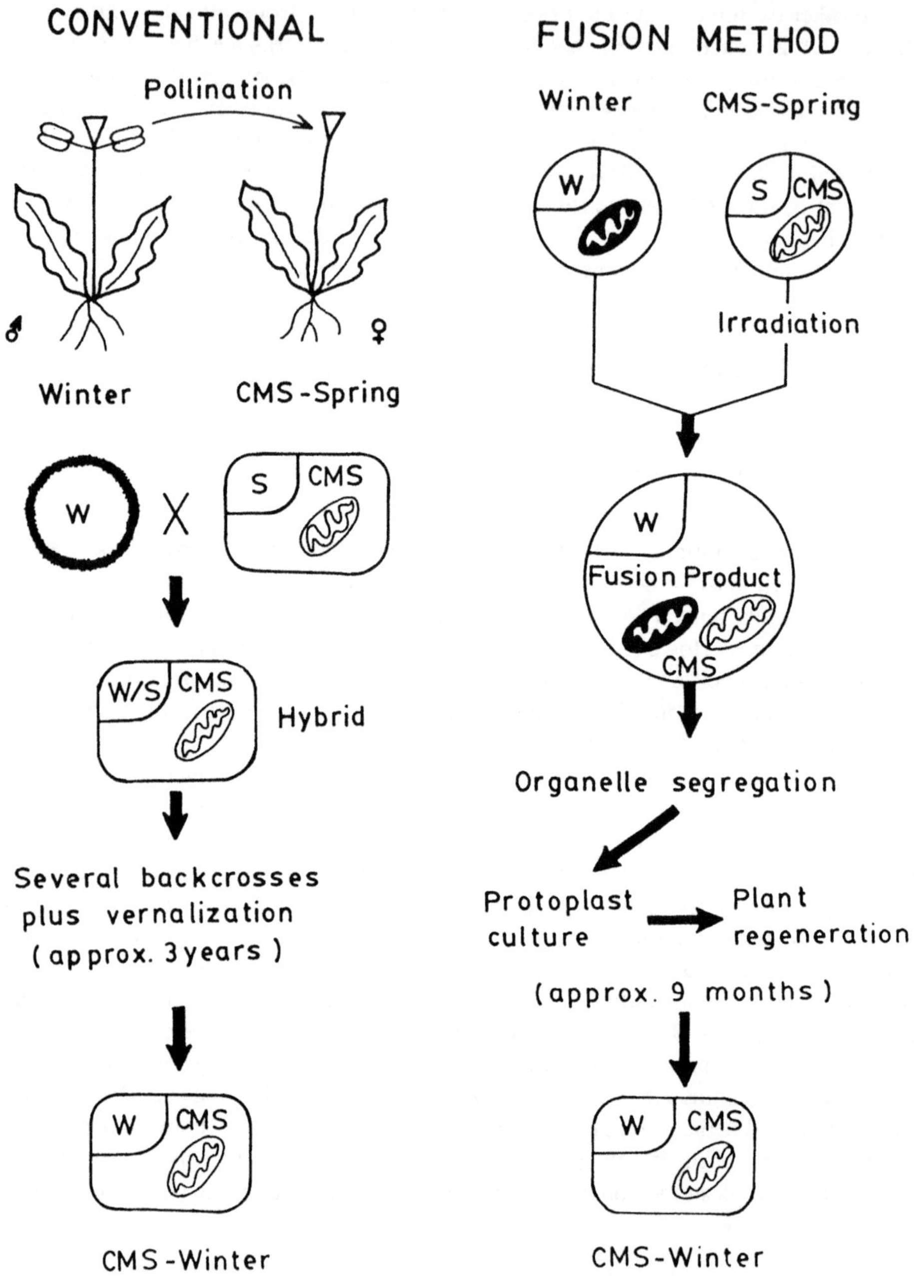

Fig. 1. Comparison between conventional and fusion mediated methods of alloplasmically transferring *cms* cytoplasm from spring to winter lines. Conventional: a pollen grain and an ovule are depicted with their nuclear and mitochondrial contents, which following fertilization form a hybrid. The final *cms*-winter product is depicted as a single cell. Fusion mediated: Single protoplasts from each parent, fuse forming a heterokaryon containing a mixture of mitochondria and a single winter nucleus. The final segregated *cms*-winter regenerant is depicted as a single cell.

Pol-type male sterility was displayed in some of the triazine-resistant progeny. However, this phenomenon was only observed in this one specific line combination. Further research suggests that maternal inheritance of mitochondria is controlled by genes active during pollen development and following fertilization (Erickson and Kemble, in prep). Although paternal inheritance has been observed recently in other plant species, the phenomenon will be of limited usefulness in *Brassica* improvement.

9.4 Cytoplasmic Manipulation by Protoplast Fusion

Whereas conventional breeding methods offer very limited possibilities for overcoming the limitations of maternal inheritance, protoplast fusion can allow the generation of quite extensive cytoplasmic heterogeniety. Protoplasts are single cells devoid of their cellulosic cell walls. They can be isolated from virtually any species and plant tissue. Protoplasts from an increasing number of species, under specific culture conditions, can be induced to regenerate a cell wall, divide to form cell colonies and regenerate whole plants. Protoplast-to-plant regeneration is routinely possible with most of the important *Brassica* species, with the exception of *B. campestris* (Glimelius 1984).

The advantage of plant protoplasts is that they can be fused together to form hybrid cells, which, in turn, can be regenerated into hybrid plants, thus potentially bypassing sexual incompatibilities that might otherwise prevent many desired interspecific and intergeneric combinations. Fusion can conceivably produce any nuclear/mitochondrial/chloroplast genome combinaton (see Fig. 2a, where *atr*^{R} and CMS cytoplasms are mixed together). Although quite diversely related species have been "somatically" hybridized (i.e. involving nuclear fusion) with some success (e.g. *Arabidopsis thaliana* × *B. campestris*, Gleba and Hoffman 1980; potato × tomato, Melchers et al. 1979 and Shepard et al. 1983), generally the phyllogenetic relationship between the two parental species cannot be too distant without chromosomal dysfunction.

In the case of cybrid production, nuclei of one fusion parent are intentionally disabled so that they cannot undergo fusion, such that (in the absence of selection pressure) the remaining chloroplast/mitochondrial mixture in the fusion product segregates randomly as the cell divides, leading to novel nuclear/ chloroplast/ mitochondrial combinations, or cybrids (Fig. 2b). Initial research in cybrid formation through protoplast fusion was developed in *Nicotiana*, as recently reviewed by Kumar and Cocking (1987).

The first application of cybrid formation in *Brassica*, in the case where the plastome or chondriome was independently transferred, was where the defective *ogu* plastome, described earlier, was substituted with the regular *B. napus* plastome, by fusing protoplasts of *ogu* CMS *B. napus* rapeseed with regular *B. napus* rapeseed, thus correcting the problem of cold-sensitive chloroplasts (Pelletier et al. 1983; Menczel et al. 1987; Jarl and Bornman 1988; Jarl et al. 1988; Kemble et al. 1988a).

Another useful cybrid organelle combination was produced when protoplasts of *ogu* CMS *B. napus* rapeseed were fused with those from a *B. napus* rapeseed line carrying the *atr*^{R} cytoplasm from *B. campestris*, producing an atrazine-tolerant CMS line (Pelletier et al. 1983). Other groups have since combined different CMS systems

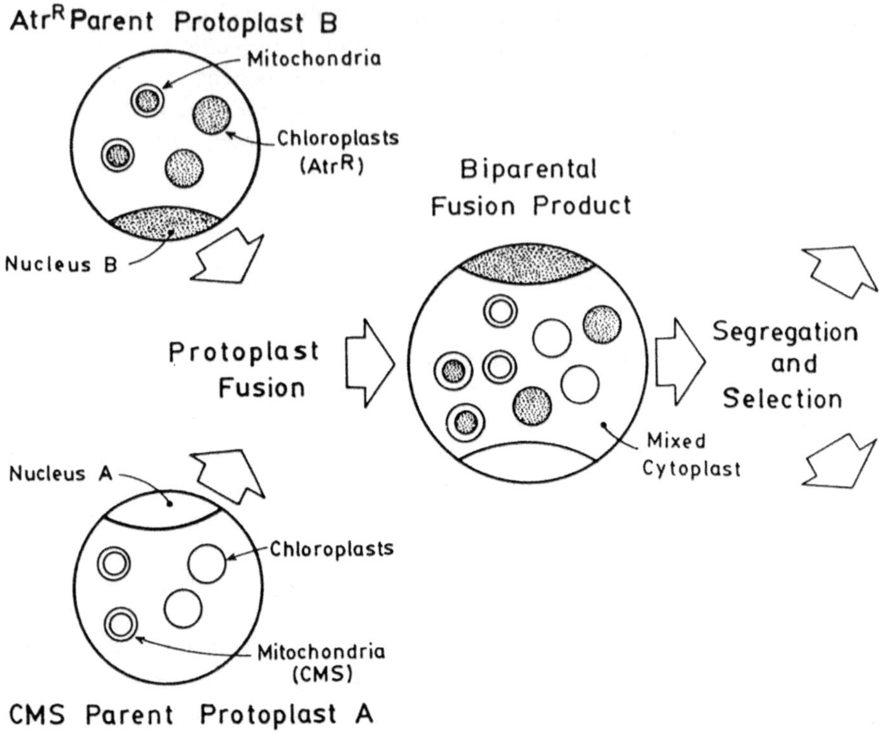

Fig. 2a. Protoplast fusion producing a heterokaryon containing both nuclei and a mixture of organelles.

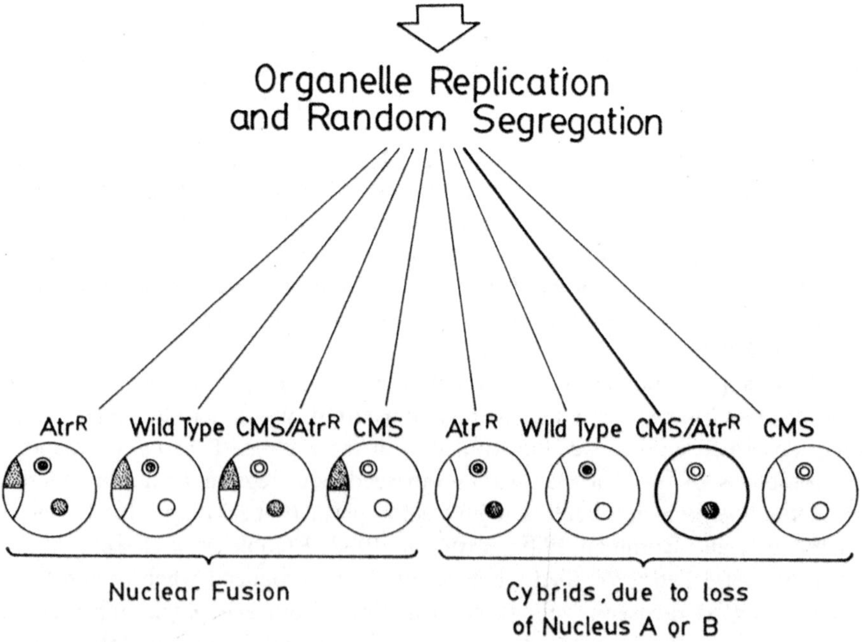

Fig. 2b. Oganelle segregation leading to combined *cms/atr*^{R} cybrids and other sorted out types

(*pol*, *nigra* and *nap*) with the *atr*^{R} plastome (Yarrow et al. 1986; Barsby et al. 1987a; Chuong et al. 1988; Christey et al. 1991) (see Fig. 2).

Cybrids with CMS/*atr*^{R} cytoplasms could be incorporated in a theoretical scheme to produce F_1 hybrids on a commercial scale, as described by Beversdorf et al. (1985). The cybrids are field-mixed with male fertile triazine-sensitive lines. Following pollination, triazines are applied to the plants, leaving only the surviving females, carrying the hybrid seed.

Protoplast fusion techniques have also been utilized to produce alloplasmic substitution–type cybrids, whereby entire cytoplasms are transferred from one *Brassica* line to another. Due to the requirement to vernalize winter rapeseed at each backcrossing stage, transferring CMS from one line to another can take approximately 3 years. Utilizing protoplast fusion techniques, backcrossing was obviated when CMS was transferred from spring CMS canola into winter lines in one step, taking only 9 months (Barsby et al. 1987b), with obvious savings in time and expenses (see Fig. 1). In another example, the *atr*^{R} cytoplasm of the canola variety Triton was transferred to winter lines via fusion (Thomzik and Hain 1988). Similarly, Jourdan et al. (1989) used fusion to transfer *atr*^{R} from *B. napus* to *B. oleracea*.

9.5 Cybrid Production Methodologies

The production of cybrids involves the following stages:

9.5.1 Protoplast Isolation and Culture

Protoplasts can be isolated routinely from all the commercially valuable *Brassica* species, and plants can be regenerated from most, though the techniques are sometimes difficult to reproduce from one laboratory to another, due to variations in genotype and growth conditions (Pelletier et al. 1983; Glimelius 1984; Chuong et al. 1985; Barsby et al. 1986; Barsby et al. 1987b; Yarrow 1990; Louden et al. 1990; Batra and Dhingra 1990; Kirti and Chopra 1990; Jaiswal et al. 1990).

9.5.2 Fusion

Fusion protocols used for *Brassica* cytoplasmic manipulations described to date involve the use of polyethylene glycol solutions as originally developed by Kao and Michayluk (1974) and Wallin et al. (1974). Solutions of calcium ions buffered at high pH are often incorporated into the process, as described by Keller and Melchers (1973). The increasingly popular technique of fusing protoplasts using electrical stimuli, first described by Zimmerman and Scheurich (1981), has clear applications for *Brassica* hybrid production.

Using current methodologies, fusion of two different populations of protoplasts produces both homokaryotic and heterokaryotic fused cells, where commonly less than 5% of protoplasts contribute to heterokaryon formation. Ideally, for cybrid formation, the nuclei within the heterokaryons do not fuse. In this case, only one parental nuclear genome is contributed to the final plant, as the other is lost during cell division and plant regeneration.

However, without some level of control, nuclear fusion can occur. The frequency varies, depending on such factors as the phylogeny of the parents, and their physiological state prior to fusion. Polyploidy or aneuploidy of the fusion product regenerants as a consequence of nuclear fusion often can be corrected by backcrossing, if female fertility is retained in the fusion products (Barsby et al. 1987a). The use of haploid parental lines can encourage formation of diploid cybrids (Chuong et al. 1988).

Although *Brassica* cybrids have been produced without nuclear fusion control (Pelletier et al. 1983; Yarrow et al. 1986; Jarl and Bornman 1988), the application of irradiation treatments to the undesired nuclear fusion partner prior to fusion inhibits nuclear fusion. Both X-ray-type irradiation (Menczel et al. 1987; Thomzik and Hain 1988) and gamma rays (Barsby et al. 1987a, b; Yarrow 1990; Yarrow et al. 1990) have been employed. Zelcer et al. (1978) were the first to develop this essentially cytoplasmic "donor-recipient" technique when transferring CMS in *Nicotiana*. In combination with donor parent irradiation, the recipient line is often treated with a metabolic inhibitor, such as iodoacetic acid (Sidorov et al. 1981), which eliminates any non-fused recipient protoplasts from the population (Barsby et al. 1987a, b; Morgan and Maliga 1987; Yarrow 1990). Heterokaryons survive through complementation of the induced deficiencies. With this technique, the proportion of fusion regenerants with the desired nuclear complement is increased. Use of enucleated protoplasts, cytoplasts, may be an attractive alternative. Maliga et al. (1982) were the first to demonstrate the transfer of streptomycin—resistant chloroplasts in *Nicotiana* by cytoplast-protoplast fusion. Later, Watanabe and Yamaguchi (1988) demonstrated the possibility of using this technique in *Brassica*. Fusion of cytoplasts isolated from cytoplasmic male sterile *Raphanus sativus* cv. Kosena with iodoacetamide (IOA)–treated protoplasts of *B. napus* resulted in male sterile *B. napus* (Sakai and Imamura 1990), thereby permitting the transfer of newly found CMS cytoplasm in *Raphanus sativus* to *B. napus* cv. Westar in a single step.

Another approach used by Yarrow et al. (1986) is to manually select and isolate individual heterokaryons using a micromanipulator. The parental protoplasts were differentially stained with fluorescent dyes prior to fusion, to aid visual identification of the heterokaryons.

Whereas the frequency of heterokaryon-derived plants in the regenerant population can be increased by selection of fusion products, currently there is no direct method for selecting a particular desired chloroplast/mitochondrial combination. The addition of triazine-type herbicides to the culture medium may be one case where a particular organelle complement (i.e. atr^{R} chloroplasts) could be selected for. Whether a heterokaryon selection procedure is used or not, it is necessary to screen the regenerant population for the desired nucleo-cytoplasmic combination.

9.5.3 Identification of Fusion Products

The presence or absence of particular organelles can often be determined by relatively simple tests. In the case of CMS, visual inspection of the flowers can determine pollen production. *Atr*^{R} can be detected by herbicide applications or by testing for photosynthetic activity using nitro-blue tetrazolium (Robertson and Earle 1987). The correction of chloroplast-defective *ogu* CMS can be established by growing the plants under cool conditions and looking for yellowing leaves (Pelletier et al. 1983; Menczel et al. 1987) or by detailed physiological studies (Jarl et al. 1988).

Unfortunately, these methods can be unreliable, particularly in the case of CMS, as this phenotype is frequently observed in fusion products due to aneuploidy. Also, some organelle types do not exhibit any obvious phenotypic traits, necessitating more sophisticated techniques of genotype identification.

The analysis of cytoplasmic DNA is the definitive identification methodology. Kemble (1987) developed a restriction fragment length polymorphism (RFLP) analysis system that can differentiate between all the main *Brassica* mt-and cpDNA types of interest. DNA is isolated specifically from either the mitochondria or chloroplasts and digested with a restriction enzyme of choice. The fragmented DNA is then electrophoresed on agarose gels and stained with ethidium bromide to exhibit the banding pattern which is analogous to a fingerprint for each of the different organelle types. Techniques of this type were used for a number of the *Brassica* cybrids reported (Pelletier et al. 1983; Chetrit et al. 1985; Vedel et al. 1986; Yarrow et al. 1986, 1990; Barsby et al. 1987a, b; Chuong et al. 1988).

An alternative method of determining the organelle complement or a regenerant by molecular means is by isolating total DNA from the plant tissue. This is digested with a restriction enzyme, electophoresed on an agarose gel and then the DNA is Southern blotted (Southern 1975) onto nitrocellulose filters. The DNA is then hybridized with radioactively labelled DNA probes that bind to the different restriction fragments. Apart from the necessity to handle radioactive material, a further disadvantage to the technique is that there are often chloroplast and mitochondrial sequences repeated in the nuclear DNA, and even chloroplast sequences in mitochondrial DNA (Lonsdale 1985; Timmins and Steele-Scott 1985), that can lead to confusing results. However, with diligent selection of probes, such Southern analysis has successfully identified cybrids (Morgan and Maliga 1987).

In most cases the organelle mixture originally present in the heterokaryon, following fusion, usually sorts out, resulting in a cell line or plant with a population of mitochondria of one type only and also a uniform chloroplast population. Occasionally, plants are regenerated that retain a mixture of different mitochondria or different chloroplasts, that are not due to chimeras. However, generally these eventually sort out in planta towards one of the parental types, which are stably inherited through a seed generation (Kemble et al. 1988a).

An additional observation following organelle DNA analysis is often a high frequency of mitochondrial DNA recombination (Table 1), albeit to date, commonly only in those reports involving the *ogu* CMS cytoplasm (Chetrit et al. 1985; Vedel et al. 1986; Morgan and Maliga 1987; Kao et al. 1992). Novel hybridization patterns of mt DNA were also observed in four out of ten plants resulting from cytoplast-

Table 1. Occurrence of mitochondrial DNA recombination following fusions involving different cytoplasmic combinations in *Brassica napus*

Parental mitochondrial combinations	No. of fusion plants analysed	No. of plants with mtDNA alternations	Reference
ogu × *nap*	4	4	{Pelletier et al. (1983) {Chetrit et al. (1985)
ogu × *ctr*	1 (5)	1 (5)	{Vedel et al. (1986)
ogu × *nap*	22	1	{Menczel et al. (1987)
ogu × *nap*	7 (29)	6 (7)	{Morgan and Maliga (1987)
pol × *nap*	100	0	{Yarrow et al. (1986)
pol × *ctr*	32	0	{Barsby et al. (1987a, b)
pol × *cam*	72	0	{Kemble et al. (1988a)
nap × *ctr*	29	0	
ogu × *nap*	2	0	
ogu × *cam*	2	0	
mau × *nap*	1 (238)	0 (0)	

cam designates the mitochondria in regular *B. campestris*; *mau* designates mitochondria in *B. maurorum*. Numbers in parentheses are totals for each research group.

protoplast fusion between CMS *R. sativus* cv. Kosena and *B. napus* cv. Westar (Sakai and Imamura 1990). In contrast, fusions largely involving the *pol* cytoplasm have not resulted in mtDNA recombinations (Kemble et al. 1988b). In an elaborate study, Landgren and Glimelius (1990) observed mtDNA rearrangements in three combinations of somatic hybrids studied. This results from the interaction between the parental mtDNA genomes in heteroplasmic state following protoplast fusion. All rearranged *B. napus* (+) *B. nigra* and *B. napus* (+) *Eruca sativa* hybrids contained hybrid-specific mtDNA fragments, while the rearranged mtDNA found in *B. campestris* (+) *B. oleracea* hybrids consisted of a mixture of parental-specific fragments, without any hybrid-specific fragments. Hybrid-specific mtDNA fragments perhaps result from intermolecular recombination between parental mtDNA genomes in heteroplasmic state after protoplast fusion. Similar to results of Kemble et al. (1988b), no mtDNA rearrangement were observed in *B. napus* plants regenerated from protoplasts. Whether the degree of mtDNA recombination is dependent on the cytoplasmic genome, the fusion methodology or culture techniques, or a combination of these factors, remains unclear.

As discussed later, such mtDNA recombinations could be beneficial in some situations, although faithful conservation of the mitochondrial genome following cybrid formation is generally desired by most plant breeders.

Although mtDNA recombination following protoplast fusion is not infrequent, recombination in the chloroplast DNA remains rare. There are no examples in *Brassica*. To date, the two reports of cpDNA rearrangements are from fusions between two *Nicotiana* species (Medgyesy et al. 1985) and between *N. tabacum* and *Solanum tuberosum* (Thanh and Medgyesy 1989).

Recombination of mtDNA as described above refers to events in the chromosomal DNA of the mitochondria. However, extrachromosomal DNA plasmids exist in some *Brassica* lines (Palmer et al. 1983; Turpen et al. 1987). Previous association of one particular plasmid with CMS has recently been refuted (Chetrit et al. 1985; Kemble et al. 1986). This plasmid, one of 11.3 kb, was found, following fusion, to occasionally disappear (Kemble et al. 1988b) and even to transfer from the mitochondria of one protoplast population to mitochondria of the other fusion partner protoplast population (Kemble et al. 1988b; Rosen et al. 1988), an interesting phenomenon which may have some application for mitochondrial transformations in the future, although for what specific goals remains unclear at present.

9.6 Cybrids in Breeding Programmes

Brassica cybrids have immediate applications in F_1 hybrid programmes based on cytoplasmic male sterility systems, and in developing lines that are resistant to the triazine herbicides. To be acceptable and useful to the plant breeders it is essential that the cybrids exhibit the minimum of aberrations or loss of female fertility as a consequence of the fusion and culture process. Tissue culture of plants in general can frequently induce variation, known as somaclonal variation (Larkin and Scowcroft 1981), depending on the cultural practices and the genotypes utilized. Somaclonal variation can be a potentially useful source of novel variation that can be exploited in some aspects in plant breeding. However, in cybrid production, which concentrates on changes in the cytoplasmic genome as opposed to the nuclear genome, somaclonal variation resulting in abnormal plant morphologies is usually unwanted. Aneuploidy is a common outcome of cybrid production. However, if some seed set is possible, albeit low, female fertility problems can be corrected in the next generation, where euploidy is usually restored (Barsby et al. 1987a). With fertility resolved, cybrids have been successfully incorporated into breeding programmes (Barsby et al. 1987c; Pelletier 1990).

Mitochondrial DNA recombination has been associated with female fertility problems in some cases (Pelletier 1990), although such recombination may be beneficial for producing improved CMS systems. Ideally, a rearranged mtDNA genome should combine the genetic material specifically responsible for the CMS trait, with those genes that allow for restoration. Additionally, genes that influence healthy nectary formation should be incorporated, since nectar production is of importance to bee activity for cross-pollination towards F_1 hybrid seed development.

Apart from improving current CMS systems, by either substituting defective chloroplasts or recombining mtDNA, fusion can assist in the formation of cybrids that introduce new CMS systems to breeding programmes. Current mitochondrial types can be substituted with those from related but sexually incompatible *Brassica* species via fusion, to test for male sterility potential. For example, mitochondria from the wild species *B. maurorum* have been transferred to *B. napus* canola for this purpose (Yarrow, unpubl.).

With further physiological and genetic understanding of mitochondrial functions in the cell, it is conceivable that there will be a future need to insert particular beneficial gene(s) into the mitochondrial genome. Observations of the mitochondrial plasmid movements following *Brassica* fusions suggest that fusion may be a mechanism for mtDNA transformation (Kemble et al. 1988b). Desired genes, perhaps, could be transferred to the mitochondria, using the *Brassica* plasmid as a vector, via liposomes fused to *Brassica* protoplasts.

9.7 Conclusions

The production of cybrids is the first example of where the biotechnology of protoplasts fusion has a direct application to the breeding of *Brassica* oilseed crops. The technology has since been transferred to parallel breeding goals in the vegetable species *B. oleracea* (Yarrow et al. 1990). Clearly this research has and will continue to contribute to the development of CMS systems, but the importance of CMS itself in the oilseed industry remains unclear at this current time, as almost all the available CMS systems suffer from one defect or another. Thus the major efforts will continue to be directed towards selective correction of such defects. Asymmetric protoplast fusion will also help in the simultaneous production of CMS lines and introgression of fertility restorer nuclear genes from cytoplasmic donor alien species.

Molecular biology approaches to pollination control currently offer novel alternatives to both CMS and self-incompatibility, such as inserting anti-sense genes into *Brassica* that specifically inactivate vital pollen development genes. Alternatively, there is progress towards inserting RNA degradation genes that are controlled by pollen tissue-specific promoters, which again perform by interfering with pollen development. These developments are beyond the scope of this chapter, but serve to illustrate other applications of biotechnology to *Brassica* breeding.

References

Ayotte RR (1986) The transfer of triazine resistance from *Brassica napus* to *B. oleracea*. Ph D thesis, Univ Guelph, Canada

Bannerot H, Boulidard L, Cauderon Y, Tempe J (1974) Cytoplasmic male sterility transfer from *Raphanus* to *Brassica*. Proc Eucarpia Crop Sect Cruciferae 25: 52–54

Bannerot H, Boulidard L, Chupeau Y (1977) Unexpected difficulties met with the radish cytoplasm *Brassica oleracea*. Cruciferae Newslett 2: 16

Barsby TL, Chuong PV, Yarrow SA, Wu SC, Coumans M, Kemble RJ, Powell AD, Beversdorf WD, Pauls KP (1987a) The combination of Polima CMS and cytoplasmic triazine resistance in *Brassica napus*. Theor Appl Genet 73: 809–814

Barsby TL, Kemble RJ, Yarrow SA (1987b) *Brassica* cybrids and their utility in plant breeding. In: Von Wetstein D, Chua NH (eds) Plant molecular biology. Nato Ad Stud Inst, Copenhagen. Series A: Life Sci Vol 140. Plenum

Barsby TL, Yarrow SA, Kemble RJ, Grant I (1987c) The transfer of cytoplasmic male sterility to winter-type oilseed rape (*Brassica napus* L.) by protoplast fusion. Plant Sci 53: 243–248

Barsby TL, Yarrow SA, Shepard JF (1986) A rapid and efficient alternative procedure for the regeneration of plants from protoplasts of *Brassica napus*. Plant Cell Rep 5: 101–103

Batra A, Dhingra M (1990) A method for the rapid isolation of hypocotyl protoplasts of *Eruca sativa*. Curr Sci 59: 277–279

Beversdorf WD, Erickson LR, Grant I (1985) Hybridization process utilizing a combination of cytoplasmic male sterility and herbicide tolerance. US Patent 4517763

Beversdorf WD, Weiss-Lerman J, Erickson LR, Souza-Machado Z (1980) Transfer of cytoplasmic inherited triazine resistance from bird's rape to cultivated oilseed rape (*Brassica campestris* and *B. napus*). Can J Genet Cytol 22: 167–172

Chetrit P, Mathieu C, Vedel F, Pelletier G, Primard C (1985) Mitochondrial DNA polymorphism induced by protoplast fusion in *Cruciferae*. Theor Appl Genet 69: 361–366

Christey MC, Makaroff CA, Earle ED (1991) Atrazine-resistant cytoplasmic male sterile-*nigra* broccoli obtained by protoplast fusion between cytoplasmic male sterile *B. oleracea* and atrazine resistant *B. campestris*. Theor Appl Genet 83: 201–208

Chuong PV, Pauls KP, Beversdorf WD (1985) A simple culture method for *Brassica* hypocotyl protoplasts. Plant Cell Rep 4: 4–6

Chuong PV, Beversdorf WD, Powell AD, Pauls KP (1988) Somatic transfer of cytoplasmic traits in *Brassica napus* L. by haploid protoplast fusion. Mol Gen Genet 211: 197–201

Erickson L, Kemble RJ (1990) Paternal inheritance of mitochondria in rapeseed (*Brassica napus*). Mol Gen Genet 218: 419–422

Gleba YY, Hoffman F (1980) "Arabidobrassica" — a novel plant obtained by protoplast fusion. Planta 149: 112–117

Glimelius K (1984) High growth rate and regeneration capacity of hypocotyl protoplasts in some Brassicaceae. Physiol Plant 61: 38–44

Jaiswal SK, Hammatt N, Bhojwani SS, Cocking EC, Davey MR (1990) Plant regeneration from cotyledon protoplasts of *Brassica carinata*. Plant Cell Organ Cult 22: 159–166

Jarl CI, Bornman CH (1988) Correction of chlorophyll-defective, male-sterile winter oilseed rape (*Brassica napus*) through organelle exchange: Phenotypic evaluation of progeny. Hereditas 108: 97–102

Jarl CI, Ljungberg UK, Bornman CH (1988) Correction of chlorophyll-defective, male-sterile winter oilseed rape (*Brassica napus*) through organelle exchange: Characterization of the chlorophyll deficiency. Physiol Plant 72: 505–510

Jourdan PS, Earle ED, Mutschler MA (1989) Atrazine-resistant cauliflower obtained by somatic hybridization between *Brassica oleracea* and *ATR-B. napus*. Theor Appl Genet 78: 271–279

Kao HM, Keller WA, Gleddie S, Brow GG (1992) Synthesis of *B. oleracea/B. napus* somatic hybrid plants with novel organelle DNA compositions. Theor Appl Genet 83: 313–320

Kao KN, Michayluk MR (1974) A method for high frequency intergeneric fusion of plant protoplasts. Planta 115: 355–367

Keller WA, Melchers G (1973) The effect of high pH and calcium on tabacco leaf protoplast fusion. Z Naturforsch 28C: 737–741

Kemble RJ (1987) A rapid single leaf, nucleic acid assay for detemining the cytoplasmic organelle complement of rapeseed and related *Brassica* species. Theor Appl Genet 73: 364–370

Kemble RJ, Barsby TL, Yarrow SA (1988a) Transformation of plant mitochondria with mitochondrial DNA plasmids via protoplast fusion. Mol Gen Genet 213: 202–205

Kemble RJ, Carlson JE, Erickson LR, Sernyk JL, Thompson DJ (1986) The *Brassica* mitochondrial DNA plasmid and large RNAs are not exclusively associated with cytoplasmic male sterility. Mol Gen Genet 205: 193–195

Kemble RJ, Yarrow SA, Wu SC, Barsby TL (1988b) Absence of mitochondrial and chloroplast DNA recombinations in *Brassica napus* plants regenerated from protoplasts, protoplast fusions and anther culture. Theor Appl Genet 75: 875–881

Kirti PB, Chopra VL (1990) Rapid plant regeneration through organogenesis and somatic embryogenesis from cultured protoplasts of *Brassica juncea*. Plant Cell Tissue Organ Cult 20: 65–68

Kumar A, Cocking EC (1987) Protoplast fusion: a novel approach to organelle genetics in higher plants. Am J Bot 74: 1289–1303

Landgren M, Glimelius K (1990) Analysis of chloroplast and mitochondrial segregation in three different combinations of somatic hybrids produced within Brassicaceae. Theor Appl Genet 80: 776–784

Larkin PJ, Scowcroft WR (1981) Somaclonal variation—a novel source of variability from cell cultures for plant improvement. Theor Appl Genet 60: 197–214

Lonsdale DM (1985) Movement of genetic material between the chloroplast and mitochondrion in higher plants. In: Hohn B and Dennis ES (eds) Genetic flux in plants. Springer, Berlin-Heidelberg New York Tokyo pp 51–60

Louden PT, Nelson RS, Ingram DS (1990) Studies of protoplast culture and plant regeneration from commercial and rapid-cycling *Brassica* species. Plant Cell Tissue Organ Cult 19: 213–224

Maliga P, Lortz H, Lazer C, Nagy F (1982) Cytoplast-protoplast fusion for interspecific chloroplast transfer in *Nicotiana.* Mol Gen Genet 185: 211–215

Matais B, Bouchard CJ (1978) Une moutarde des oiseaux (*Brassica rapa* L.) resistance à l'atrazine. Phytoprotection 59: 117–119

Medgyesy P, Fejes E, Maliga P (1985) Interspecific chloroplast recombination in a *Nicotiana* somatic hybrid. Proc Nat Acad Sci USA 82: 6960–6964

Melchers G, Sacristan MD, Holder AA (1979) Somatic hybrid plants of potato and tomato regenerated from fused protoplasts. Carlsberg Res Commun 43: 203–218

Menczel L, Morgan A, Brown S, Maliga P (1987) Fusion-mediated combination of Ogura-type cytoplasmic male sterility with *Brassica napus* plastids using X-irradiated cms protoplasts. Plant Cell Rep 6: 98–101

Morgan A, Maliga P (1987) Rapid chloroplast segregation and recombination of mitochondrial DNA in *Brassica* cybrids. Mol Gen Genet 209: 240–246

Ogura H (1968) Studies of the new male sterility in Japanese radish, with special reference to the utilization of this sterility towards the practical raising of hybrid seed. Mem Fac Agric Kagoshima Univ, 6: 39–78

Palmer JD, Shields CR, Cohen DB, Orton TJ (1983) An unusual mitochondrial DNA plasmid in the genus *Brassica.* Nature (London) 301: 725–728

Pelletier G (1990) Cybrids in oilseed *Brassica* crops through protoplast fusion. In: Bajaj YPS (ed) Biotechnology in agriculture and forestry, vol 10, Legumes and oilseed crops. I. Springer Berlin Heidelberg New York Tokyo pp 418–433

Pelletier G, Primard C, Vedel F, Chetrit P, Remy R, Rouselle P, Renard M (1983) Intergeneric cytoplasmic hybridization in Curciferae by protoplast fusion. Mol Gen Genet 191: 244–250

Prakash S, Chopra VL (1990) Male sterility caused by cytoplasm of *Brassica oxyrrhina* in *B. campestris* and *B. juncea.* Theor Appl Genet 79: 285–287

Reith M, Straus NA (1987) Nucleotide sequence of the chloroplast gene responsible for triazine resistance in canola. Theor Appl Genet 73: 357–363

Robertson D, Earle ED (1987) Nitro-blue tetrazolium: a stain for photosynthesis in protoplasts. Plant Cell Rep 6: 70–73

Rosen B, Hallden C, Heneen WK (1988) Diploid *Brassica napus* somatic hybrids: Characterization of nuclear and organellar DNA. Theor Appl Genet 76: 197–203

Sakai T, Imamura J (1990) Intergeneric transfer of cytoplasmic male sterility between *Raphanus sativus* (CMS line) and *Brassica napus* through cytoplast-protoplast fusion. Theor Appl Genet 80: 421–427

Shepard JF, Bidney DL, Barsby TL, Kemble RJ (1983) Genetic transfer in plants through interspecific protoplast fusion. Science 219: 683–688

Sidorov VA, Menczel L, Nagy F, Maliga P (1981) Chloroplast transfer in *Nicotiana* based on metabolic complementation between irradiated and iodoacetate treated protoplasts. Planta 152: 341–345

Southern EM (1975) Detection of specific sequences among DNA fragments separated by gel electrophoresis. J Mol Biol 98: 502–527

Thanh ND, Medgyesy P (1989) Limited chloroplast gene transfer via recombination overcomes plastome-genome incompatibility between *Nicotiana tabacum* and *Solanum tuberosum.* Plant Mol Biol 12: 87–93

Thomzik JE, Hain R (1988) Transfer and segregation of triazine-tolerant chloroplasts in *Brassica napus* L. Theor Appl Genet 76: 165–171

Timmins JN, Steele-Scott N (1985) Movement of genetic information between the chloroplast and nucleus. In: Hohn B, Dennis ES (eds). Genetic flux in plants. Springer, Berlin Heidelberg New York Tokyo pp 61–78

Turpen T, Garger SJ, Marks MD, Grill LK (1987) Molecular cloning and physical characterization of a *Brassica* linear mitochondiral plasmid. Mol Gen Genet 209: 227–233

Vedel F, Chetrit P, Mathieu C, Pelletier G, Primard C (1986) Several different mitochondrial DNA regions are involved in intergenomic recombination in *Brassica napus* cybrid plants. Curr Genet 11: 17–24

Wallin AK, Glimelius K, Eriksson T (1974) The induction of aggregation and fusion of *Daucus carota* protoplasts by polyethylene glycol. Z Pflanzenphysiol 74: 64–80

Watanabe M, Yamaguchi H (1988) The methods for isolation of cytoplasts in several crop plants. Jpn J Breed 38: 43–52

Yarrow SA (1990) Production of cybrids in Rapeseed (*Brassica napus*). In: Pollard JW, Walker JM (eds) Methods in molecular biology Vol 6: Plant cell cult, Humana, Clifton, New Jersey

Yarrow SA, Burnett LA, Wildeman RP, Kemble RJ (1990) The transfer of Polima cytoplasmic male sterility from oilseed rape (*Brassica napus*) to broccoli (*B. oleracea*) by protoplast fusion. Plant Cell Rep 185–188

Yarrow SA, Wu SC, Barsby TL, Kemble RJ, Shepard JF (1986) The introduction of cms mitochondria to triazine tolerant *Brassica napus* L. var. Regent, by micromanipulation of individual heterokaryons. Plant Cell Rep 5: 415–418

Zelcer A, Aviv D, Galun E (1978) Interspecific transfer of cytoplasmic male sterility by fusion between protoplasts of normal *Nicotiana sylvestris* and X-ray irradiated protoplasts of male sterile *N. tabacum*. Z Pflanzenphysiol 90: 397–407

Zimmerman U, Scheurich P (1981) High frequency fusion of plant protoplasts by electric fields. Planta 151: 26–32

CHAPTER 10

Transformation and Foreign Gene Expression

M.M. Moloney and *L.A. Holbrook*

10.1 Introduction

Genetic modification by plant transformation has advanced more rapidly in *Brassica* spp. than in most other oilseeds. This is because of the availability of a number of tissue culture methods for whole-plant regeneration via organogenesis (Kartha et al. 1974; Stringham 1977) and embryogenesis (Keller and Armstrong 1978; Crouch and Sussex 1981). Single-cell systems, including root and shoot protoplasts (Xu et al. 1982; Newell et al. 1984) and microspores (Chuong et al. 1988), have demonstrated the totipotency of cells from many tissues and species of *Brassica*. Although these manipulations are frequently less efficient than analogous work in model systems like tobacco or petunia, *Brassica* nevertheless is one of the most readily manipulated genera among the major food crops. This genus comprises a large number of oilseed and vegetable crops. The application of genetic engineering in these crops for agronomic improvement has been widely discussed (Holbrook et al. 1988; Battey et al. 1989). However, it may still take several years to realize some of the most important goals, including the biochemical modification of *Brassica* seed oils and the development of disease resistance. To a large extent, progress in this area will depend primarily on continued commitment to fundamental work in lipid biochemistry and plant-pathogen interactions.

To date, *Brassica* species have not been widely used in more basic studies of plant gene expression, although there are a few notable exceptions to this (Crouch et al. 1985; Radke et al. 1988; Nasrallah et al. 1987). In part, this has been due to the relative difficulty associated with *Brassica* transformation as compared, for example, to tobacco or petunia. Improvements in *Brassica* transformation will greatly assist in the use of the genus for fundamental studies where it would have some singular advantages. These advantages include a relatively short life cycle (in the case of annual Brassicaceae), seeds of moderate size which can be handled easily even during their early stages of development, flowers of manipulable proportions, a relatively small genome and broad scope for tissue culture procedures.

Clearly, then, the ability to perform genetic transformations in members of the *Brassica* genus will have many consequences for fundamental and applied research. In this chapter recent progress in the transformation of this genus with special reference to the oilseed brassicas and the nature of current studies in gene expression using *Brassica* spp. as a model will be discussed.

10.2 Gene Transfer

10.2.1 *Agrobacterium*-Mediated Gene Transfer

Since the initial reports in the early 1980s on the use of *Agrobacterium tumefaciens* as a vector for plants (Horsch et al. 1984; DeBlock et al. 1984), there has been great interest in the use of such a system for *Brassica* transformation. It has been known for many years that the *Brassica* genus may act as a host for *Agrobacterium* infection (DeCleene and DeLey 1976). More detailed studies were initiated by Holbrook and Miki (1985) and by Ooms et al. (1985a). Holbrook and Miki (1985) established, using five *Brassica* species including four members of the evolutionary triangle of U (1935) and *Brassica hirta*, that all these species were susceptible to *A. tumefaciens* infection, as indicated by tumour induction. Significant differences were also found in the numbers of tumours induced by different strains of *A. tumefaciens* in *B. napus*. Among the strains tested, it was found that those harbouring nopaline Ti plasmids (pTiC58, T37) were much more effective in initiating tumours than pTi $B_6$806, A6S and B6-T. These experiments confirmed the transformed phenotype by the use of opine biosynthesis assays on extracts of axenically grown tumours. Using the nopaline strain, C58, teratomous tissue was observed from crown galls grown on *B. napus*, *B. oleracea* and *B. hirta*. These teratomous shoots were invariably abnormal and did not give rise to rooted plants. Attempts to graft them were also unsuccessful.

Ooms et al. (1985a) exploited *Agrobacterium rhizogenes* rather than using *A. tumefaciens*. This approach was based on the fact that *A. rhizogenes* also transfers plasmid DNA to host plants (White et al. 1982) and that many workers had succeeded in regenerating whole plants from the "hairy roots" which are the phenotype of *A. rhizogenes* infections in most hosts (Chilton et al. 1982; Tepfer 1984; Ooms et al. 1985b). Ooms et al. (1985a) could regenerate whole plants from an axenic hairy-root culture obtained after infection of *B. napus* with *A. rhizogenes* strain LBA 9402. These plants had several abnormalities, including excessive root development, curly leaves and the inability to set seed. Subsequent experiments by Guerche et al. (1987a) yielded viable plants using this method and both single and multiple T-DNA insertions were found. These plants set seed, but the progeny exhibited a phenotype characteristic of transformations mediated by *A. rhizogenes*. This included abnormalities such as drastically reduced apical dominance and curly leaves (Ackermann 1977; Tepfer 1984). Apart from establishing the ability to transform *Brassica* species using *A. tumefaciens* or *A. rhizogenes*, these studies also demonstrated that the oncogenes of *Agrobacterium* had severe effects on the phenotype of transformed tissue similar to those observed in solanaceous species.

The application of "disarmed" *Agrobacterium* vectors has now solved the problem of abnormal plants or tissues encountered by the initial work on *Brassica* transformation. Experiments using co-integrate vectors (Fraley et al. 1985) and binary vectors (Bevan 1984; Horsch and Klee 1986) showed the possibilities of regenerating transformed tissue. The primary focus of attention for much of this work has been *B. napus*.

Whereas selection of transformed tissue after infection with wild-type Ti or Ri vectors is normally dependent on a morphological phenotype, disarmed vectors also introduce a selection step usually involving drug resistance. Pua et al. (1987) reported the recovery of transformed *B. napus* plants using a binary vector pMON 809 (Eichholtz et al. 1987) containing a gene encoding methotrexate-resistant mouse dihydrofolate reductase (DHFR). These workers co-cultivated the binary *Agrobacterium* vector with longitudinally cut stem sections from 4-5-week-old shoot cultures. Selection was performed using 0.01 mg/l methotrexate (MTX), which is approximately tenfold higher than the lowest concentration found to inhibit completely the shoot regeneration in controls. Using this selection, about 10% of the explants yielded transformed plants. Transformation was confirmed by a leaf-disc callus growth assay on MTX (0.01 mg/l) and Northern blots of putative transformants using a mouse DHFR probe. Pua et al. (1987) also attempted transformations using a binary vector containing a neomycin phosphotransferase (NPT II) gene specifying kanamycin resistance. Using an identical transformation protocol, but substituting kanamycin for MTX, they were unable to recover transformed shoots.

Using NPT II as a selectable marker, Fry et al. (1987) were able to produce large numbers of confirmed stable transformed *B. napus* plants. They used both co-integrate (Fraley et al. 1985) and binary (Beven 1984; Horsch and Klee 1986) vectors in their study. They also made use of transverse sections of floral stems (Stringham 1977) cut as 5-mm discs. These cut surfaces were infected with an overnight culture of the appropriate *Agrobacterium* strain for 5 min and then blotted to absorb excess bacteria. Co-cultivation took place for 48-72 h in the presence of a feeder layer of tobacco cells (Horsch et al. 1985), after which the infected sections were placed on regeneration medium supplemented with carbenicillin (50 mg/l) to eliminate residual *Agrobacterium* and kanamycin (100 mg/l) for selection. This method gave numerous transformed shoots that could be rooted readily and resulted in fertile whole plants. Both octopine and nopaline strains were used by Fry et al. (1987), and it was found in agreement with the earlier studies of Holbrook and Miki (1985) that nopaline strains were more virulent on *B. napus* than octopine strains. Using this method, Fry et al. (1987) achieved an average of 10% of explants giving rise to at least one transformed plant. In the best experiments reported, they obtained transformed shoots from 21% of the explants. The selection conditions (100 mg/l kanamycin) used by these workers employs significantly higher kanamycin concentrations than most other workers have used for the Brassicaceae. This may result in the selection of transformants expressing the introduced gene at higher levels. Some of the shoots obtained by this method did not survive a secondary selection on 100 mg/l of kanamycin. However, it should be noted that untreated explants did not show any shoot regeneration when cultured on 100 mg/l kanamycin. Thus, it is possible that the secondary screen eliminates a number of bona fide transformants which express NPT II at marginal to low levels.

Several other laboratories have also been successful in exploiting NPT II as a selectable marker for *B. napus* transformation. Charest et al. (1988) performed *Agrobacterium*-mediated transformation experiments using thin cell layer explants as the infection target. These thin cell layers (from floral stems) were reported by Klimazewska and Keller (1985) to be highly regenerable. Up to 87% of the unselected explants produced shoots. In these explants, meristemoids appeared to initiate from periclinal divisions in the parenchyma four or five cell layers below the epidermal surface. This is usually accompanied by a small amount of callus formation. This proved to be a viable target tissue for transformation using *Agrobacterium*, and Charest et al. (1988) obtained, in several experiments, confirmed transformants from about 2% of the explants. These workers also attempted transformation and selection using chloramphenicol acetyl transferase (CAT) as a selectable marker gene. The use of CAT was not very effective for selection of transformed plants. In fact, it has been shown that *Brassica* spp. exhibit endogenous CAT-like activity (Balazs and Bonneville 1987). This may confound the selection procedure as a high number of "escapes" are possible. It should be noted that because of this, CAT is also not recommended as a reporter protein for studies involving transient expression in *Brassica* cells. Other enzymes such as NPT II (Moloney et al. 1989) or β-glucuronidase (Jefferson 1987; Moloney et al., unpubl.) are, however, useful as quantifiable markers in such studies using *Brassica* spp. Charest et al. (1988) also investigated a variety of regimes using kanamycin as selection agent. Using their transformation system, in contrast to that of Fry et al. (1987), it was necessary to select at quite low kanamycin concentrations (15-20 mg/l) in order to obtain transgenic plants. Of the plants selected, very few were resistant to subsequent selection on higher concentrations of kanamycin (80 mg/l). Phenotypically normal and fertile *B. napus* transformants were also obtained following co-cultivation of stem epidermal explants with an *Agrobacterium tumefaciens* strain containing a disarmed octopine pMON 316, a cointegrate vector carrying the gene for kanamycin resistance and the scorable marker nopaline synthatase (Misra 1990). Kanamycin resistance could be transferred to the progeny of self-fertile plants.

Moloney et al. (1989) reported a new procedure for *Agrobacterium*-mediated transformation of *Brassica* spp. This work exploits the high regeneration rates obtained from cotyledonary petioles of *Brassica* spp. Sharma et al. (1990) developed this system for high frequency regeneration in *Brassica juncea*. This system routinely produces shoots on >80% of the explants with *B. juncea* (Sharma 1987; Sharma et al. 1990), *B. napus* (Moloney et al. 1989) and *B. oleracea* (Szarka and Moloney, unpubl.). The regeneration arises from cells in the petiole cut end, and occurs after about 14 to 21 days in culture (Sharma 1987). The petiolar cut surface has proven to be a susceptible target for *Agrobacterium* transformation. Moloney et al. (1989), using NPT II as a selectable marker, recovered transformed plants from more than 50% of the explants of *B. napus* cv. Westar. These transformants were selected on a regeneration medium supplemented with 15 mg/l kanamycin. Homo- and heterozygous progeny carrying the introduced sequences were resistant to higher concentrations of kanamycin (30 mg/l). Numerous multiple insertions were reported and demonstrated by Southern blotting and Mendelian segregation ratios. Noteworthy features of this method include the rapidity of shoot development and the minimal callus formation. The latter property may be a reason for relatively few "escapes" being produced in this system. This method has also been applied to other members of the *Brassica* genus including *B. campestris*, *B. oleracea* and *B. juncea* (Moloney et al., unpubl.) with positive results. Culture conditions, in particular growth substance concentrations, vary between species for optimal transformation and regeneration rates.

10.2.2 Direct DNA Uptake

DNA uptake into plant cells may be induced by controlling the permeability of the plasmalemma. This is comparable with the classical transformation method used for prokaryotes, which requires competent cells and rigorous ionic conditions. Such methods have been applied successfully in animals, where naked DNA is precipitated onto the cell surface in the presence of calcium phosphate. Treatments with soluble polymers (e.g. polyethylene glycol) which induce changes in membrane permeability and cellular turgidity result in the absorption into the cell of some of the DNA from the cell surface.

For plant cells, this system is dependent on the removal of cell walls which are an efficient barrier to macromolecule uptake. The initial reports successfully applying this technique to plant cells were those of Davey et al. (1980), Draper et al. (1982) and Krens et al. (1982). These original studies exploited the isolated Ti plasmid from *Agrobacterium*, but it became clear that the fragments of the entire Ti plasmid could become integrated, including DNA outside the T-region. With this came the recognition that any plasmid DNA (non-Ti-derived) might be suitable for direct uptake and integration experiments. The use of this technique has recently been reviewed by Davey et al. (1989). *Brassica* species have been among the primary targets for transformation experiments using direct DNA uptake. This can, in part, be attributed to the relative ease with which *Brassica* protoplasts may be obtained and cultured.

Potrykus et al. (1985) demonstrated that several plant species including *B. campestris* var. *rapa* could be stably transformed using direct DNA uptake. They did not, however, regenerate *B. campestris* plants in these experiments. The same group (Paszkowski et al. 1986) demonstrated, using this technique, that an engineered viral genome could be introduced into *Brassica campestris* cells and that the viral genome itself could be used as a vector for foreign genes. This experiment is discussed later in this chapter. The rates of recovery of stably transformed colonies were between 10^{-4} and 10^{-5}.

The development of electroporation as a technique for plant cell transformation (Fromm et al. 1985; Shillito et al. 1985) was a significant technical advance for studies in direct DNA uptake. Using this technique it was demonstrated that both transient expression (Fromm et al. 1985) and stable integration (Fromm et al. 1986; Shillito et al. 1985) of foreign DNA was possible. This technique was exploited for transformation and regeneration of stably transformed *B. napus* plants by Guerche et al. (1987b). These workers introduced NPT II into leaf protoplasts of *B. napus*. They recovered transformed colonies using the agarose embedding technique (Shillito et al. 1983) and were able to obtain whole regenerated plants from these cultures. The expression of NPT II was stable and segregated normally through meiosis.

A novel way of encouraging direct DNA uptake by plant cells is the use of a laser microbeam. Weber et al. (1988) demonstrated such DNA uptake by *B. napus* cells using a UV laser microbeam. This technique, although not yet shown to result in expression of integrated DNA, has a singular advantage. It may be used on intact cells and does not require the use of protoplasts, which frequently show low regeneration rates. Recently, Weber et al. (1989) also demonstrated the possibility of introduction of DNA into isolated organelles such as chloroplasts. Using *B. napus* chloroplasts, they demonstrated generation of pores, uptake of DNA and resealing of the pore. The pore was produced using a microbeam from UV laser. If stable integration and expression of the introduced DNA is proven, this technique promises broader application in plants.

10.2.3 Microinjection

The introduction by injection into single cells of a diverse array of compounds including proteins, nucleic acids, antibodies and fluorescent probes has been a widely used tool in the cell biology experiments. This injection, using fine glass needles, has been particularly adapted to studies on in vitro mRNA translation and protein processing using the *Xenopus* oocyte system. For example, plant storage protein mRNAs have been introduced and translated in these oocytes (for method and references, see Kawata et al. 1988). Another major application has been the production of transgenic animals by DNA injection into fertilized eggs, i.e. zygotes, most often using mice (Gordon et al. 1980; Wagner et al. 1981; Palmiter et al. 1982). However, transgenic rabbits, sheep and pigs have also been produced (Hammer et al. 1985).

Microinjection has been more widely used in animal cells than in plant cells. There are a variety of reasons for this, including: lack of cell walls, monolayer growth in tissue culture, large nucleus to cytoplasm volume ratio, lack of large vacuoles, and an easily visualized nucleus, all of which are frequently features of cultured animal cells. Despite the obstacles inherent to such work with plant cells, several interesting studies have been undertaken. The advantages of attempting plant cell transformation by microinjection include: high frequency transformation reported to be as high as 14 to 26% of injected vectors such as *Agrobacterium*; and the possibility to effect organellar injection to alter cytoplasmic genomes (Crossway et al. 1986; Reich et al. 1986). All plant cell microinjection experiments have originally been carried out with protoplasts (Steinbiss and Stabel 1983; Lawrence and Davies 1985; Morikawa and Yamada 1985).

The first report of *Brassica* protoplast injection indicated up to 66% of the transformed cells (Miki et al. 1986). This account of plasmid injection into nuclei of *Brassica* protoplasts was carried out with a non-regenerable suspension culture line, therefore whole plants were not obtained to permit phenotypic descriptions or foreign gene inheritance studies. Extensive efforts have been made to micro-inject *B. napus* microspores even with their sporopollenin walls. This is technically feasible and progressing, but as yet is without proven results (Miki, pers. commun.).

An interesting recent development in *Brassica* microinjection studies is that of Neuhaus et al. (1987). Their approach was to inject DNA into most or all of 12 cells of an intact early-stage microspore-derived embryoid. If the individual embryoids were allowed to develop and germinate this would yield a chimera, with certain tissues derived from successful transformations forming a mosaic of transformed cells throughout the plant. To circumvent this, they subjected the target embryoids to further tissue culture to induce secondary embryogenesis. Upon induction of secondary embryogenesis from the cotyledonary stage primary embryoid, up to one half of the regenerated plantlets revealed integrated DNA sequences of the original injected plasmid. Growth on selection medium was not required because of the high frequency transformation, and only screening by NPT II activity was needed. This method would alleviate the need to produce and regenerate protoplasts, does not have the technical difficulty of holding or immobilizing individual protoplasts, does not require a selectable marker like antibiotic resistance, and could be applied to cereal crops. It would appear to require only embryogenic cell systems starting from single isolated cells, such as microspores, in culture. The evaluation of the inheritability of such injected DNA into *Brassica* embryoids, copy number, reorganization, and phenotypic or mutagenic effects has not yet been fully reported.

Significantly, most foreign DNA uptake and integration into both plant and animal cells that has been analyzed appears to be randomly inserted. However, an important goal of gene transfer is to replace a functionally defective gene, or alternatively, to insert an in vitro mutagenized gene to displace the normal allele. This gene targeting would require double reciprocal recombination. Such homologous recombination or partial replacement, as in gene conversion, appears to be rare in most reported gene transfer experiments to date. Only by microinjection have estimates of gene replacement even been made and were reported to be in the range of one in 10^3 cells or 0.1% (Thomas et al. 1986). Continuing efforts are being made into elucidating the biochemical components of site-specific recombination and techniques for selecting such events (Dorin et al. 1989). It might be speculated that only by micro-injection could a gene of interest, perhaps nucleotide primers, and an associated package of enzymes like polymerases, helicases, ligases be directly placed into the nucleus of a cell. If such a mixture could, in fact, increase the likelihood of homologus recombination, this would place microinjection in a unique light vis-a-vis other DNA delivery systems.

10.2.4 Viral Vectors

Along with *Agrobacterium*, the other biological method for DNA transfer into cells of *Brassica* spp. is by viral infection. The most extensively studied virus that was considered to have potential for use as a transformation vector is cauliflower mosaic virus (CaMV), a natural pathogen of the Brassicaceae. This 8-kb double-stranded DNA virus has two regions (ORF II and VII) not required for infection when the virus is rubbed onto healthy leaves with an abrasive compound such as carborundum paste. There is a systemic spread of the virus, yielding a high copy number in infected cells. Replication is mediated by reverse transcription and therefore CaMV is retroviral-like. Its use as a vector is, therefore, similar in concept to that of retroviruses in animal transformation. Furthermore, this virus provides one of the most widely used constitutive promoters for genes used in plant transformation experiments, i.e. the 35S promoter sequence.

The initial work to evalute CaMV as a potential vector in *Brassica* was performed by Brisson et al. (1984). The non-essential ORF II (encoding the aphid transmission factor) could be replaced by short sequences (<500 bp) without interfering with the viral life cycle. These workers used the small MTX-resistant DHFR gene ligated into this region. Upon infecting *B. napus* leaves with the isolated vector, infection symptoms developed and isolated mature viral particles revealed the presence of the DHFR gene which was properly maintained in the viral genome. Western analysis demonstrated the expression of the DHFR gene in producing the enzyme. The activity of this enzyme was also tested and found to be resistant to methotrexate inhibition in an in vitro assay. This was the first demonstration of expression of a foreign gene transferred into *Brassica* using a viral vector.

Since ORF II is very limiting in size for DNA insertion, a later study by Paszkowski et al. (1986) inserted the selectable marker NPT II into region ORF VI which normally accommodates the inclusion body protein gene. They then infected *B. campestris* var. *rapa* both in planta and by direct DNA uptake into protoplasts. Infection, replication and resistance to kanamycin only developed if the NPT II-containing vector was complemented by a wild-type vector. The protoplast system revealed genomically integrated selectable marker and functional expression of kanamycin resistance. This was one of the first published stable genetic transformation of Brassicaceae. However, the protoplast system used was not regenerable to plants, therefore inheritance studies could not be carried out.

One further application of CaMV infection for gene transfer in *Brassica* was the replacement of ORF II by an animal gene, Chinese hamster metallothionein (Lefebvre et al. 1987). A high level of the protein was produced after systemic spread of the inoculated plasmid. Excised leaves tested in vitro showed enhanced cadmium tolerance in the infected plants relative to controls. This was not a stably integrated gene and therefore the phenotype was not heritable. Interestingly, this was one of the first examples of an animal gene functioning in *Brassica*.

For reasons including tissue culturability and economic significance, it appears that most experimental effort towards genetic transformation of the Brassicaceae has been afforded to *B. napus*. However, with improvements in tissue culture, and improved efficacy of vector constructs, many other species of *Brassica* are now being used in gene transfer experiments. One approach involving many species has been that of using *A. rhizogenes*. A recent review by Tepfer et al. (1989) lists a number of *Brassica* species which have now been transformed with *A. rhizogenes* including *B. napus*, *B. oleracea*, *B. hirta*, *B. chinensis* and *B. pekinensis*.

A. tumefaciens-mediated transformation of *B. juncea* was demonstrated (Mathews et al. 1985) using wild-type T-DNA. This produced callus from which some nopaline-positive shoots developed that would not root. A similar oncogenic transformation of *B. oleracea* producing shoots and plants has been reported (Srivastava et al. 1988). Protoplasts of *B. campestris* were transformed by an engineered cauliflower mosaic virus genome (Paszkowski et al. 1986) and by non-oncogenic pTi plasmids (Ohlsson and Eriksson 1988), but in neither case were plants regenerated. Sacristan et al. (1989) transformed *B. nigra* with hygromycin resistance for use in protoplast fusion experiments. De Block et al. (1989) recently applied the technique of hypocotyl transformation (Radke et al. 1988) to *B. oleracea*. This procedure yielded fertile plants on >30% of the explants. Despite a number of successes in the transformation of other *Brassica* species, there is clearly much work to be done to extrapolate methods originally designed for use with *B. napus* to other members of the Brassicaceae.

10.3 Factors Affecting Recovery of Transformants

10.3.1 Target Cells and Tissues

Cells and tissues used as transformation targets may affect the efficiency of a transformation procedure. In most circumstances, the desired product is a regenerated plant expressing the introduced gene(s). Most systems for introduction of foreign genes result in only a fraction of the target cells being stably transformed. The regenerability of these cells is frequently the limiting factor in recovery of transgenic plants. For this reason, tissues such as epidermal thin layers (Charest et al. 1988), floral stem explants (Fry et al. 1987) and cotyledonary petioles (Moloney et al. 1989) have proven quite effective targets. All these explant systems, in the absence of selective agents, undergo shoot regeneration on 70-90% of the explants. It is noteworthy that, in spite of the obvious success of leaf discs as transformation targets among the Solanaceae (Horsch et al. 1985), leaf disc transformation has not been successful in most *Brassica* species studied. This is not simply a result of inefficient transformation, as it is quite easy to obtain tumourous callus using wild-type *Agrobacterium* and *Brassica* leaf discs. It is most likely due to the low levels of shoot regeneration from *Brassica* leaf species, which involves an intermediate callus phase of more than one month (Stringham 1979).

For non-*Agrobacterium* system, cellular targets are more limited. Direct-DNA uptake into cells normally requires removal of cell walls and production of protoplast cultures. In general, the rates of recovery of transgenic *Brassica* plants from protoplasts subjected to electroporation of PEG-induced DNA uptake has been very low. Guerche et al. (1987b) recovered two transformed plants from 4×10^6 electroporated protoplasts. Several factors may be helpful in recovery of transformed colonies or calli from protoplasts. These include the use of the agarose embedding technique (Shillito et al. 1983) which reduces sample handling considerably and yields high plating efficiency. Secondly, certain pre-treatments of protoplasts may increase their susceptibility to transformation by direct DNA uptake. Meyer et al. (1985) showed that synchronized tobacco protoplasts underwent greatly elevated levels of transformation when DNA uptake occurred during S-phase or mitosis. This was confirmed by Okada et al. (1986), and it was proposed that the absence of a nuclear membrane favours recombination of exogenous DNA with chromosomal DNA. Low doses of ionizing radiation prior to DNA uptake appear to promote integration of foreign DNA. This has promoted speculation that DNA repair systems are involved in the incorporation of exogenous DNA into the plant genome (Kohler et al. 1989).

10.3.2 Strains and Host Range of Biological Vectors

Relatively little quantitative work has been performed to define host ranges of biological vectors suitable for transformation of *Brassica* spp. Most experiments have been on the virulence of *Agrobacterium* strains in the transformation of *B. napus* and

B. juncea. The conclusions may not necessarilly be applicable to other members of the *Brassica* genus, and therefore further work may be required to determine the ideal vectors for important *Brassica* species such as *B. campestris* and *B. oleracea.* Holbrook and Miki (1985) investigated the efficiency of transformation of octopine and nopaline strains of *Agrobacterium* in a tumorigenesis assay on five *Brassica* species. Among the species investigated, a differential sensitivity to nopaline (pTi C58) and octopine (pTi $B_6$806) strains of *Agrobacterium* was found with *B. napus.* In fact, octopine strains failed to give rise to tumours in infected plants, whereas >95% of the plants infected with nopaline strains gave tumours. Interestingly, this dramatic difference in rates of tumorigenesis was not paralleled in other *Brassica* species. Three other *Brassica* species, *B. hirta*, *B. campestris* and *B. nigra*, all gave approximately equivalent rates of tumorigenesis with octopine or nopaline strains. Fry et al. (1987), using co-integrate and binary vectors, found that it was possible to recover transgenic *B. napus* plants using either nopaline or octopine strains. However, these authors noted that there was a reduced transformation efficiency with octopine strains.

A more recent study (Charest et al. 1989b) confirmed these results using several cultivars of *B. napus.* Even the cultivar Tower, which was the most susceptible to the octopine strain, TT2, yielded tumours on only 46% of the plants infected, whereas with a nopaline strain, C58, over 90% yielded tumours. A criticism of tumorigenesis assays is that they may merely represent differential responsiveness to hormonal accumulation specified by the oncogenes of the Ti plasmids used, which may vary widely between different *Agrobacterium* strains. In order to circumvent this problem, Charest et al. (1989b) used binary vectors (Bevan 1984) with a variety of virulence regions to derive the transfer of the oncogenic T-DNA from pAL 1050 originally obtained from pTiAch5 (an octopine strain). A tumorigenesis assay performed after infection of *Brassica* spp. was thus a measure of virulence rather than susceptibility to tumour formation with a particular T-DNA. These workers demonstrated a contribution from both the genotype of the bacterial chromosome and the virulence plasmid. These results confirmed that a nopaline chromosomal background as well as a nopaline Ti-plasmid virulence region act together to yield a high frequency of gene transfer to *B. napus.* With *B. juncea*, nopaline strains were also highly efficient in gene transfer although differences between nopaline and octopine strains were primarily determined by the virulence plasmid. Those obtained from nopaline strains uniformly gave tumours on all infection sites, whereas those obtained from octopine strains (e.g. LBA 4404) yielded tumours on 16% of the infection sites. In general, it could be concluded that *B. juncea* is susceptible to a wider range of *Agrobacterium* genotypes than *B. napus.*

Agrobacterium vectors derived from succinamopine strains such as A281 (Hood et al. 1986) have also been used to transform *Brassica* spp. Radke et al. (1988) showed that T-DNA mobilized by a virulence plasmid derived from a disarmed A281 (EHA101; Hood et al. 1986) was transferred to *B. napus* hypocotyl explants, albeit at low frequencies. The transformations reported here were determined by the expression of a selectable marker rather than by tumorigenesis. Moloney et al. (1989) used a similar succinamopine strain for obtaining a high degree of transformation (>50%). The use of octopine strains and their virulence plasmids resulted in much lower transformation efficiencies in *B. napus* even using the highly susceptible cut surfaces of cotyledonary petioles.

While much work remains to determine virulence of *Agrobacterium* strains for the *Brassica* genus, it is clear that nopaline and succinamopine strains of *Agrobacterium tumefaciens* are efficient in gene transfer. Octopine strains and the associated vir regions of their Ti-plasmids are relatively inefficient for gene transfer to *B. napus*, but appear to function quite well with certain *Brassica* species such as *B. nigra*.

The possibility that the application of "inducers" of virulence such as acetosyringone (Stachel et al. 1985) might improve transformation frequencies in *Brassica* spp. has not been widely tested. Moloney et al. (1989) found no enhancement of transformation efficiency of *B. napus* with succinamopine strains of *Agrobacterium* Charest et al. (1989b) found a significant enhancement of transformation frequency when octopine strains were treated for 2 h with acetosyringone prior to infection.

Studies on the related crucifer *Arabidopsis thaliana* have produced conflicting results with respect to the use of phenolic inducers of virulence. Sheikholeslam and Weeks (1987) reported an increase in transformation frequencies of leaf explants from 2 to 3% (untreated) to 50-60% on overnight pre-treatment of the bacteria with acetosyringone. Schmidt and Willmitzer (1988), however, obtained transformation rates of up to 80% of treated explants without exploiting acetosyringone, while using the same strain of *Agrobacterium* as Sheikholeslam and Weeks (1987). Their best experiments involved the use of tobacco cell feeder-layers (Horsch et al. 1985) although this was not essential. Even without feeder layers, frequencies close to 80% were obtained in some experiments.

10.3.3 Selectable and Screenable Markers

Most transformation methods, with the notable exception of microinjection, require the use of a selective agent for the efficient recovery of transformants. With *Brassica* spp., several selectable markers have been exploited with variable success. Aminoglycoside antibiotics such as kanamycin and geneticin (G418) have been widely used (Fry et al. 1987; Radke et al. 1988; Charest et al. 1988). NPT II appears to rescue transformed cells in the presence of 15-100 mg/l kanamycin. Among *Brassica* species, kanamycin appears to be a suitable antibiotic for selections. However, it should be noted that there is differential sensitivity of *Brassica* species and tissues to a given concentration of the antibiotic. Also, in our experience, there are probably some secondary toxicosis effects where dying cells not resistant to kanamycin can be detrimental to nearby cells which actually contain and express the resistance gene. For this reason, selection on kanamycin prior to the formation of large amounts of callus tissue appears preferable.

Other aminoglycoside antibiotics such as hygromycin have not yet been widely used in *Brassica* transformation. It is clear, however, that hygromycin selection may be used for the recovery of transformants in some *Brassica* species. Sacristan et al. (1989) used hygromycin-resistant *Brassica nigra* protoplasts in order to aid in the selection of heterokaryons in a protoplast fusion experiment. This resulted in transfer and expression of hygromycin phosphotransferase into *B. napus* host cells in asymmetric fusions.

Chloramphenicol acetyltransferase (CAT) has found only limited utility as a selectable marker in plants (De Block et al. 1984). It is generally less efficient than the aminoglycoside antibiotics and often permits an unacceptable number of "escapes". It has, however, proven to be an excellent reporter enzyme and has been used widely in monocots and dicots (Hauptmann et al. 1987). In several *Brassica* spp., CAT-like activity has been detected (Balazs and Bonneville 1987). This poses problems for the use of CAT either for selection or as a reporter in *Brassica* spp. Charest et al. (1989a) found evidence for inhibition of bacterial CAT activity in extracts of *B. napus*. These extracts inhibited CAT activity of extracts of transgenic tobacco down to a level equivalent to the background CAT-like activity normally found in *Brassica* extracts. These findings certainly limit the possibilities for the use of CAT in gene expression studies in *Brassica*. Balazs and Bonneville (1987) did, however, find that the *Brassica* CAT-like activity was quite heat-labile, whereas the CAT activity deriving from a gene on the bacterial transposon Tn9 was not. Some workers have suggested that a 10-min pre-treatment at 65 °C eliminates most of the endogenous CAT activity.

While genes of bacterial origin have been used successfully as selectable markers in plant transformation, there is also one case of a mammalian gene which has been used. Mutant mouse DHFR specifying resistance to MTX was used in *Petunia hybrida* (Eichholtz et al. 1987) and *B. napus* (Pua et al. 1987). Transformed callus and plants of *B. napus* grew on 0.005-0.01 mg/l of methotrexate. This concentration of methotrexate is very toxic to wild-type *Brassica* plants and tissue cultures.

The difficulties in using selectable markers such as antibiotic resistance for recovery of transformed *Brassica* plants might be avoided by the use of screenable markers such as β-glucuronidase (Jefferson 1987). Several laboratories have now confirmed that it is possible to use the uidA gene from *E. coli* (encoding β-glucuronidase or GUS) as a screenable marker and reporter protein (Moloney et al., unpubl.; Kridl et al. 1989; Facciotti et al. 1989). The advantages of this method may be the elimination of large amounts of dying tissue inherent to a selection procedure, which may have deleterious consequences for the entire culture. The recent report by Gould and Smith (1989) involving non-destructive testing for GUS activity in culture may offer an alternative route involving rapid visual screening under UV light for 4-methylumbelliferone fluorescence.

10.4 Gene Expression in Transgenic Plants

The interest among plant molecular biologists in elucidating the regulation of gene expression has been greatly assisted by the use of transgenic plants. Many *Brassica* species lend themselves readily to such studies for reasons outlined earlier. In addition, heterologous gene expression in *Brassica* has the potential to produce novel phenotypes of agronomic value. There are now a few examples that address questions of regulation of gene expression or enhanced agronomic value of *Brassica* spp.

Radke et al. (1988) used a tagged napin gene (Scofield and Crouch 1987) reintroduced into *B. napus* by *Agrobacterium*-mediated transformation to evaluate seed-specific gene expression. The construct used contained a 278-bp fragment of bacterial DHFR, which was utilized as expression tag. This was inserted 10 bp after the stop codon of a napin-coding sequence with 300 bp of flanking DNA 5′ to the initiation codon. This transcriptionally tagged chimeric sequence was ligated into an *Agrobacterium* binary plasmid with a NPT II selection marker. Transgenic rapeseed plants obtained using this vector were found to express the introduced gene correctly, i.e. only in the developing seeds. The expression of the introduced gene was distinguished from the native napin genes using the DHFR tag as a probe in Northern blots. From this it may be concluded that within the 300 bases 5′ to the initiation codon there is sufficient information to regulate tissue specificity of napin expression. It is not yet known how much of the 5′ end is required for expression which is quantitatively similar to the native storage protein gene.

Recently, seed-specific expression using the system described by Radke et al. (1988) has been exploited in a very novel way. Vanderkerckhove et al. (1989) introduced into *B. napus* and *Arabidopsis thaliana* a chimeric gene comprising the gene encoding 2S albumin (napin-like) from *Arabidopsis* with its own 3′ and 5′ flanking sequences and with an oligonucleotide inserted which encoded the pentapeptide "Leu-enkephalin". This sequence was incorporated in-frame and was flanked by lysine codons to permit tryptic digestion and purification of the neuropeptide from the 2S protein. It was found that the chimeric gene expressed correctly in developing seeds of *Arabidopsis* and *B. napus*. By this technique it was possible to obtain up to 200 nmol of the neuropeptide per gram of seed. This experiment illustrates the potential for seed-specific expression of non-plant polypeptides of commercial interest in plants and shows the advantages of using *Brassica* spp. for such applications.

Transgenic *B. napus* plants have been produced for the purposes of obtaining herbicide resistance. Several herbicide resistance genes are currently being investigated for application to *Brassica* spp. These include: 5-enolpyruvyl shikimate-3-phosphate synthase (EPSP synthase), acetolactate synthase (ALS) and phosphinothricine acetyltransferase (PAT). High-level expression or expression of mutant forms of EPSP synthase can give rise to glyphosate (RoundUp®) resistance (Stalker et al. 1985; Shah et al. 1986) in plants. This work, which originated in transgenic solanaceous plants, has been successfully extrapolated to *Brassica napus* (Metz and Fedele 1989) and field trials have commenced.

Experiments to evaluate the use of mutated forms of ALS for production of plants resistant to sulfonyl ureas (DuPont) and imidazolinones (American Cyanamid) have also been undertaken recently in *B. napus*. Bedbrook et al. (1987) reported the introduction of a mutant tobacco ALS gene with its own regulatory sequences into *B. napus*. This was found, in some transformants, to provide some protection against applied sulfonyl ureas although the tolerance was too low to be of agronomic value. Recently, Miki et al. (1990) could recover the transgenic plants of *B. napus* cv. Profit by direct selection for resistance to chlorsulfuron at field level under greenhouse conditions. This indicated the feasibility of employing extremely simple T-DNA such as a mutant gene AHAS (enzyme acetohydroxy acid synthase) flanked by T-DNA borders for further experiments. The transgenic plants, which expressed the *Arabidopsis* AHAS wild type or a mutant gene csr-1-1, could resist sulfonylurea chlorsulfuron. This will help in the control of broad leaf crucifer weeds. Wiersma et al. (1989) isolated members of the *B. napus* ALS gene family and introduced specific mutations into them. These mutations of the *Brassica* ALS gene were shown to give a low level of herbicide tolerance in transgenic plants.

De Block et al. (1989) performed transformations on *B. napus* and *B. oleracea* using the *bar* gene (De Block et al. 1987) which encodes PAT. Transformants were selected using NPT II and there was generally a correlation between kanamycin resistance and resistance to phosphinothricine (PPT) or to the herbicide Basta® (active ingredient PPT) applied at the equivalent of 8 l/ha.

An interesting example of foreign gene expression for the production of a novel phenotype in *B. napus* is the use of mammalian metallothionein as a chelation agent of heavy metals. Although the first report of this was using viral-based vectors (Lefebvre et al. 1987), it was not possible using this method to evaluate the heritability of the metallothionein gene or to test the phenotype in a whole plant. Misra and Gedamu (1989), therefore, used a human metallothionein II in a stable transformation experiment involving *Agrobacterium* transformation of epidermal strips as described by Charest et al. (1988). This resulted in stably transformed *B. napus* plants which showed quite high levels of resistance to 0.1 nM $CdCl_2$, which is generally very inhibitory of *B. napus* seedling growth. There is the potential in this approach of retaining heavy metals by chelation in roots, thus minimizing accumulation in leaves or seeds. Such a phenotype could be a valuable property in soils with unusually high levels of heavy metal residues, e.g. due to mining pollution. It may also be possible to use such recombinant phenotypes in land reclamation as a tool in the removal of heavy metal pollution.

A novel exploitation of transgenic phenotypes of *Brassica* spp. was recently reported by Sacristan et al. (1989). These workers used a hygromycin-resistant line of *B. nigra* to produce protoplasts for fusion experiments. Hygromycin-resistant *B. nigra* protoplasts were rendered inviable by irradiation and then fused with viable *B. napus* protoplasts. Recombinant *B. napus* cells expressing genes derived from the *B. nigra* genome were recovered by hygromycin selection. This procedure will be particularly valuable for the transfer of complex genetic traits between sexually incompatible species. In the case in question, the *B. nigra* lines used were resistant to blackleg (*Phoma lingam*) and clubroot (*Plasmodiophora brassicaceae*), both important diseases of the Brassicaceae.

10.5 *Arabidopsis* Research and *Brassica* Oilseeds

A review of *Brassica* transformation and useful genes for isolation would be deficient without acknowledgement of the potential benefits of studying the crucifer *Arabidopsis thaliana*, a member of the Brassicaceae. There exist several reviews of *Arabidopsis* and its potential for studies in molecular genetics (Meyerowitz 1987; Meyerowitz 1989) and plant morphogenesis (Haughn and Somerville 1988). Included in the list of properties that make this crucifer particularly appealing for experiments are its small nuclear genome, a 4- to 5-week generation time, the minimal amount of dispersed repetitive DNA, and ease of mutagenesis and selection or screening.

Chemical mutagenesis of seed has led to numerous interesting mutations such as herbicide resistance (Haughn and Somerville 1986) and leaf membrane fatty acyl composition (Browse et al. 1985). A general review of *Arabidopsis* mutants has been presented by Estelle and Somerville (1986), and one of embryo-lethal mutants by Meinke (1986). An interesting new aspect to previous mutagenesis studies is the recent demonstration of *Arabidopsis* insertional mutagenesis by T-DNA integration (Feldmann et al. 1989). The compact genome with little repetitive DNA makes this methodology of gene conversion and thereby tagging of practical value. These same workers also used an innovative method of transforming germinating seeds of *Arabidopsis* by cocultivation with *Agrobacterium* (Feldmann and Marks 1987).

Improvements in tissue culture methodology for plant regeneration from explants such as cotyledons (Patton and Meinke 1988) and from suspension cultures (Gleddie 1989) will greatly assist in experiments on transformation for mutant complementation. There have been reports of successful *Arabidopsis* transformation using leaf explants (Lloyd et al. 1986; Sheikholeslam and Weeks 1987). However, the highest frequencies reported involve the use of cotyledonary explants (Schmidt and Willmitzer 1988) as target tissue for *Agrobacterium*-mediated transformation.

Apart from basic studies in gene regulation, particularly of fatty acid and triacylglycerol metabolism within *Arabidopsis* itself, major efforts are being launched to "walk" the chromosomes for intensive linkage mapping studies. Many unique DNA sequences, of known coding function or not, may provide very useful probes for RFLP mapping in *Brassica* spp. Furthermore, it is very likely that isolated *Arabidopsis* genes will have sufficiently high sequences homology to crop plant brassicas to enable isolation of *Brassica* homologues from genomic libraries. An example of this is the isolation of ALS genes of *B. napus* using an *Arabidopsis* ALS mutant probe (Wiersma et al. 1989).

10.6 Conclusions

Since the development of transformation methods for plants involving *Agrobacterium* and direct DNA delivery, the application of these methods to *Brassica* species has been a major focus of numerous laboratories. The *Brassica* genus will probably be among the first major food crops to produce results at the field level using recombinant foreign genes. The Brassicaceae have not proven to be quite as amenable to transformation methods as the Solanaceae, but nevertheless progress at both the fundamental and applied levels has been made.

Although some particular refinements of technique are required for efficient gene transfer in *Brassica* spp., all established methods of gene transfer using *Agrobacterium*, PEG-mediated DNA uptake, electroporation and viral-based vectors appear to work in *Brassica* spp. Areas which warrant more attention include: further analysis of host-range relationships for *Agrobacterium*-mediated transformation of the Brassicaceae, more evaluation of appropriate selectable markers, which might include using herbicide resistance genes for selection in culture and improvement in regeneration rates for some of the more recalcitrant *Brassica* species such as *B. campestris*. The majority of studies has been focussed on the oilseed *B. napus*, but there remains a great deal of work to be performed on other members of the genus. There is currently great agronomic interest in the use of other *Brassica* oilseeds such as *B. juncea* and

B. campestris. These species have received inadequate attention relative to their agricultural importance.

Acknowledgements. The authors wish to thank several colleagues and collaborators for helpful discussions especially Kiran Sharma, John Hachey and Steven Szarka (University of Calgary) and Dr. Brian Miki (Agriculture Canada, Ottawa). We also thank Erin Moloney for typing, editing and formatting of the manuscript.

References

Ackermann C (1977) Pflanzen aus *Agrobacterium rhizogenes* tumoren an *Nicotiana tabacum*. Plant Sci Lett 8: 23–30

Balazs E, Bonneville JM (1987) Chloramphenicol acetyl transferase activity in *Brassica* spp. Plant Sci 50: 65–68

Battey JF, Schmidt KM, Ohlrogge JB (1989) Genetic engineering for plant oils: potentials and limitations. Trend Biotechnol 7: 122–126

Bedbrook JR, Chaleff RS, Falco SC, Mazur BJ, Yadav NS (1987) Nucleic acid fragment encoding herbicide resistance plant acetolactate synthase. Eur Pat Appl 87307384 S

Bevan M (1984) Binary *Agrobacterium* vectors for plant transformation. Nucl Acids Res 12: 8711–8716

Brisson N, Paszkowski J, Penswick JR, Gronenborn B, Potrykus I, Hohn T (1984) Expression of a bacterial gene in plants by using a viral vector. Nature (London) 310: 511–514

Browse J, McCourt P, Somerville CR (1985) A mutant of *Arabidopsis* lacking a chloroplast-specific lipid. Science 227: 763–765

Charest PJ, Holbrook LA, Gabard J, Iyer VN, Miki BL (1988) *Agrobacterium*-mediated transformation of thin cell layer explants from *Brassica napus* L. Theor Appl Genet 75: 438–445

Charest PJ, Iyer VN, Miki BL (1989a) Factors affecting the use of chloramphenicol acetyltransferase as a marker for *Brassica* genetic transformation. Plant Cell Rep 7: 628–631

Charest PJ, Iyer VN, Miki BL (1989b) Virulence of *Agrobacterium tumefaciens* strains with *Brassica napus* and *Brassica juncea*. Plant Cell Rep 8: 303–306

Chilton MD, Tepfer D, Petit A, David CC, Delbert F, Tempe J (1982) *Agrobacterium rhizogenes* inserts T-DNA into plant roots. Nature (London) 295: 432–434

Chuong PV, Deslauriers C, Kott LS, Beversdorf WD, (1988) Effects of donor genotype and bud sampling on microspore culture of *Brassica napus* L. Can J Bot 66: 1653–1657

Crossway A, Oakes JV, Irvine JM, Ward B, Knauf VC, Shewmaker CK (1986) Integration of foreign DNA following microinjection of tobacco mesophyll protoplasts. Mol Gen Genet 202: 179–185

Crouch ML, Sussex IM (1981) Storage protein synthesis in *Brassica napus*. Planta 153: 64–74

Crouch ML, Tenbarge K, Simon A, Finkelstein R, Scofield SR, Solberg L (1985) Storage protein mRNA levels can be regulated by abscisic acid in *Brassica* embryos. In: VanVloten-Doting L, Groot GSP, Hall TC (eds) Molecular form and function of the plant genome. Plenum New York

Czernilofsky AP, Hain R, Herrera-Estrella L, Lortz H, Goyvaerts E, Baker BJ, Schell J (1986) Fate of selectable marker DNA integrated into the genome of *Nicotiana tabacum*. DNA 5: 101–113

Davey MR, Cocking EC, Freeman J, Pearce N, Tudor I (1980) Transformation of *Petunia* protoplasts by isolated *Agrobacterium* plasmids. Plant Sci Lett 18: 307–313

Davey MR, Rech EL, Mulligan BJ (1989) Direct DNA transfer to plant cells. Plant Mol Biol 13: 273–285

DeBlock M, Botterman J, Vandewiele M, Dockx J, Thoen C, Gossele V, Movva NR, Thompson C, Van Montagu M, Leemans J (1987) Engineering herbicide resistance in plants by expressing a detoxifying enzyme. EMBO J 6: 2513–2518

DeBlock M, De Brouwer D, Tenning P (1989) Transformation of *Brassica napus* and *Brassica oleracea* using *Agrobacterium tumefaciens* and the expression of the bar and neo genes in transgenic plants. Plant Physiol 91: 694–701

DeBlock M, Herrera-Estrella L, Van Montagu M, Schell J, Zambryski P (1984) Expression of foreign genes in regenerated plants and their progeny. EMBO J 3: 1681–1684

DeCleene M, DeLey J (1976) The host range of crown gall. Bot Rev 42: 389–408

Dorin JR, Inglis JD, Porteous DJ (1989) Selection for precise chromosomal targetting of a dominant marker by homologous recombination. Science 243: 1357–1360

Draper J, Davey MR, Freeman JP, Cocking EC, Cox BJ (1982) Ti-plasmid homologous sequences present in tissues from *Agrobacterium* plasmid transformed *Petunia* protoplasts. Plant Cell Physiol 23: 451–458

Eichholtz DA, Rogers SG, Horsch RB, Klee J, O'Connell KM, Fraley RT (1987) Expression of mouse dihydrofolate reductase gene confers methotrexate resistance in transgenic petunia plants. Somat Cell Mol Genet 13: 67–76

Estelle MA, Somerville CR (1986) The mutants of *Arabidopsis*. Trends in Genet 2: 89–93

Facciotti D, Turner J, Justicia R, Radke SE (1989) Somatic regeneration and recovery of transgenic plants from *B. campestris*. In: Proc 5th Cruciferae Genet Workshop 1989, Univ Calif, Gen Resources Conserv Prog Rep 4, p 48

Feldmann KA, Marks MD (1987) *Agrobacterium*-mediated transformation of germinating seeds of *Arabidopsis thaliana* : a non-tissue culture approach. Mol Gen Genet 208: 1–9

Feldmann KA, Marks MD, Christianson ML, Quatrano RS (1989) A dwarf mutant of *Arabidopsis* generated by T-DNA insertion mutagenesis. Science 243: 1351–1354

Food and Agriculture Organization (FAO) (1988) Commodity rev Outlook 1986-87, FAO Econ Soc Dev Ser 43, FAO Rome, Italy pp 41–47

Fraley RT, Rogers SG, Horsch RB, Eichholtz DA, Flick JS, Fink CL, Hoffmann NL, Sanders PR (1985) The SEV system: a new disarmed Ti plasmid vector system for plant transformation. Bio Technol 3: 629–635

Fromm M, Taylor LP, Walbot V (1985) Expression of genes transferred into monocot and dicot plant cells by electroporation. Proc Nat Acad Sci USA 82: 5824–5828

Fromm M, Taylor LP, Walbot V (1986) Stable transformation of maize after gene transfer by electroporation. Nature (London) 319: 791–793

Fry JE, Barnason AR, Horsch RB (1987) Transformation of *Brassica napus* with *Agrobacterium tumefaciens*-based vectors. Plant Cell Rep 6: 321–325

Gleddie S (1989) Plant regeneration from cell suspension cultures of *Arabidopsis thaliana* Leynh. Plant Cell Rep 8: 1–5

Gordon JW, Scangos GA, Plotkin DJ, Barbosa JA, Ruddle FH (1980) Genetic transformation of mouse embryos by microinjection of purified DNA. Proc Nat Acad Sci USA 77: 7380–7384

Gould JH, Smith RH (1989) A non-destructive assay for GUS in the media of plant tissue cultures. Plant Mol Biol Rep 7: 209–216

Graessmann A, Graessmann M, Mueller C (1980) Microinjection of early SV40 DNA fragments and T antigen. Methods Enzymol 6: 816–825

Guerche L, Charbonnier M, Jouanin L, Tourneur C, Paszkowski J, Pelletier G (1987a) Direct gene transfer by electroporation in *Brassica napus*. Plant Sci 52: 111–116

Guerche L, Jouanin L, Tepfer D, Pelletier G (1987b) Genetic transformation of oilseed rape (*B. napus*) by the Ri T-DNA of *Agrobacterium rhizogenes* and analysis of inheritance of the transformed phenotype. Mol Gen Genet 206: 382–386

Hammer RE, Pursel VG, Rexroad CA Jr, Wall RJ, Bolt DJ, Ebert KM, Palmiter RD, Brinster RL (1985) Production of transgenic rabbits, sheep and pigs by microinjection. Nature (London) 315: 680–683

Haughn GW, Somerville CR (1986) Sulfonylurea-resistant mutants of *Arabidopsis thaliana*. Mol Gen Genet 201: 430–434

Haughn GW, Somerville CR (1988) Genetic control of morphogenesis in Arabidopsis. Dev Genet 9: 73–89

Hauptmann RM, Ozias-Akins P, Vasil V, Tabaeizadeh Z, Rogers SG, Horsch RB, Vasil IK, Fraley RT (1987) Transient expression of electroporated DNA in monocotyledonous and dicotyledonous species. Plant Cell Rep 6: 265–270

Holbrook LA, Miki BL (1985) *Brassica* crown gall tumorigenesis and in vitro culture of transformed tissue. Plant Cell Rep 4: 329–332

Holbrook LA, Scowcroft WR, Moloney MM, Schmiemann MG (1988) Towards directed fatty acid modification in oilseeds.In: Gavora J, Gerson DF, Luong J, Storer A, Woodley JH (eds) Biotechnology Research and Applications. Barkng UK; Elsevier Appl Sci London, pp 33–42

Hood EE, Helmer GL, Fraley RT, Chilton MD (1986) The hypervirulence of *Agrobacterium tumefaciens* A 281 is encoded in a region of pTi Bo 542 outside of T-DNA. J Bacteriol 168: 1291–1301

Horsch RB, Fraley RT, Rogers SG, Sanders PR, Lloyd AM, Hoffmann W (1984) Inheritance of functional foreign genes in plants. Science 223: 496–499

Horsch RB, Fry JE, Hoffmann NL, Wallroth M, Eichholtz D, Rogers SG, Fraley RT (1985) A simple and general method for transferring genes to plants. Science 227: 1229–1231

Horsch RB, Klee HJ (1986) Rapid assay of foreign gene expression of leaf-discs transformed by *Agrobacterium tumefaciens* : role of T-DNA borders in the transfer process. Proc Nat Acad Sci USA 83: 4428–4432

Jefferson RA (1987) Assaying chimaeric genes in plants: the GUS gene fusion system. Plant Mol Biol Rep 5: 387–405

Kartha KK, Gamborg OL, Constable F (1974) In vitro plant formation from stem explants of rape (*Brassica napus* cv. Zephyr). Physiol Plant 31: 217–220

Kawata EE, Galili G, Smith LD, Larkins BA (1988) Translation in Xenopus oocytes of mRNAs transcribed in vitro. In: Gelvin SB, Schilperoort RA, Verma DPS (eds) Plant molecular biology manual B7. Kluwer, Dordrecht, Netherlands, pp 1–22

Keller WA, Armstrong KC (1978) High-frequency production of microspore-derived plants from *Brassica napus* anther cultures. Z Pflanzenzuecht 80: 100–108

Klimazewska K, Keller WA (1985) High frequency plant regeneration from thin cell-layer explants of *Brassica napus*. Plant Cell Tissue Organ Cult 4: 183–197

Kohler F, Cardon G, Pohlmann M, Gill R, Scheider O (1989) Enhancement of transformation rates in higher plants by low dose irradiation: are DNA repair systems involved in the incorporation of exogenous DNA into the plant genome ? Plant Mol Biol 12: 189–199

Krens FA, Molendijk L, Wullems GJ, Schilperoort RA (1982) In vitro transformation of plant protoplasts with Ti-plasmid. Nature (London) 296: 72–74

Kridl JC, McCarter DW, Knutzon DS, Scherer DE, Knauf VC (1989) Characterization of *Brassica* seed-specific promoters and their expression in transformed rapeseed. In: Proc 5th Cruciferae Genet Workshop 1989 Univ Calif, Gen Resources Conserv Prog Rep #4 pp 20–21

Lawrence WA, Davies DR (1985) A method for the microinjection and culture of protoplasts at very low densities. Plant Cell Rep 4: 33–35

Lefebvre DD, Miki BL, Laliberte JF (1987) Mammalian metallothionein function in plants. Bio Technol 5: 1053–1056

Lloyd AM, Barnason AR, Rogers SG, Byrne MC, Fraley RT, Horsch RB (1986) Transformation of *Arabidopsis thaliana* with *Agrobacterium tumefaciens*. Science 234: 464–466

Mathews VH, Bhatia CR, Mitra R, Krishna TG, Rao PS (1985) Regeneration of shoots from *Brassica juncea* (L.) Czern and Coss cells transformed by *Agrobacterium tumefaciens* and expression of nopaline dehydrogenase genes. Plant Sci 39: 49–54

Meinke DW (1986) Embryo-lethal mutants and the study of plant embryo development. Oxford Surv Plant Mol Cell Biol 3: 123–165

Metz S, Fedele J (1989) Molecular biology and transformation of *Brassica* spp. In Proc 5th Cruciferae Genet Workshop 1989, Univ Calif Genet Resources Conserv Prog Rep #4 p 27

Meyer P, Walgenbach E, Bussmann K, Hombrecher G, Saedler H (1985) Synchronized tobacco protoplasts are efficiently transformed by DNA. Mol Gen Genet 201: 513–518

Meyerowitz EM (1987) *Arabidopsis thaliana*. Ann Rev Genet 21: 93–111

Meyerowitz EM (1989) *Arabidopsis*, a useful weed. Cell 56: 263–269

Miki BL, Labb'e L, Hattori J, Ouellet T, Gabard J, Sunohara G, Charest PJ, Iyer VN (1990) Transformation of *B. napus* canola cultivars with *Arabidopsis thaliana* acetohydroxy acid synthase gene and analysis of herbicide resistance. Theor Appl Genet 80: 449–458

Miki BL, Reich TJ, Simmonds D, Charest PJ, Holbrook L, Iyer VN, Keller W (1986) The use of mechanical and biological mechanisms for the delivery of genes into plant cells of crop species. Abstr 3rd Inter Symp Mol Genet Plant-Microbe Ineract. Montreal, Canada, p 20

Misra S (1990) Transformation of *Brassica napus* L. with a disarmed octopine plasmid of *Agrobacterium tumefaciens* molecular analysis and inheritance of transformed phenotype. J Exp Bot 20: 147–156

Misra S, Gedamu L (1989) Heavy metal tolerant transgenic *Brassica napus* L. and *Nicotiana tabacum* plants. Theor Appl Genet 78: 161–168

Moloney MM, Walker JM, Sharma KK (1989) High efficiency transformation of *Brassica napus* using *Agrobacterium* vectors. Plant Cell Rep 8: 238–242

Morikawa H, Yamada Y (1985) Capillary microinjection into protoplasts and intranuclear localization of injected materials. Plant Cell Physiol 26: 229–236

Nasrallah JB, Kao TH, Chen CH, Goldberg ML, Nasrallah ME (1987) Amino-acid sequences of glycoproteins encoded by three alleles of the S locus of *Brassica oleracea*. Nature (London) 326: 617–619

Neuhaus G, Spangenberg G, Mittelsten Scheid O, Schweiger HG (1987) Transgenic rapeseed plants obtained by the microinjection of DNA into microspore-derived embryoids. Theor Appl Genet 75: 30–36

Newell CA, Rhoads ML, Bidney DL (1984) Cytogenetic analysis of plants regenerated from tissue explants and mesophyll protoplasts of winter rape, *Brassica napus* L. Can J Genet Cytol 26: 752–761

Ohlsonn M, Eriksonn T (1988) Transformation of *Brassica campestris* protoplasts with *Agrobacterium tumefaciens*. Hereditas 108: 173–177

Okada K, Takebe I, Nagata T (1986) Expression and integration of genes introduced into highly synchronized plant protoplasts. Mol Gen Genet 205: 398–403

Ooms G, Bains A, Burell M, Karp A, Twell D, Wilcox E (1985a) Genetic manipulation in cultivars of oilseed rape (*Brassica napus*) using *Agrobacterium*. Theor Appl Genet 71: 325–329

Ooms G, Karp A, Burrell MM, Twell D, Roberts J (1985b) Genetic modification of potato development using Ri T-DNA. Theor Appl Genet 70: 440–446

Palmiter RD, Chen HY, Brinster RL (1982) Differential regulation of metallothionein-thymidine kinase fusion genes in transgenic mice and their offspring. Cell 29: 701–710

Paszkowski J, Pisan B, Shillito RD, Hohn T, Hohn B, Potrykus I (1986) Genetic transformation of *Brassica campestris* var. *rapa* protoplasts with an engineered cauliflower mosaic virus genome. Plant Mol Biol 6: 303–312

Paszkowski J, Shillito RD, Saul M, Mandak VB, Hohn T, Hohn B, Potrykus I (1984) Direct gene transfer to plants. EMBO J 3: 2717–2722

Patton DA, Meinke DW (1988) High-frequency plant regeneration from cultured cotyledons of *Arabidopsis thaliana*. Plants Cell Rep 17: 233–237

Potrykus I, Paszkowski J, Saul MW, Petruska J, Shillito RD (1985) Molecular and general genetics of a hybrid foreign gene introduced into tobacco by direct gene transfer. Mol Gen Genet 199: 169–177

Pua EC, Mehra-Palta A, Nagy F, Chua NH (1987) Transgenic plants of *Brassica napus*. Bio Technol 5: 815–817

Radke SE, Andrews BM, Moloney MM, Crouch ML, Kridl JC, Knauf VC (1988) Transformation of *Brassica napus* using *Agrobacterium tumefaciens* : developmentally regulated expression of a reintroduced napin gene. Theor Appl Genet 75: 685–694

Reich TJ, Iyer VN, Miki BL (1986) Efficient transformation of alfalfa protoplasts by the intranuclear microinjection of Ti plasmids. Bio Technol 4: 1001–1004

Sacristán MD, Gerdemann-Knorck M, Schieder O (1989) Incorporation of hygromycin resistance in *Brassica nigra* and its transfer to *B. napus* through asymmetric protoplast fusion. Theor Appl Genet 78: 583–586

Schmidt R, Willmitzer L (1988) High efficiency *Agrobacterium tumefaciens*-mediated transformation of *Arabidopsis thaliana* leaf and cotyledon explants. Plant Cell Rep 7: 583–586

Scofield SR, Crouch ML (1987) Nucleotide sequence of a member of the napin storage protein family from *B. napus*. J Biol Chem 262: 12202–12208

Shah DM, Horsch RB, Klee HJ, Kishore GM, Winter JM, Tumer NE, Hironaka CM, Sanders PR, Gasser CS, Aykent S, Siegel NR, Rogers SG, Fraley RT (1986) Engineering herbicide tolerance in transgenic plants. Science 233: 478–481

Sharma KK (1987) Control of organ differentiation from somatic tissues and pollen embryogenesis in anther cultures of *B juncea*. Ph D theses Dep Bot Univ Delhi, Delhi, India

Sharma KK, Bhojwani SS, Thorpe TA (1990) Factors affecting high frequency differentiation of shoots and roots from cotyledon explants of *B juncea* (L.) Czern. Plant Sci 66: 247–254

Sheikholeslam SN, Weeks DP (1987) Acetosyringone promotes high-efficiency transformation of *Arabidopsis thaliana* explants by *Agrobacterium tumefaciens*. Plant Mol Biol 8: 291–298

Shillito RD, Paszkowski J, Potrykus I (1983) Agarose plating and a bead-type culture technique enable and stimulate development of protoplast-derived colonies in number of plant species. Plant Cell Rep 2: 244–247

Shillito RD, Saul MW, Paszkowski J, Muller M, Potrykus I (1985) High efficiency direct gene transfer to plants. Bio Technol 3: 1099–1103

Srivastava V, Reddy AS, Guha-Mukherjee S (1988) Transformation and regeneration of *Brassica oleracea* mediated by an oncogenic *Agrobacterium tumefaciens*. Plant Cell Rep 7: 504–507

Stachel SE, Messens E, Van Montague M, Zambryski P (1985) Identification of the signal molecules produced by wounded plant cells that activate T-DNA transfer in *Agrobacterium tumefaciens*. Nature (London) 318: 624–629

Stalker DM, Hiatt WR, Comai L (1985) A single amino-acid substitution in the enzyme 5-enolpyruvylshikimate-3-phosphate synthase confers resistance to the herbicide glyphosate. J Biol Chem 260: 4724–4728

Steinbiss HH, Stabel P (1983) Protoplast derived tobacco cells can survive capillary micoinjection of the fluorescent dye Lucifer Yellow. Protoplasma 116: 223–232

Stringham GR (1977) Regeneration in stem explants of haploid rapeseed (*Brassica napus* L.). Plant Sci Lett 9: 115–119

Stringham GR (1979) Regeneration in leaf callus cultures of haploid rapeseed (*Brassica napus* L.). Z Pflanzenphysiol 92: 459–462

Tepfer DL (1984) Transformation of several species of higher plants by *Agrobacterium rhizogenes*—sexual transmission of the transformed genotype and phenotype. Cell 37: 9159–9167

Tepfer DL, Metzger L, Prost R (1989) Uses of roots transformed by *Agrobacterium rhizogenes* in rhizophere research: applications in studies of cadmium assimilation from sewage sludge. Plant Mol Biol 13: 295–302

Thomas KR, Folger KR, Capecchi MR (1986) High frequency targetting of genes to specific sites in the mammalian genome. Cell 44: 419–428

UN (1935) Genome analysis in *Brassica* with special reference to experimental formation of *B. napus* and peculiar mode of fertilization. Jpn J Bot 7: 389–452

Vanderkerkhove J, Van Damme J, Van Lijsbettens M, Botterman J, De Block M, Vandewiele M, De Clerq A, Leemans J, Van Montagu M, Krebers E (1989) Enkephalins produced in transgenic plants using modified 2S seed storage proteins. Bio Technol 7: 829–832

Wanger TE, Hoppe PC, Jollick JD, Scholl DR, Hodinka RL, Gault JB (1981) Microinjection of a rabbit β-globin gene into zygotes and its subsequent expression in adult mice and their offspring. Proc Nat Acad Sci USA 78: 6376–6380

Weber G, Monajembashi S, Greulich KO, Wolfrum J (1988) Genetic manipulation of plant cells and organelles with a laser microbeam. Plant Cell Tissue Organ Cult 12: 219–222

Weber G, Monajembashi S, Greulich KO, Wolfrum J (1989) Uptake of DNA in chloroplasts of *Brassica napus* L. facilitated by a UV-laser microbeam. Eur J Cell Biol 49: 73–79

White FF, Ghodossi G, Gordon MP, Nester EW (1982) Tumor induction by *Agrobacterium rhizogenes* involves the transfer of plasmid DNA to the plant genome. Proc Nat Acad Sci USA 79: 3193–3197

Wiersma PA, Schmiemann MG, Condie JA, Crosby WL, Moloney MM (1989) Isolation, expression and phylogenetic inheritance of an acetolactate synthase gene from *Brassica napus*. Mol Gen Genet 219: 413–420

Xu ZH, Davey MR, Cocking EC (1982) Plant regeneration from root protoplasts of *Brassica*. Plant Sci Lett 24: 117–121

CHAPTER 11

Status of Isoenzyme Research

Marian L. Thorpe and *Louise H. Duke*

11.1 Introduction

Studies of isoenzymes of *Brassica* species were initiated in 1967 (Vaughan and Waite 1967 a, b). Initially, the research was concerned mainly with the taxonomy of *Brassica* species (Vaughan and Waite 1967a, b, 1968; Vaughan and Denford 1968); subsequently, however, the use of isoenzymes as genetic markers for cultivar identification, purity analysis (Chen and Tong 1985; Thorpe et al. 1987; Duke 1988) and for identification of hybrids received greater attention (Wills et al. 1979; Kato and Tokumasu 1979; Thorpe and Beversdorf 1988; Toriyama et al. 1987). Isoenzymes have also been used to study the effectiveness of plant transformation (Mathews et al. 1985) and tissue culture techniques (Cardy 1986; Charne et al. 1988). The association between isoenzymes and agronomic traits has been reported by a few researchers (Thukral et al. 1985). A limited amount of research has been done on the genetics of specific isoenzymes (Arus and Orton 1983; Quiros et al. 1987); results from most isoenzyme research are reported as zymograms. Each of the above-mentioned applications of isoenzyme research to the breeding of oilseed *Brassica* species will be covered in detail in this chapter.

11.2 Isoenzymes: A Definition

Isoenzymes are multiple molecular forms of an enzyme which catalyze the same biochemical reaction. These multiple forms may occur in a single organism or in different members of the same species (Harris and Hopkinson 1976). Multiple enzyme forms may result from different loci or from different alleles at a locus (Gottlieb 1982). Isoenzymes can vary in size, shape, structure, isoelectric point, amino acid composition, cellular location, substrate specificity, temperature and pH optima. Post-translational modification of enzyme subunits also may influence isoenzyme expression (Harris and Hopkinson 1976).

Isoenzymes can be identified by a variety of techniques. This chapter, however, will be restricted to the isoenzyme studies which use electrophoretic techniques: starch gel electrophoresis, polyacrylamide electrophoresis, or isoelectric focussing.

11.3 Taxonomic Studies

Early investigations into isoenzyme expression in *Brassica* species were undertaken to provide further evidence for the taxonomic origin of various amphiploid *Brassica* species. A series of studies (Vaughan and Waite 1967a, b; Vaughan and Denford 1968; Vaughan et al. 1970) analyzed seed proteins, including isoenzymes, to provide information on the phylogenetic relationships among elementary diploid and amphiploids *Brassica* species. Results correlated well with the accepted taxonomy (UN 1935).

Vaughan et al. (1970) resynthesized *B. napus* and studied the expression of five isoenzymes in the resynthesized hybrid, natural amphiploid *B. napus*, and in the parental diploid species. In general, isoenzyme bands in the hybrids comprised bands from both parents; natural *B. napus* had fewer isoenzyme bands than the resynthesized hybrid. The loss of isoenzyme bands in the natural *B. napus* could be attributed to the selection for desirable agronomic characteristics. In a similar study, Coulthart and Denford (1982) used phosphoglucomutase (PGM) to confirm the hybrid origin of *B. napus*. As in the previous study, most *B. napus* expressed isozymes from both parental species.

Denford (1975) used isoenzymes to divide $n = 10$ members of the genus *Brassica* into three groups: Indo-European (*B. rapa*), Chinese (*B. chinensis*) and Mediterranean (*B. tournefortii*). In a subsequent study (Denford and Vaughan 1977), 11 isoenzyme systems were studied in the same taxonomic group. Phylogenetic relationships among the allies of *B. campestris* were calculated according to the percentage similarity of the isoenzyme patterns. Three basic groups were postulated: *B. campestris* including *rapa*, *sarson* and *toria* (Indo-European); *B. chinensis*, including *pekinensis* and *perviridis* (Chinese), and *B. tournefortii* (Mediterranean).

Taxonomic studies in *Brassica* species, based on isoenzyme variability, may be of limited use due to the amount of intraspecific variation in isoenzyme pattern. Vaughan et al. (1970) noted that bands in some resynthesized hybrids were unique to certain individual crosses. Wills et al. (1979) reported that beta-galactosidase (beta-GAL) and beta-glucosidase (beta-GLU) expession in *B. oleracea* differed from that reported by Vaughan and Waite (1967a). These differences may have been due to the use of single-seed extracts (Wills et al. 1979) in place of bulk seed extracts (Vaughan and Waite 1967a). Wills et al. (1979) cautioned that further knowledge of intraspecific variation in isoenzyme expression was necessary before taxonomic studies based on interspecific variability were definitive.

11.4 Cultivar Purity

Isoenzymes may be useful to *Brassica* breeders as a tool for measuring heterogeneity within the breeding material, varifying the uniformity of seed stocks (Morgan 1984), confirming homozygosity, or testing for genetic drift within the breeder seed. Isoenzymes also may be used for varietal characterization prior to application for Plant Breeders' Rights (Bailey 1983). The use of isoenzymes in cultivar identification has been reviewed extensively (Cooke 1984; Nielsen 1985).

Kumar and Gupta (1985) reported that peroxidase (PER) isoenzyme patterns could be used to characterize different genotypes of Indian mustard (*B. juncea*) based on differences in the number, placement and intensity of the bands. Chen and Tong (1985), also working with *B. juncea*, indicated that both PER and acid phosphatase

(ACP) were useful in separating morphotypes of this species. Cardy (1986) found considerable variation among and within cultivars of *B. napus* and *B. campestris*. This was confirmed by Thorpe et al. (1987), who studied variability in 22 enzymes of *B. napus* and *B. campestris*. Duke (1988) was able to separate seed lots of 25 cultivars and breeding lines of *B. napus* on the basis of glucose phosphate isomerase (GPI) and diaphorase (DIA) isoenzyme ratios.

11.5 Hybrid Determination

11.5.1 Intraspecific Hybrids

Isoenzyme electrophoresis as a tool for determining F_1 hybrid purity, by detection of self- or sib-contaminants, was first utilized in *Brassica* species by Wills et al. (1979) for hybrids of *B. oleracea*. The technique can be applied to any *Brassica* species in which there is sufficient isoenzyme variability within breeding material and parental lines. Chen and Tong (1985) could distinguish hybrid *B. juncea* from parental lines by PER zymograms; Li (1986) reported distinct estrase (EST) patterns in F_1 hybrids of *B. napus*. GPI and DIA have also been used to identify hybrids of *B. napus* (Thorpe and Beversdorf 1988).

When the parents of a hybrid are genetically fixed for different alleles of the isoenzyme under study, identification of contaminants due to maternal self- or sib-pollination is simple, providing the isoenzyme is inherited in a codominant manner. However, outcrossing rates among oilseed *Brassica* species are variable, and each parental line may have more than one allele at any isoenzyme locus. In this case, the proportion of seed which results from self- or sib-pollination in the maternal line can be estimated using formulae developed by Samaniego and Arus (1983) for *B. oleracea* hybrids.

If hybrid purity is being estimated in a commercial hybrid *Brassica* breeding programme, large numbers of seed may need to be tested. Suurs (1986) described a large-scale electrophoretic system designed for use in a hybrid *B. oleracea* programme. The system does not require sophisticated apparatus and allows one person to handle 768 samples in a day.

11.5.2 Interspecific Hybrids

Iwasaki (1983) used PER isoenzymes to determine interspecific hybrids of *Brassica* at the seedling stage. Schenk and Wolf (1986) reported that somatic hybrids between *B. oleracea* and *B. campestris* had EST and phosphorylase (PHO) bands from each parent. Terada et al. (1987) used GPI and leucine aminopeptidase (LAP), respectively, to distinguish somatic hybrids between *B. oleracea* and *B. campestris*.

When isoenzymes are used to distinguish somatic hybrids, it is important that the specific plants used as parents are analyzed before the hybrid is made. Terada et al. (1987) found a two-banded LAP zymogram in the somatic hybrid between

B. oleracea and *B. campestris* comprising a single slow-migrating band from *B. campestris* and a single fast-migrating band from *B. oleracea*. Neither of these bands is species-specific. Arus and Orton (1983) and Orton and Browers (1985) reported that populations of *B. oleracea* have slow, fast and two-banded zymograms for LAP; the same three isoenzyme phenotypes have been found in *B. campestris* (Thorpe, unpubl.). Unless specific parental phenotypes are known, misidentification of escapes as hybrids is possible.

Kanazawa et al. (1981) used EST and PER isoenzymes to monitor the elimination of C genome chromosomes in *B. oleracea* × *B. campestris* backcrossed three times to *B. campestris*. Isoenzyme variability in the backcrosses was related to the number of C genome chromosomes, and the frequency of C genome-specific isoenzymes declined as the number of C genome chromosomes was reduced by backcrossing.

11.5.3 Intergeneric Hybrids

Isoenzyme analysis also can be used to identify intergeneric hybrids. *Brassicoraphanus*, the amphidiploid between *B. japonica* and *Raphanus sativus*, was analyzed for EST and PER isoenzymes by Kato and Tokumasu (1979). They found that banding patterns in the amphidiploid comprised bands from both parental species, plus additional bands which may have been the product of subunit reassociation. Gleba and Hoffman (1980) used protoplast fusion between cells of *Arabidopsis thaliana* and *B. campestris* to produce *Arabidobrassica*. EST isoenzymes from both parental species were present in the hybrid. Similarly, Toriyama et al. (1987) studied isoenzyme expression in somatic hybrids between *Sinapis turgida* and *B. oleracea* and *S. turgida* and *B. nigra*. They reported that *S. turgida* × *B. oleracea* hybrids could be distinguished on the basis of both ACP and PER isoenzymes, while the hybrids with *B. nigra* had distinct PER isoenzymes only.

Isoenzymes may also be used as an indicator of vector-based transfer of foreign genetic material. Mathews et al. (1985) transformed cells of *B. juncea* using a strain of *Agrobacterium tumefaciens* which contained nopaline. Regenerated shoots and leaves were analyzed by electrophoresis; successfully transformed regenerants contained nopaline dehydrogenase (NDH) isoenzymes.

11.6 Anther and Microspore Culture

Tissue culture techniques such as anther or microspore culture are expected to result in homozygous progeny; diploid regenerants, arising either through spontaneous chromosome doubling or through chemical treatment, should be homozygous for isoenzyme expression. Isoenzymes, therefore, can be used to monitor the origin of plants derived from tissue culture.

Orton and Browers (1985) surveyed regenerants of *B. oleracea* from anther culture for PGM, GPI and LAP isoenzymes. Of 762 plantlets, one was heterozygous for PGM alleles, indicating sporophytic origin. Segregation ratios for PGM-1, PGM-2 and LAP alleles, expected to exhibit codominant, independent Mendelian segregation, were distorted, with an over-representation of "fast" alleles (Orton and Browers 1985).

Cardy (1986) surveyed 28 anther-derived diploid lines of *B. napus* obtained from cultures of F_1 hybrids. Two lines in this study were putatively heterozygous for genes controlling DIA and GPI isoenzymes. These lines possibly resulted from somatic cells or from unreduced microspores. In a similar study, Charne et al. (1988) used GPI and DIA to detect heterozygosity in microspore-derived spontaneous diploids of *B. napus*. As in Orton and Browers' (1985) study in *B. oleracea*, low levels of heterozygosity, coupled with distortion of expected segregation ratios, were reported.

11.7 Association with Agronomic Traits

Although the potential of isoenzymes as genetic markers in breeding programmes has been realized (Coulthart and Denford 1982), little work has been published in this area. Thukral et al. (1985) studied PER isoenzymes in *B. juncea*, *B. napus*, *B. carinata* and *B. tournefortii*. They reported that band number and activity were higher in drought-tolerant genotypes, indicating that PER isoenzymes could perhaps be used to screen breeding material for this trait.

Lui et al. (1983) screened fertile and cytoplasmic male sterile (CMS) lines of *B. oleracea* for RuBP carboxylase by isoelectric focussing. Expression of three large subunits differed between fertile and CMS lines, indicating that RuBP carboxylase has potential as a marker to identify CMS germplasm. The presence of new bands (PER) in male sterile lines and their complete absence in restored hybrids could be taken as evidence of some inhibitory activity in *B. juncea* (Banga et al. 1984).

Other agronomic traits of interest of *Brassica* breeders include triazine resistance, erucic acid and glucosinolate content, and winter vs. spring growth habit. Duke (1988) screened 25 lines of *B. napus* for 14 enzymes and found no association between these traits and isoenzyme expression.

11.8 Genetic Control of Isoenzyme Expression

Genetic control of a number of isoenzymes has been studied in *B. oleracea*, *B. napus* and *B. campestris*. Following the convention of Arus and Shields (1983), isoenzymes of *Brassica* are recorded as patterns within a number of zones on the gel. Isoenzymes generally are negatively charged and migrate toward the anodal end of the gel; zone 1 is therefore the most anodal gel zone, zone 2 the second most anodal, etc. For reference, the zymogram of ACP within zone 1 would be referred to as ACP-1. The designation of a section of a complete zymogram as a specific zone is at the discretion of the researcher, but generally zymograms can easily be divided into two or more discrete zones.

Research into specific enzymes will be discussed using the following criteria: genetic control and subcellular location, and zymogram descriptions based on zonal designation. The enzymes discussed here are those which are used most commonly in isoenzyme studies in oilseed *Brassica* species.

11.8.1 Acid Phosphatase (ACP; EC 3.1.3.2)

Wills et al. (1979) studied ACP in *B. oleracea* and characterized two loci that are active in seed, and one locus active in seedlings and older plants. Thorpe et al. (1987) published data which showed three zones of activity in seedling tissue of both *B. campestris* and *B. napus*. In *B. campestris*, ACP-1 had 16 patterns, ACP-2 had 11 patterns and ACP-3 had 12 patterns. Thorpe et al. (1987) reported eight zymograms in *B. napus* for ACP-1; Duke (1988) reported one additional ACP-1 pattern. ACP-2 was found to have seven zymograms and ACP-3 had 11 (Thorpe et al. 1987).

Chen and Tong (1985) reported variation for ACP among cultivars in *B. juncea* and suggested its use as a genetic marker in an Indian mustard breeding programme.

11.8.2 Aconitate Hydratase (ACO; EC 4.2.1.3)

Cardy (1986) observed six patterns in each of *B. napus* and *B. campestris*. Thorpe et al. (1987) published 16 zymograms for ACO in *B. campestris*; in *B. napus*, ACO could be separated into three zones of activity. ACO-1 was single banded and monomorphic; ACO-2 had three four-banded patterns which varied on the basis of mobility. ACO-3 had four single-banded patterns and one double-banded pattern. Two additional four-banded patterns in ACO-2 have also been reported (Duke 1988).

11.8.3 Alcohol Dehydrogenase (ADH; EC 1.1.1.1)

Truco (1986) reported that ADH in *B. campestris* was monomorphic; in contrast, Thorpe et al. (1987) observed 11 zymograms varying in band number and mobility. They also reported two single-banded patterns which varied in mobility in *B. napus*. Duke (1988) identified a third single-banded pattern which was slightly anodal to the other patterns. Variation in migratory distance of one-quarter to one-half band width was interpreted as resulting from differences in conductivity across the gel and not from genetic variation (Duke 1988). Most *B. napus* analyzed had one ADH zymogram; the other two patterns were rarely observed.

11.8.4 Diaphorase (DIA; EC 1.6.4.3)

Thorpe et al. (1987) reported three DIA zymograms within a single zone in *B. campestris*: one three-banded patterns and two with four bands. In *B. napus*, eight zymograms within a single zone were observed: seven four-banded patterns and one six banded pattern.

11.8.5 Esterase (EST; EC 3.1.1.1; 3.1.1.2)

Esterases have been the subject of numerous studies in plant enzyme polymorphism. Kumar and Gupta (1985) reported developmental differences in EST patterns in *B. juncea*, with five to eight anodal isoenzyme bands observed. Denford (1975) reported three anodal bands in *B. campestris* and four anodal bands in *B. nigra*.

Thorpe et al. (1987) reported two substrate-specific esterases in *B. campestris*: EST-A had three patterns while EST-B had four. In *B. napus*, EST appeared not to be substrate-specific.

11.8.6 Beta-Galactosidase (beta-GAL; EC 3.2.1.23)

Vaughan and Waite (1967a) reported six anodal and two cathodal isoenzyme bands of beta-GAL in *B. campestris* and one to three cathodal bands in *B. nigra*. In a subsequent study (Vaughan and Waite 1967b), both cathodal and anodal bands were reported in *B. campestris* (nine bands); *B. napus* (four bands), and *B. juncea* (eight bands). *B. carinata* had four cathodal isoenzyme bands; *B. nigra* three. Denford and Vaughan (1977) reported two zones of activity in *B. campestris*.

11.8.7 Beta-Glucosidase (beta-GLU; EC 3.2.1.21)

Vaughan and Waite (1967a) observed one cathodal and one anodal isoenzyme band of beta-GLU in *B. nigra*, and two anodal zymograms in *B. campestris*. Denford (1975) reported two zones of activity in *Brassica* species and found this enzyme useful in distinguishing among species complexes.

11.8.8 Glucose-6-Phosphate Dehydrogenase (G6PDH; EC 1.1.1.49)

Two zones of activity in both *B. campestris* and *B. napus* have been recorded (Thorpe et al. 1987). G6PDH-1 had three double-banded zymograms in *B. campestris*; G6PDH-2 had two double-banded patterns. In *B. napus*, two triple-banded patterns were observed for G6PDH-1; three single-banded, one double-banded, and one triple-banded patterns were reported for G6PDH-2 (Duke 1988).

11.8.9 Glucose Phosphate Isomerase (GPI; EC 5.3.1.9)

Two zones of activity have been reported for *B. oleracea* (Weeden and Gottlieb 1982), *B. campestris* (Hodgkin, pers. commun.) and *B. napus* (Arus 1984). The most anodal zone, GPI-1, is produced by plastid isoenzymes, while GPI-2 is a cytosolic isoenzyme (Weeden and Gottlieb 1982).

Thorpe et al. (1987) reported three zones of activity in *B. campestris*. GPI-1 had five double-banded and four triple-banded patterns; GPI-2 had ten zymograms varying in band number (two to five) and migratory distance. Truco (1986) found GPI-2 to be polymorphic in *B. campestris*. GPI-3, which was frequently poorly resolved, had 11 patterns of one to three bands.

In *B. napus*, Thorpe et al. (1987) and Duke (1988) recorded three zones of activity. GPI-1 had five triple-banded patterns which varied on the basis of migratory distance. Thorpe et al. (1987) reported 11 patterns with one to five bands in GPI-2; Duke (1988) observed another three patterns. Subsequent to this research, several more GPI-2 patterns have been observed in *B. napus*; some of these are undoubtedly hybrid zymograms (Thorpe, unpubl.). Arus (1984) reported duplicated loci at GPI-2, with at least one locus polymorphic. Two single-banded and one double-banded pattern were observed in GPI-3 (Thorpe et al. 1987).

11.8.10 Glutamate Dehydrogenase (GDH; EC 1.4.1.3)

Vaughan et al. (1970) observed differences in GDH banding patterns between *B. napus* and *B. campestris* but no intraspecific variation was recorded. However, Denford and Vaughan (1977) reported intraspecific variation in *B. campestris*. Thorpe et al. (1987) showed five zymograms in *B. campestris* varying in band number and mobility within one zone of activity. *B. napus* had two zones of GDH activity. GDH-1 was single-banded and monomorphic; GDH-2 had one four-banded and three six-banded patterns (Thorpe et al. 1987).

11.8.11 Glutamate Oxaloacetate Transaminase (GOT; EC 2.6.1.1)

Arus and Orton (1983) suggested that three loci control the expression of GOT in *B. oleracea*; Arus (1984) also reported three loci control of GOT expression in *B. napus*, but observed no variation. Cardy (1986) found variability in *B. napus* but none in *B. campestris*. Thorpe et al. (1987) reported 14 GOT patterns in *B. campestris* and four in *B. napus*.

Resolution in GOT zymograms, especially on starch gels, is often poor. Truco and Arus (1987) stated that GOT is a dimeric enzyme, and the poor resolution may be caused by the formation of interlocus heterodimers and the overlapping of regions of activity of products of different loci.

11.8.12 Isocitrate Dehydrogenase (IDH; EC 1.1.1.41)

Three zones of activity were reported in both *B. napus* and *B. campestris* (Thorpe et al. 1987). In *B. napus*, one double-banded and five triple-banded zone 1 patterns were observed; zone 2 was monomorphic for a single-banded pattern. Duke (1988) reported a third zone of activity cathodal to zone 2; zone 3 had four zymograms.

Three zones were present in *B. campestris*. Thirteen patterns were observed in zone 1: six double-banded, six triple-banded, and one pattern probably hybrid, with six bands (Thorpe et al. 1987). Double-banded zymograms in this zone may be artefacts resulting from poor staining and not due to actual phenotypes. Zone 2 in *B. campestris* had five patterns; zone three had 13 of which some may again be staining artefacts (Thorpe et al. 1987).

11.8.13 Leucine Aminopeptidase (LAP; EC 3.4.11.1)

Denford and Vaughan (1977) found two- and three-banded patterns within a single zone in *B. campestris*; Truco (1986) also suggested *B. campestris* to be polymorphic for LAP. Terada et al. (1987) reported a single-banded phenotype in *B. campestris*. Thorpe et al. (1987) observed two zones of activity in both *B. campestris* and *B. napus*. Zone 2 is monomorphic in both species; in *B. campestris* three zone 1 patterns were recorded: two double-banded and one triple-banded. More recently, two single-banded phenotypes have been found (Thorpe, unpubl.). In *B. napus*, Thorpe et al. (1987) reported three zone 1 patterns, confirming Arus's (1984) report of LAP-1 in *B. napus* as exhibiting fixed heterozygosity and polymorphism. Duke (1988) observed five zymograms in LAP-1.

11.8.14 Malate Dehydrogenase (MDH; EC 1.1.1.37)

Cardy (1986) and Thorpe et al. (1987) reported variation for MDH in both *B. napus* and *B. campestris*. Quiros (1987) found five isoenzymes for MDH in both *B. nigra* and *B. campestris*; one isoenzyme was found in plastids, while the others were cytosolic. Goonewardena and Wilson (1979) reported one cytoplasmic and two mitochondrial MDHs in *B. napus*; of the mitochondrial isoenzymes, one was in the matrix and one was membrane-bound within the matrix.

Analysis of MDH zymograms may be difficult; MDH is known to be subject to post-translational modification (Tyson et al. 1985) and intergenic dimers may be formed (Doebley et al. 1986). Non-replicable variability has been recorded (Duke 1988).

11.8.15 Myrosinase (MYR; EC 3.2.3.1)

Myrosinases (glucosinolases), which degrade glucosinolates, were first examined by Vaughan and Denford (1968), with limited success. MacGibbon and Allison (1970) separated myrosinases of *B. napus* and *B. campestris* using acrylamide electrophoresis. Root myrosinase isoenzymes were identical in both species, but leaf, petiole, and seed zymograms could be used to distinguish *B. napus* from *B. campestris*.

Hoffman et al. (1982) studied myrosinase expression in *B. napus* lines derived from anther culture. A five-banded pattern was seen when sinigrin was used as the substrate; glucoiberin produced a six-banded phenotype. No differences were seen in isoenzyme expression in roots, leaves, or seeds.

G. Seguin Swartz (pers. commun.) is currently investigating the inheritance of MYR in *B. nigra*. Preliminary results using seed have indicated two zones of activity with three bands in one zone and one to three bands in the second zone.

11.8.16 6-Phosphogluconate Dehydrogenase (6-PGD; EC 1.1.1.44)

Quiros (1987) reported two zones of activity for 6-PGD in both *B. nigra* and *B. campestris*. The most anodal zone (6-PGD-1) is believed to be located in the plastids; this zone was triple-banded, but the bands differed in migratory distance between species. The less anodal cytosolic zone (6-PGD-2) had five bands in *B. nigra* and one band in *B. campestris*. A subsequent study (Quiros et al. 1987) suggested three-banded zymograms in zone 1 of 6-PGD in both *B. oleracea* and *B. campestris* and a six-banded phenotype in *B. napus*. It was proposed that both parental species contain duplicated loci: 6-pgd-1 ol and 6-pgd-1'ol, and 6-pgd-l cp and 6-pgd-1'cp. Within a species, the duplicated loci interact to form heterodimers, resulting in separate, three-banded phenotypes at zone 1 in both species. The hybrid *B. napus* displays the five bands from the parental species (most anodal band is common in both species) plus an interlocus heterodimeric band formed by polypeptides from the 6-pgd-1'ol and 6-pgd-1'cp isoenzymes (Quiros et al. 1987).

Thorpe et al. (1987) reported two zones of activity within both *B. campestris* and *B. napus*. Variability was observed in both regions and in both species.

11.8.17 Phosphoglucomutase (PGM; EC 2.7.5.1)

Coulthart (1979) and Arus and Orton (1983) proposed that PGM is coded for by two nuclear loci in *Brassica* species; Arus and Orton (1983) observed that these loci behave independently and are codominantly inherited. Arus (1984) reported fixed heterozygosity at PGM-1 and PGM-2 in *B. napus*, and polymorphism at PGM-2. Truco (1986) suggested that PGM-1 is polymorphic in *B. campestris*, whereas PGM-2 is monomorphic.

Preliminary studies by Cardy (1986) indicated isoenzyme activity within three zones in both *B. campestris* and *B. napus*. Thorpe et al. (1987) reported two zones of activity in *B. campestris* and three zones in *B. napus*, all of which exhibited phenotypic variation (Duke 1988).

11.8.18 Peroxidase (PER; EC 1.11.1.7)

Many isoenzyme studies in *Brassica* species have involved PER, but results are difficult to compare because of differences in tissue and gel type used. Denford and Vaughan (1977) reported no intraspecific variability in *B. campestris*; Yadava et al. (1979) observed no differences in PER zymograms of *B. napus*, *B. campestris* and *B. oleracea*. However, intraspecific variation has been recorded in *B. napus* (Vaughan and Waite 1967a; Iwasaki 1983), and in *B. campestris*, *B. carinata* and *B. juncea* (Iwasaki 1983). Genotype differences in anodal PER isoenzymes were recorded for *B. juncea* by Kumar and Gupta (1985); cathodal PER patterns differed with developmental stage.

Thorpe et al. (1987) reported one anodal zone of activity for PER in *B. campestris* and *B. napus*; single- and double-banded phenotypes were observed in both species. Duke (1988) also reported a polymorphic cathodal PER in *B. napus*.

11.8.19 Shikimate Dehydrogenase (SKDH; EC 1.1.1.25)

Thorpe et al. (1987) reported intraspecific variation for SKDH in both *B. napus* and *B. campestris*. Duke (1988) suggested that two zones of activity are present in both species; SKDH-1 in *B. campestris* and SKDH-2 in *B. napus* stain poorly on starch gels. The other zones stain well and exhibit phenotypic variation.

11.9 Conclusions

The primary use of isoenzymes within a breeding programme is as molecular markers. Isoenzymes closely linked to specific quantitative traits may be easier to screen and select for than the trait itself. Development-dependent expression of isoenzymes may act as a marker for the initiation or cessation of specific physiological activities and could be used in conjunction with genetic engineering for development-specific control of the transcription of inserted foreign genetic material. Isoenzymes may also be used as inexpensive markers simply to monitor successful crosses and line purity. Regardless of the use within a specific programme, a greater knowledge of the genetics of isoenzymes is needed; specifically, the number of loci controlling isoenzyme expression, the potential linkages among isoenzyme loci and other traits, and the chromosomal location of these genes all need to be determined.

Acknowledgements. The authors would like to thank Dr. W.D. Beversdorf for his encouragement and support, and Dr. D.J. Hume for his editorial review of this manuscript. Thanks are also due to Maria Sousadias and Lyne Renaud for their expert and invaluable assistance.

References

Arus P (1984) *Brassica oleracea* and *Brassica napus* isozymes. Cruciferae Newslett 9: 64–65

Arus P, Orton TJ (1983) Inheritance and linkage relationships of isozyme loci in *Brassica oleracea*. J Hered 74: 405–412

Arus P, Shields CR (1983) Cole crops (*Brassica oleracea*). In: Tanksley SD and Orton TJ (eds) Isozymes in plant genetics and breeding, Part B, Elsevier, Amsterdam. pp 339–350

Bailey DC (1983) Isozymic variation and plant breeders' rights. In: Tanksley SD and Orton TJ (eds) Isozymes in plant genetics and breeding, Part B, Elsevier, Amsterdam. pp 425–440

Banga SS, Labana KS, Banga SK (1984) Male sterility in Indian mustard [*Brassica juncea* (L.) Coss]—a biochemical characterization. Theor Appl Genet 67: 515–519

Cardy BJ (1986) Production of anther-derived doubled haploids for breeding oilseed rape (*Brassica napus* L.) Ph D theses, Univ Guelph, Guelph

Charne DJ, Thorpe ML, Beversdorf WD (1988) Isozyme electrophoresis as a tool for detecting heterozygosity in microspore-derived spontaneous diploids of rapeseed. (Abstr) Can J Plant Sci 68: 831

Chen S, Tong N (1985) A preliminary study of different morphotypes in mustard (*Brassica juncea*). Cruciferae Newslett 10: 37

Cooke RJ (1984) The characterisation and identification of crop cultivars by electrophoresis. Electrophoresis 5: 59–72

Coulthart MB (1979) A comparative electrophoretic study of an allopolyploid complex in *Brassica*. MSc theses, Univ Alberta, Edmonton

Coulthart MB, Denford KE (1982) Isozyme studies in *Brassica*. I. Electrophoretic techniques for leaf enzymes and comparison of *B. napus*, *B. campestris* and *B. oleracea* using phosphoglucomutase. Can J Plant Sci 62: 621–630

Denford KE (1975) Isoenzyme studies in members of the genus *Brassica*. Bot Notiser 128: 455–462

Denford KE, Vaughan JG (1977) A comparative study of certain seed isoenzymes in the ten chromosome complex of *Brassica campestris* and its allies. Ann Bot 41: 411–418

Doebley J, Morden CW, Schertz KF (1986) A gene modifying mitochondrial malate dehydrogenase in *Sorghum*. Biochem Genet 24: 813–819

Duke LH (1988) Cultivar identification of rapeseed (*Brassica napus*) using isoenzyme analysis. MSc theses, Univ Guelph, Guelph

Gleba YY, Hoffman F (1980) "Arabidobrassica", a novel plant obtained by protoplast fusion. Planta 149: 112–117

Goonewardena H, Wilson BS (1979) The oxidation of malate by isolated turnip (*Brassica napus* L) mitochondria II. The malate oxidizing enzymes, number, and location. J Exp Bot 118: 877–887

Gottlieb LD (1982) Conservation and duplication of isozymes in plants. Science 216: 373–379

Harris H, Hopkinson DA (1976) Handbook of enzyme electrophoresis in human genetics. North Holland, New York

Hoffman F, Thomas E, Wenzel G (1982) Anther culture as a breeding tool in rape. Theor Appl Genet 61: 225–232

Iwasaki F (1983) Zymogram analysis in *Brassica* species. Jpn J Breed 33: 171–177

Kanazawa K, Eguchi H, Iwasa S, Uemoto S (1981) Electrophoretic study on esterases and peroxidases in strains backcrossed with pollen of chinese cabbage with reference to nuclear substitution. In: Chi Cabbage: Proc first int symp pp 377–383

Kato M, Tokumasu S (1979) An electrophoretic study of esterase and peroxidase isozymes in *Brassicoraphanus*. Euphytica 28: 339–349

Kumar R, Gupta VP (1985) Isozyme studies in Indian mustard (*Brassica juncea*). Theor Appl Genet 79: 107–110

Li D (1986) Success in large scale extension of breeding male sterile, maintainer, and restorer lines of swede rape (*Brassica napus* L.). (Abstr). Sci Agric Sinica 4: 94

Lui ZC, Li JG, Luo HX, Chen FT (1983) Studies on ribulose bisphosphate carboxylase and male sterility. Acta Genet Sinica 10: 36–42

MacGibbon DB, Allison RM (1970) A method for the separation and detection of plant glucosinolates (myrosinases). Phytochem 9: 541–544

Mathews VH, Bhatia CR, Mitra R, Krishna TG, Rao PS (1985) Regeneration of shoots from *Brassica juncea* (Linn.) Czern and Coss cells transformed by *Agrobacterium tumefaciens* and expression of nopaline dehydrogenase genes. Plant Sci 39: 49–54

Morgan AG (1984) Breeding for crop quality. M Phil degree seminar. Nat Inst Agric Bot Cambridge

Nielsen G (1985) The use of isozymes as probes to identify and label plant varieties and cultivars. In: Rattazzi MC, Scandalios JG, White GS (eds) Isozymes: Current Topics in Biological and Medical Research, Vol 12 Liss New York pp 1–32

Orton TJ, Browers MA (1985) Segregation of genetic markers among plants regenerated from cultured anthers of broccoli (*Brassica oleracea* var. *italica*). Theor Appl Genet 69: 637–643

Quiros CF (1987) Duplicated isozyme loci and cellular compartmentalization of their products in *Brassica*. Cruciferae Newslett 12: 24

Quiros CF, Ochoa D, Kinian SF, Douches D (1987) Analysis of the *Brassica oleracea* genome by the generation of *B. campestris-oleracea* chromosome addition lines: characterization by isozymes and rDNA genes. Theor Appl Genet 74: 758–766

Samaniego FJ, Arus P (1983) On estimating the sib proportion in seed purity determinations. Biometrics 39: 563–572

Scheck HR, Wolf G (1986) Characterization of somatic *Brassica napus* hybrids by polyacrylamide electrophoresis. Plant Breed 97: 72–74

Suurs LCJM (1986) Routine large scale electrophoresis for plant breeding. An example with F_1 seeds of Brussels sprouts (*Brassica oleracea* var. *gemmifera*). Euphytica 36: 147–151

Terada R, Yamashita Y, Nishibayashi S, Shimamoto K (1987) Somatic hybrids between *Brassica oleracea* and *B. campestris*: selection by the use of iodoacetamide inactivation and regeneration ability. Theor Appl Genet 73: 379–384

Thorpe ML, Beversdorf WD (1988) Application of starch-gel electrophoresis to hybrid purity testing in *Brassica napus*. Can J Plant Sci 68: 839

Thorpe ML, Duke LH, Beversdorf WD (1987) Procedures for the detection of isozymes of rapeseed (*Brassica campestris* and *B. napus*) by starch gel electrophoresis. Univ Guelph Tech Bull OAC 887, Univ Guelph, Guelph

Thukral SK, Behl RK, Kumar R (1985) Water stress effects on some important physiomorphological attributes in oilseed *Brassica*. Ann Biol 85: 209–215

Toriyama K, Kameya T, Hinata K (1987) Selection of a universal hybridizer in *Sinapis turgida* Del and regeneration of plantlets from somatic hybrids with *Brassica* species. Planta 170: 308–313

Truco MJ (1986) Tesi de lliceniatura. Fac Biol, Univ Barcelona, Barcelona

Truco MJ, Arus P (1987) Comparative study of the isozymes of *Brassica campestris*, *B. oleracea*, and *B. napus*. Cruciferae Newslett 12: 18–19

Tyson H, Fieldes MA, Cheung C, Starobin J (1985) Isozyme relative mobility (Rm) changes related to leaf position: apparently smooth Rm trends and some implications. Biochem Genet 23: 641–654

UN (1935) Genome analysis in *Brassica* with special reference to the experimental formation of *B. napus* and peculiar mode of fertilization. Jpn J Bot 7: 389–452

Vaughan JG, Denford KE (1968) An acrylamide gel electrophoretic study of the seed proteins of *Brassica* and *Sinapis* species with special reference to their taxonomic value. J Exp Bot 19: 724–732

Vaughan JG, Denford KE, Gordon EL (1970) A study of the seed proteins of synthesized *Brassica napus* with respect to its parents. J Exp Bot 21: 892–898

Vaughan JG, Waite A (1967a) Comparative electrophoretic studies of the seed proteins of certain species of *Brassica* and *Sinapis*. J Exp Bot 18: 100–109

Vaughan JG, Waite A (1967b) Comparative electrophoretic studies of the seed proteins of certain amphidiploid species of *Brassica*. J Exp Bot 18: 269–276

Weeden NF, Gottlieb LD (1982) Dissociation, reassociation, and purification of plastid and cytosolic phosphoglucose isomerase isozymes. Plant Physiol 69: 717–723

Wills AB, Fyfe SK, Wiseman EM (1979) Testing F_1 hybrids of *Brassica oleracea* for sibs by seed isoenzyme analysis. Ann Appl Biol 91: 263–270

Yadava JS, Chowdhury JB, Kakkar SN, Nainawatee HS (1979) Comparative electrophoretic studies of proteins and enzymes of some *Brassica* species. Theor Appl Genet 54: 89–91

CHAPTER 12

Disease Resistance

G.S. Saharan

12.1 Introduction

Oilseed brassicas are subject to attack by a number of pathogenic diseases (Table 1), not including several non-pathogenic disorders. The yield losses vary, depending upon the type and nature of pathogen association, time and severity of the attack, resistance level of the cultivars grown and prevailing environmental conditions. These diseases limit oilseed brassica production over a wide geographical area all over the world. Even in normal years, some regions may have substantial yield reductions due to a particular pathogen, e.g., *Sclerotinia* attack in northeastern Saskatchewan (Canada) in 1973 and in Rajasthan (India) during 1988-89.

Pathogens of oilseed brassicas vary greatly in the way and rate they grow, multiply, and spread. In order to restrict the damage from pathogens, plants employ a wide range of defense mechanisms, e.g. avoidance, resistance and tolerance. The first mechanism operates before parasitic contact between host and pathogen, whereas the two other mechanisms operate after the pathogen has made contact with the host. Not only does resistance occur in a variety of mechanisms, it also varies in intensity, from complete to almost imperceptible. Selection for resistance implies measurements of plants' resistance. Ideally, it can be done by measuring the growth and development of the pathogen; the more they are reduced the more resistant the host plant is. This is generally not possible, because in most cases the pathogen is either not or only partially visible. However, one can evaluate the pathogen's effects on the host in the form of symptoms. The information generated so far on different aspects of disease resistance against few important diseases in oilseed brassica crops, i.e. disease scoring procedures, germplasm screening techniques, occurrence of pathotypes, sources of resistance, genetics of host-parasite interaction, biochemical basis of resistance and incorporation of resistance, is presented.

12.2 *Alternaria* Blight

This is the most destructive and widespread disease of brassicas wherever these crops are grown. Although the literature reveals its occurrence in only 12 countries (Table 1) of the world, its presence in the rest of the world cannot be ruled out. Estimates of yield losses due to this disease vary between 10 and 70% in different species of oilseed brassicas grown in different countries (Kolte 1985; Saharan and Chand 1988).

Table 1. Distribution of diseases of oilseed brassicas (Kolte 1985; Saharan and Chand 1988)

S. No.	Name of disease	Pathogen	Geographical distribution
1	2	3	4
1.	*Alternaria* blight	*Alternaria brassicae* *A. brassicicola* *A. raphani*	Canada, England, Ethiopia, France, Germany, Holland, India, Poland, Sri Lanka, Spain, Sweden, Trinidad, UK
2.	Downy mildew	*Peronospora parasitica*	Canada, China, Ethiopia, France, Germany, Hong Kong, India Japan, Pakistan, Phillipines, Soviet Union, Sweden, UK
3.	White rust	*Albugo candida*	Brazil, Canada, Ethiopia, France, Germany, India, Japan Pakistan, Palestine, Romania, Soviet Union, Turkey
4.	Sclerotinia rot	*Sclerotinia sclerotiorum*	Brazil, Canada, China, Denmark, Ethiopia, Finland, France, Germany, Sweden, UK
5.	Blackleg or stem canker	*Leptosphaeria maculans* (*Phoma lingam*)	Australia, Canada, Germany, Ethiopia, France, Holland, Kenya, New Zealand, Sweden, United Kingdom
6.	Clubroot	*Plasmodiophora brassicae*	Germany, India, Malaysia, New Zealand, Poland, Sweden, UK, USA
7.	Light leaf spot	*Pyrenopeziza brassicae* (*Cylindrosporium concentricum*)	France, Germany, Sweden, UK
8.	Phyllody	Mycoplasma-like organism	India
9.	Bacterial rot	*Xanthomonas campestris* pv. *campestris*	Brazil, Canada, Ethiopia, Germany, Sweden, USA
10.	Broom rape	*Orobanche aegyptiaca* *O. cernua*	India

1	2	3	4
11.	Powdery mildew	*Erysiphe cruciferarum*	France, Germany, India, Japan, Sweden, Turkey, UK, USA
12.	Fusarium wilt	*Fusarium oxysporum* f. sp. *conglutinans*	India, USA
13.	Root gall smut	*Urocystis brassicae*	India
14.	Bacterial stalk rot	*Erwinia carotovora*	India
15.	Mosaic	Turnip mosaic virus	Canada, China, Germany, Hungary, India, Soviet Union, Trinidad, UK, USA
16.	Damping off	*Pythium aphanidermatum*	India
		P. butleri	India
		P. debaryanum	Canada, Germany, India, Phillipines
17.	Root rot	*Alternaria alternata*	Canada, India
		Rhizoctonia solani	Canada, India, Phillipines, USA
		Sclerotium rolfsii (*Pellicularia rolfsii*)	India, USA
		(*Macrophomina phaseoli*)	India
18.	Leaf spot	*Alternaria alternata*	Canada
		A. longipes	India
		A. napiformae	India
		Cercospora cheiranthi var. *brassica*	India
		C. brassicicola	Sri Lanka
		Cercosporella brassicae	Canada
		Leptosphaerulina brassicae	India
		Mycosphaerella brassicicola	Canada, Ethiopia
		Myrothecium roridum	India
		Pseudo-cercosporella capsellae	Canada

Fig. 1. *Alternaria* infection on *Brassica* leaves (**a**) and pods (**b**)

12.2.1 Disease-Scoring Scale

In India, oilseed workers score *Brassica* germplasm on a 0-5 scale. The germplasm lines which fall in the 0-2 point scale at leaf and/or pod stages are considered as resistant (Anonymous 1980-1989; Kadian and Saharan 1983). The disease index is computed by the formula:

$$\text{Disease index (\%)} = \frac{(\text{numerical rating} \times \text{sample frequency})}{\begin{pmatrix}\text{number of samples}\\ \text{examined}\end{pmatrix} \times \begin{pmatrix}\text{maximum rating score}\\ \text{point on the scale}\end{pmatrix}} \times 100$$

Mayee and Datar (1986) suggested the use of a 0-9 scale based on the growth stages of the crop. Recently, Conn et al. (1990) prepared keys for the assessment of disease on both leaves and siliqua.

12.2.2 Germplasm Screening Techniques for Testing Resistance

1. Screening at the Cotyledonary Stage. The test is an adaptation of the technique described for *Phoma* (Williams 1985). In this, the seedlings at the cotyledonary stage are inoculated with *A. brassicae* spores taken from a 16-day culture on V8 medium at a 5×10^4 concentration of spores per ml. Two calibrated 10-μl droplets are placed on the cotyledons of 8-day-old seedlings. The seedlings are raised in the growth chamber at 18 °C, 80% relative humidity and 14 h light. Plants were scored 48 and 96 h later, as follows (Brun et al. 1987).

0 = No apparent infection
1 = 1 to 2 spots
2 = 2 to 7 distinct spots
3 = 8 to 15 distinct spots
4 = Extended spotting, but the whole surface of droplet location not completely covered
5 = Whole droplet location completely covered

The studies conducted by Brun et al. (1987) demonstrated that the seedling test is highly correlated with adult plant resistance and hence is of tremendous interest for early and non-destructive testing in segregating generations.

2. *Detached Leaf Method.* The resistance to *Alternaria* blight can also be measured by the size of lesions caused by *A. brassicae* on detached wounded leaves (Bansal et al. 1990). A fully expanded fourth leaf, detached from ca. 18-day-old seedling, is placed in a clear plastic container lined with water-soaked paper towels. The leaves are punctured on both sides of the mid-rib with a needle inoculator, which creates a circular wound of ca. 3 mm in diameter (Gugel et al. 1990). A 25 μl drop of spore suspension, obtained from disease isolates cultured on V8 agar supplemented with Rose Bengal (0.4 mg/l), is then applied to each wound (50 μl per plant). Sealed plastic containers containing inoculated leaves are then incubated at room temperature under continuous light. Four days after inoculation, lesion diameter, including the chlorotic zone, is recorded.

3. *Greenhouse Method for Testing Young Brassica Plants.* In this technique, the greenhouse-grown plants are inoculated at the three-leaf stage (Grontoft and O'Connor 1990). The inoculum is isolated from *Alternaria*-infected plants and grown on PDA for 3 weeks at 18 °C with 12 h black light. Inoculation is performed using small agar discs (10 mm) cut from just behind the growth front with a corkborer. The discs are placed with the infected side against the upper side of the second leaf over a 4 mm hole. A second hole of similar size is cut on the other side of the mid-vein. Immediately after inoculation, each plant is covered with plastic to enhance the relative humidity and increase fungal growth. The resistance level is determined by measuring the diameter of necrotic lesions on leaves.

4. *Adult Plant Screening Under Natural Conditions.* In India, *Brassica* germplasm is screened at the multilocational uniform disease nursery trials of The All India Co-ordinated Research Project on Oilseeds (Anonymous 1980-1989). Promising *Brassica* germplasm lines claimed to be tolerant or resistant to disease are planted at different centres in the country, following a uniform procedure of sowing. A highly susceptible cultivar, preferably from the *B. campestris* group, is interplanted as an infector row

after every fifth test entry and all around the area covering the screening material. To ensure acute disease pressure, artificial inoculation is carried out with a spore suspension of *A. brassicae* prepared from culture and/or from previously infected leaves. Such inoculations are made three times during the crop season. High humidity is ensured by frequent irrigation (Anonymous 1980-1989).

12.2.3 Physiological Specialization

Very little information has been published on the occurrence of pathotypes in *A. brassicae*. Saharan and Kadian (1983) reported three races of *A. brassicae*. All three races are virulent on rapeseed and mustard crop groups. Race 1, designated RM 1, was isolated from rapeseed and mustard crops; race 2, called RM 2, was isolated from *B. campestris* var. *yellow* and *brown sarson*; race 3, known by the name of V3, was isolated from vegetable crops like radish, cabbage and cauliflower. The degrees of virulence in different species of *Alternaria* infecting *Brassica* crops were reported by Kolte (1987).

12.2.4 Sources of Resistance

It is believed that digenomic *Brassica* species, namely *B. napus* and *B. juncea*, have a better degree of resistance than the monogenomics like *B. campestris*. *Sinapis alba* and *B. carinata* lines have shown a higher level of resistance to this disease (Anonymous 1980-1989; Bhander and Maini 1965; Stankova 1975; Rai et al. 1977; Saharan et al. 1981; Saharan 1984; Kolte 1987; Bansal et al. 1990). Some of the resistant lines in *B. napus* reported in Czechoslovakia (Stankova 1975) are Weibull 541, 521 and Vuindsok CH/63-2; in Poland (Rozej 1976) Norde and Vestal; in Spain (Romero-Munoz and Jimenez Diaz 1980) Midas and Primor; in Canada (Skoropad and Tewari 1977) Midas and Tower. The *B. juncea* lines RC 781 and Lethbridge are reported as resistant in India (Saharan et al. 1981; Tripathi et al. 1980) and Canada (Tewari and Skoropad 1976). In India, *Brassica* genotypes, namely CSR 43, CSR 142, CSR 142-2, CSR 343, CSR 448, CSR 622, CSR 741, Gulivar, KRV-Tall, Midas, PHR 1, RC 781, TMV 2, Tower and YRT 3 were found to have consistent field resistance after testing for many years at different locations under uniform disease nursery trials (Anonymous 1980-1989; Kolte 1985, 1987; Saharan and Chand 1988). In France, Carine, a strain of *Sinapis alba*, showed only 7% disease incidence against 83% in a susceptible *B. napus* cultivar Brutor following artificial inoculation under controlled environmental conditions. Somatic hybrids between *B. napus* and *S. alba* strain Carine have already been developed (Primard et al. 1988). Preliminary results have revealed that none of the somatic hybrids was as tolerant to *Alternaria brassicae* as the *S. alba* parent. The resistance appeared to be semi-dominant and thus difficult to introduce into *B. napus*. Nevertheless, efforts are underway at Rennes (France) to improve the pathological tests and to utilize the chromosome addition lines of *S. alba* in the background of *B. napus*. (H Brun, pers. commun.).

12.2.5 The Genetics of Host-Parasite Interaction

Information on the genetics of host-parasite interaction is rather limited. In the *B. juncea* line RC 781, resistance to *A. brassicae* is reported to be governed by a single dominant gene (Tripathi et al. 1980). However, Saharan and Kadian (1983b) analyzed different components of horizontal resistance in *Brassica* genotypes against *A. brassicae*. The large amount of differences for number of lesions, size of lesions, latent period, sporulation capability and infection rate indicated a high order of horizontal resistance in *B. napus* genotype Tower and in *B. juncea* genotype RC 781. In two other *B. juncea* lines CSR 142 and CSR 448, moderate levels of horizontal resistance were recorded. Kolte (1987) suggested that the size of the lesions and the amount of sporulation should be considered as important factors in determining the level of resistance in *Brassica* species to *Alternaria* blight disease.

12.2.6 Biochemical Basis of Resistance

In some *B. napus* genotypes, resistance to *A. brassicae* has been associated with a high deposition of epicuticular wax which is water-repellent and thus reduces conidial retention. The epicuticular wax is organized into two layers—distal and proximal. The distal layer is fluffy, consisting of wax crystals, while the proximal layer has plate-like crystals. In resistant genotypes, the distal layer is thicker, while the proximal layer is almost similar (Skoropad and Tewari 1977; Tewari and Skoropad 1976). In resistant genotypes like RC 781, a higher amount of phenols and a lower amount of sugars and nitrogen than in susceptible genotypes like Prakash has been estimated (Gupta et al. 1984).

12.2.7 Incorporation of Resistance

Attempts to obtain resistance in agronomically superior genotypes of *B. juncea* through mutagenesis by exposing the seeds to gamma rays have been made with little success (Verma and Rai 1980). The transfer of resistance from *B. napus*, *B. carinata* and *S. alba* to *B. juncea* through interspecific hybridization is underway. Both pedigree selection and backcross methods are being followed to incorporate resistance into agronomically superior genotypes.

12.3 Downy Mildew

Reports of its occurrence either alone or in association with white rust have been made from 13 countries in the world (Table 1). From India, Bains and Jhooty (1979) reported that *B. juncea* plants yield 37 to 47% and 17 to 32% less siliquae and seed respectively, due to the mixed infection of downy mildew and white rust.

12.3.1 Disease-Scoring Scale

Downy mildew resistance has been rated on a scale of 1 to 5 by Ebrahimi et al. (1976) in *B. juncea* lines, where 1 indicates no sporulation, 2 = very sparse sporulation and 5 = heavy sporulation. A similar scoring scale of 1 to 4 points with slight modification has been used by Dickinson and Greenhalgh (1977). In India, oilseed workers use the same (0-5) scoring scale for assessment of downy mildew and white rust infections on leaf as well as on inflorescence (Anonymous 1980-1989). However, Mayee and Datar (1986) suggest the use of a 0-9 scale.

12.3.2 Germplasm Screening Techniques for Testing Resistance

Same as described later for white rust.

12.3.3 Physiological Specialization

There is well-documented evidence that isolates of *P. parasitica* differ in the range of cruciferous hosts that they can infect. Specialization of parasitism may be exhibited at the generic, specific and lower taxonomic levels of the host. At the generic level, Wang (1944) and Chang et al. (1964) recognized separate pathogenic varieties or formae speciales of the fungus on various species of *Brassica* (*P. parasitica* var. f. sp. *brassicae*), *Raphanus* (var. f. sp. *raphani*), and *Capsella* (var. f. sp. *capsellae*). Gaumann (1926) also recognized three formae speciales: one, *P. parasitica* f. sp. *brassicae* infected and sporulated on various *Brassica* spp. (e.g. *B. oleracea*, *B. napus*, *B. rapa*), a second (f. sp. *sinapidis*) infected *Sinapis arvensis* and *S. alba*, and a third (f. sp. *raphani*) infected *Raphanus sativus* and *R. raphanistrum*. Physiological specialization at the generic level of the host has also been reported by Hiura and Kanegae (1934) and by Dzhanuzakov (1962) on *Brassica* and *Sinapis*. At the specific level of the host Wang (1944) separated his *P. parasitica* var. *brassicae* into six specialized forms which differed in their ability to infect various *Brassica* spp. Chang et al. (1964) divided them into three subforms. Natti et al. (1967) indicated the existence of separate races 1 and 2 of the fungus. According to Sheriff and Lucas (1987), isolates of *P. parasitica* from *Brassica* are, in general, specifically adapted to their species of origin. Differential host resistance to homologous isolates of *P. parasitica* has been identified in *B. campestris*, *B. napus* and *B. oleracea* (Moss et al. 1988).

12.3.4 Sources of Resistance

Most of the reports on screening are based on natural infections. Jonsson (1966) reported Hg-vestal winter rape as the resistant variety to this disease. Ebrahimi et al. (1976) reported the P1 340207, P1 340218 and P1 347618 accession numbers of

B. juncea as resistant. Dixon (1975) showed that among the cultivars on the UK National list, Eurora, Janetski and Primor are highly resistant. Under Indian conditions all the lines of *B. carinata*, *B. napus* and *B. alba* are free from downy mildew infections. Other *Brassica* lines reported to be resistant at multilocational testing centres are CSR 43, CSR 73, CSR 343, Chamba 1, 2, Blaze, Domo, Candle, Gullivar, Metapolka, Lethbridge, EC 126743, PHR 1, RC 781, Tower, TMV 2 and Zem 1 (Anonymous 1980-1989; Saharan 1984; Kolte 1987). The oilseed rape cultivar Cressor is reported to be resistant to 14 isolates of *P. parasitica* derived from crops of *B. napus* in the UK (Lucas et al. 1988).

12.3.5 The Genetics of Host-Parasite Interaction

In *B. napus*, resistance in the oilseed rape cultivar Cressor is controlled by a single dominant gene (Lucas et al. 1988).

12.4 White Rust

Reports of white rust occurrence on oilseed brassicas have been made from 12 countries in the world (Table 1). Staghead formation (systemic infection) due to white rust and/or mixed infection of white rust and downy mildew (Fig. 2) causes losses in yield of 23-60% (Petrie and Vanterpool 1974; Saharan et al. 1984).

12.4.1 Disease-Scoring Scale

For the assessment of symptoms of white rust on mature plants of crucifers, a procedure was reported by Pound and Williams (1963). This procedure was modified/refined and was given a range of 0-5 and/or a 0-9 scale at different places by different workers (Anonymous 1980-1989). Mayee and Datar (1986) suggested the use of a 0-9 scale based on growth stages of the crop for leaf as well as staghead infections separately. Lakra and Saharan (1988a) developed a descriptive scale of 0-5 for assessing both phases of white rust development on mustard individually and/or combined. Fox and Williams (1984) suggested the use of a 0-9 visual scale in selecting for components of partial resistance to white rust disease.

12.4.2 Germplasm Screening Techniques for Testing Resistance

For the evaluation of host germplasm, procedures for growing, inoculating and incubating the inoculated plants are available (Pound and Williams 1963; Verma et al. 1975; Verma and Petrie 1978; 1980; Fox and Williams 1984; Saharan et al. 1988; Lakra and Saharan 1989).

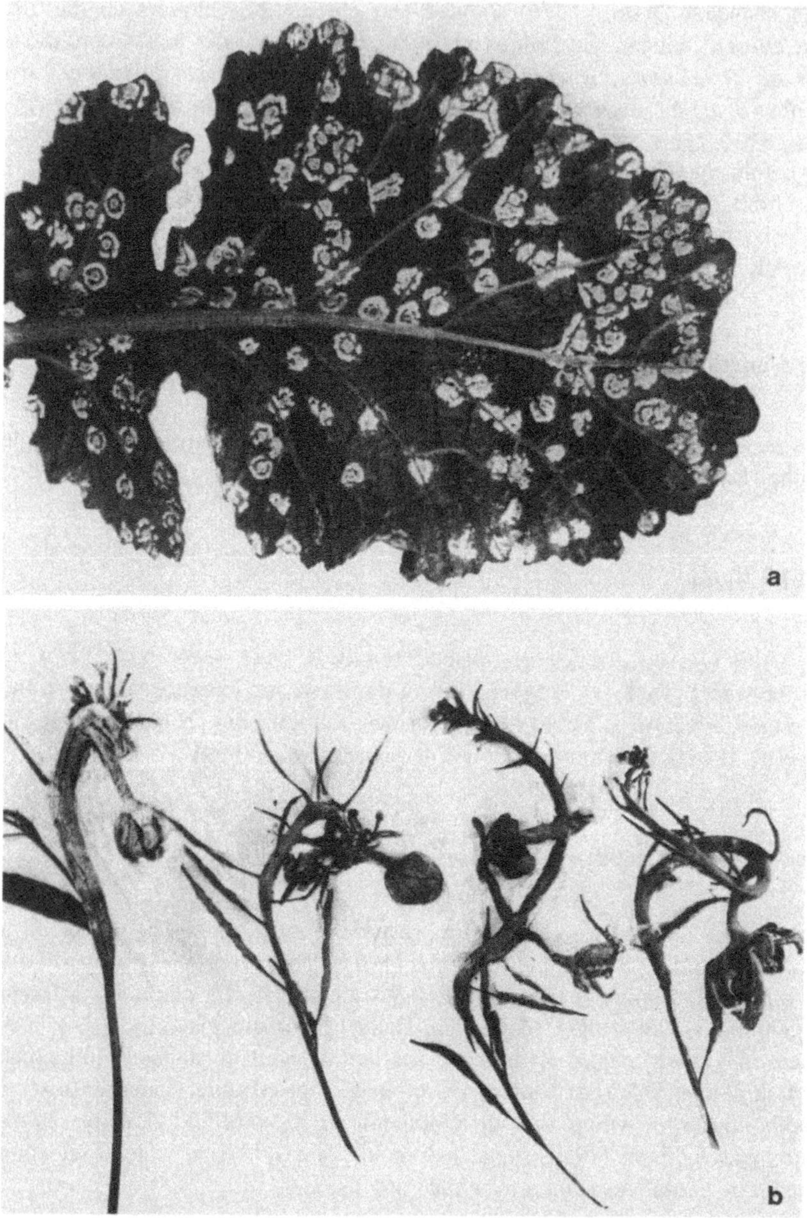

Fig. 2. **a**) White rust pustules on *B. juncea* leaf
b) Floral malformation resulting from mixed infection of downy mildew and white rust in *B. juncea*

1. Under Laboratory Conditions on the Detached Leaf. Verma and Petrie (1978) used 12-14-day-old *B. campestris* detached leaves placed in Petri dishes containing 20-25 ml of sterilized medium with 0.5 ppm benzyladenine and 0.8% agar. Leaves are placed in the dishes with their lower surface on the medium, usually within 15 min of detachment, leaves are drop-inoculated with a zoospore suspension (100,000-150,000/ml) derived from sporangia of *A. candida* and kept under 100% RH for 72 h with a day-night temperature of 21 and 16 °C, respectively. The observations are recorded on intact plants and detached leaves 10-12 days after inoculation. These are based on the percentage of infected leaves. The results of the detached plant technique are comparable with those when intact plants are used as host. This technique is a useful mean of screening breeding lines/germplasm.

2. Under Greenhouse Conditions in Pots. At the time of sowing, 5 g of oosporic material per pot is added in the soil. After emergence of the seedlings, sporangial spray inoculations are made at the seedling, branching and flower stages, respectively. High humidity is maintained by frequent irrigation (Lakra and Saharan 1989). Singh and Singh (1983) initially suggested the sporangial spray inoculation technique.

3. Under Field Conditions in a Diseased Plot. A white rust-diseased plot is prepared by adding hypertrophied malformed portions of the infected plants into the soil continuously for 3-5 years. At the time of sowing of germplasm lines, oosporic material is broadcast as well as mixed with the seed being sown. Frequent irrigation is carried out to maintain high humidity. Sporangial spray inoculations are made at seedling, branching and flowering stages of the crop growth. Genotypes are usually grown in paired rows in between the highly susceptible infector rows of plants. To obtain floral infection, the main shoot of the test row genotype grown between the infector rows is plucked 6 inches from the tip. This allows sporangial inoculum to fall on the test row plants from higher infector row plants (Saharan et al. 1988).

4. Induction of Staghead. Verma and Petrie (1980) suggested two possible ways by which staghead infection can arise: (1) an early infection of the young seedlings starting with oospore-infested seeds and progressing systematically throughout the development of the plant, and (2) infection of young flower buds by zoospores arising from wind-borne zoosporangia; this was later confirmed by Saharan et al. (1988) and Lakra and Saharan (1988b). According to Verma and Petrie (1980), oospores of *A. candida* (race 1) mixed with seeds of turnip rape, prior to sowing, result in a high infection rate of plants in both phases (leaf-staghead). Over 55% staghead infections can be induced by flower bud inoculation. However, Lakra and Saharan (1988b) could not initiate staghead infection on *B. juncea* by sporangial flower bud inoculations, but were able to obtain more than 66% staghead when 5 g/pot of oosporic inoculum was added with the seed at sowing time, combined with subsequent sporangial spray inoculation at the seedling, branching and flowering stages of the crop. Systemic infection was found to be directly correlated with host age. With increase in age there is a sharp decline in the ability to induce the systemic infection.

12.4.3 Physiological Specialization

Physiological specialization has long been noted in *A. candida*. It was Eberhardt (1904) who recognized the specialized groupings of *Albugo*, one attacking *Capsella*, *Lepidium* and *Arabis*, and the other attacking *Brassica*, *Sinapis* and *Diplotaxis*, but he was hesitant to use the phrase "biological forms" for this fungus. Later, Melhus (1911) suggested the existence of specialization in *A. candida*. Savulescu and Rayas (1930) distinguished eight morphological forms within the species of *A. candida*. Later, Savulescu (1946) divided *A. candida* into ten varieties on the basis of host specialization and morphology. Hiura (1930) in Japan distinguished three biological forms of *A. candida*, the first on *Raphanus sativus* the second on *B. juncea* and the third on *B. campestris* ssp. *chinensis*. Napper (1933) described 20 races of *A. candida* in the UK. Biga (1955) recognized two morphological taxa: *A. candida macrospora* and *A. candida microspora*, as proposed by Togashi and Shibaskaki (1934), but renamed *A. candida microspora* and *A. candida candida*.

It is clear that each of the above authors was hesitant in describing specialized races of *A. candida*. It was Pound and Williams (1963) who identified six races of *A. candida*; race 1 from *Raphanus sativus* var. Early Scarlet Globe, race 2 from *B. juncea* var. Southern Giant Curled, race 3 from *Armoracia rusticana* var. Common, race 4 from *Capsella bursa-pastoris*, race 5 from *Sisymbrium officinale*, and race 6 from *Rorippa islandica*. Later, race 7 from *B. campestris* turnip or Polish rapeseed and race 8 from *B. nigra* were added by Verma et al. (1975) and Delwiche and Williams (1977) respectively.

In India, Singh and Bhardwaj (1984) identified nine races from four host species, viz., *B. juncea*, *B. campestris* var. *toria*, *B. campestris* var. *brown sarson* and *B. campestris* ssp. *pekinensis* when tested on a differential set of 12 *Brassica* species. Lakra and Saharan (1988a) identified five races of *A. candida* on the basis of their reaction on a set of 16 host differentials. They have identified two distinct races from *B. juncea* which are different from previous records. One, (race 2) attacking *B. nigra*, *B. juncea* and *B. campestris* var. *brown sarson* and the other (race 3), infecting only *B. juncea* and *B. campestris* var. *toria*. The concept of races in *A. candida*, as proposed by Pound and Williams (1963), was based on species relationship. But recent studies have clearly demonstrated the need of including cultivars of *Brassica* crops in a set of host differentials to distinguish isolates of a pathogen within what is at present accepted as a race (Burdyukova 1980; Pidskalny and Rimmer 1985; Lakra and Saharan 1988a). There is an urgent need to standardize host differentials, keeping in mind the homogeneity and purity of species and varieties.

12.4.4 Sources of Resistance

Delwiche and Williams (1974) reported that all accessions in *B. napus* and most cultivars in *B. oleracea* are resistant to *A. candida*. According to Fan et al. (1983), the resistance in *B. napus* cv. Regent is governed by three genes, AC-7-1, AC-7-2, and AC-7-3. In India, Saharan et al. (1988) reported cvs. Tower 1, 2, 3, 4, Gullivar, Midas, Norin, Regent, H-715 and HNS-1 of *B. napus* as resistant to *A. candida*.

In *B. juncea* a number of resistant sources to *A. candida* have been reported, e.g. P.T. 347618 (Ebrahimi et al. 1976), T4 (Parui and Bandyopadhyay 1973), YRT3, Domo, Domo 4, Lethbridge, EC 126746, EC 126747, EC 126741, EC 126743, EC 126743-1, EC 126126. EC 129121, RC 781, RC 1001, RC 1401, RC 1405, RC 1408, RC 1424, RC 1425, RC 1449, RH 8541, RH 8542, RH 8543, RH 8544, RH 8545, RH 8546, RW-81-59, PHR-1, CSR-43, Metapolka, Zem, Newton, Blaze, Stoke, Purbiraya, etc. (Saharan 1984; Saharan et al. 1988; Lakra and Saharan 1989; Kolte 1987).

In *B. campestris* var. *yellow sarson* type 6 (Kolte and Tewari 1980), Tobin (Kolte 1987) and PYS-6 (Kolte and Tewari 1980). Whereas in *B. campestris* cvs. BSH-1 and BS-15 of *B. campestris* var. *brown sarson* were resistant to white rust pathogen (Kolte and Tewari 1980; Lakra and Saharan 1989).

All accessions of *B. alba* and HC 1, 2, 3, 4 and 5 of *B. carinata* were reported to be resistant to *A. candida* in India by Saharan et al. (1988).

12.4.5 The Genetics of Host-Parasite Interaction

The information on the genetics of host-parasite interaction in white rust has centred around the level of specificity between races of pathogens and genotypes of related host species. Even within the domain of race-cultivar specificity, the studies have been one-sided, since no genetic information has been published on the causal organism. Interest in such studies was originally aroused by the work of Hougas et al. (1952), who indicated the genetic control of resistance to white rust in horse radish. The exhaustive work of Williams and Pound (1963) clearly demonstrated that resistance to white rust in radish var. China Rose Winter (CRW) and Round Black Spanish (RBS) is controlled by a single dominant gene. Histological studies revealed that resistance in CRW is manifested as a hypersensitive reaction which may be modified to a sporulation-tolerant reaction by environmentally controlled minor genes. Humaydan and Williams (1976), while studying the inheritance in radish of resistance to *A. candida* (race 1) proposed to change the gene designation R into the more descriptive symbol AC-1, derived from the initials and race number of *A. candida*. The resistant reaction to *A. candida* (race 1) in *Raphanus sativus* var. *caudatus* is controlled by a single dominant gene AC-1. The resistant gene AC-1, and the gene Pi controlling pink pigmentation, were found to be linked with a recombination value of 3.28%. Bonnet (1981) found that white rust resistance in radish variety Rubigo 2 is controlled by one dominant gene.

Amongst *Brassica* species, monogenic dominant resistance to *A. candida* (race 2) has been found in *B. nigra*, *B. campestris*, *B. carinata*, and *B. juncea* (Delwiche and Williams 1974, 1976; Ebrahimi et al. 1976; Thukral and Singh 1986; Tiwari et al. 1988). A single dominant gene, AC-2, controlling resistance in *B. nigra* to *A. candida* (race 2) was identified by Delwiche and Williams in 1981. In a study to select quantitatively inherited resistance to *A. candida* (race 2) in *B. campestris*, CGS-1, Edwards and Williams (1982) found that variability for reaction to *A. candida* race 2 among susceptible *B. campestris*, DHW-Aaa-1 was due to quantitative genetic regulation, and suggested that rapid progress in resistance could be made via mass selection when starting with a susceptible base population.

According to Fan et al. (1983), Canadian cultivars of *B. napus* are resistant to white rust, while many cultivars of this species grown in China are susceptible. In India also some of the *B. napus* accessions are showing staghead infection. The inheritance of white rust resistance in *B. napus* cv. Regent is conditioned by independent dominant genes at three loci, designated AC-7-1, AC-7-2, and AC-7-3. The resistance is conferred by dominance at any one of the three loci, while plants with recessive alleles at all loci are susceptible.

12.4.6 Components of Partial Resistance

Partial resistance could be the result of low infection frequency, low spore production, a long latent period, and/or a short infectious period. According to Fox and Williams (1984), spore production by *A. candida* on *Brassica campestris* is highly correlated ($r = 0.93$) with a visual white rust infection phenotype (IP) rating scale. Plants are rated on a 0-9 scale according to the amount of leaf necrosis or area covered by white rust pustules. Means of spore production on plants rated as 1, 3, 5 and 7 are significantly different from each other ($P = 0.05$). These results support the use of a visual scale in selecting for components of partial resistance to the white rust disease.

12.4.7 Incorporation of Resistance

The results of interspecific hybridization between *B. juncea* and *B. carinata* envisaged the easy transfer of white rust resistance into high seed yield background through pedigree selection from an unadapted species, *B. carinata*, to the adapted species, *B. juncea* (Singh and Singh 1987; Singh et al. 1988). The resistance, which is monogenic, with complete dominance in *B. carinata* L., could be partially introgressed into *B. juncea* L. cultivars by selecting disease-free plants in advanced-segregating generations grown under heavy disease pressure, and their repeated backcrossing to *B. juncea* L. cultivars (Singh and Singh 1988). Some of the C-genome chromosome substitution lines of *B. juncea* developed through hybridizing *B. juncea* and *B. napus* are practically free from white rust infection (Banga 1988).

In a recent study involving inheritance of resistance to *A. candida* race 2 in mustard, Tiwari et al. (1988) found that resistance is dominant, monogenic, controlled by nuclear genes and that it could be easily transferred to adapted susceptible genotypes via backcrossing. Edwards and Williams (1987), through development of rapid-cycling *Brassica* populations, have opened unusual potential for resolving many problems relating to host-parasite relations, and breeding for disease resistance. Their preliminary studies indicated considerable isozyme variations among individuals of each population which, when inoculated with several pathogens, will show a wide range of plant-to-plant variation in the levels of resistance and susceptibility. This will be of great help to plant breeders involved in genetic resistance to plant diseases. These authors also claim that they are in the process of constructing gene pools of both major and minor genes for resistance to various crucifer pathogens, which will be of immense value to plant breeders seeking a source of resistance.

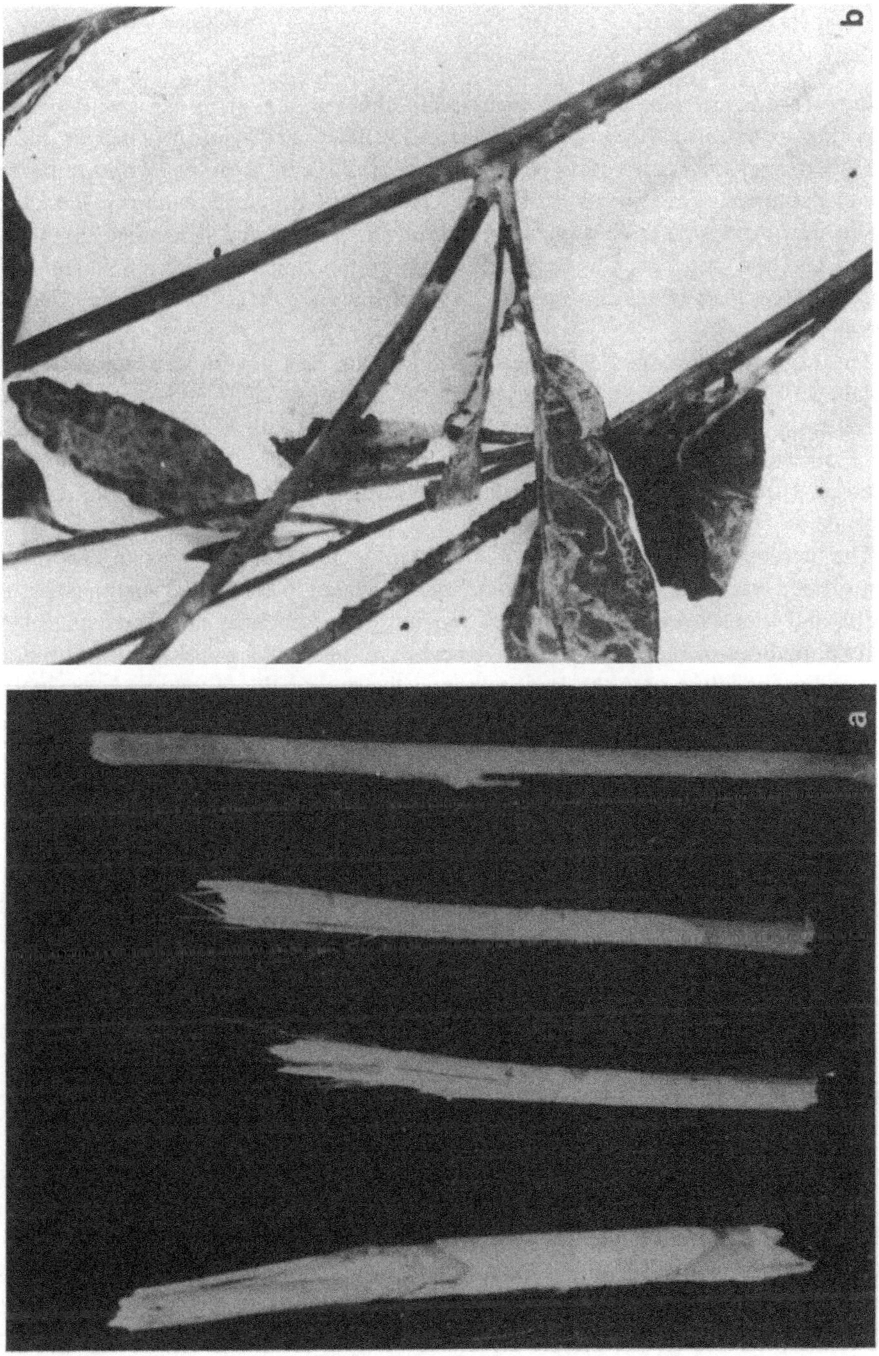

Fig. 3. **a**) *Sclerotinia* rot infection on *B. juncea* stem **b**) Powdery mildew infection on *B. juncea* stem

12.5 *Sclerotinia* Rot

Sclerotinia rot is the most common disease in temperate regions of the world, though it was first reported on oilseed *Brassica* crops by Shaw and Ajrekar (1915) in India. Since, then, frequent occurrences of the disease in a severe form have been reported from 11 countries in the world (Table 1). There may be great variation in losses in yield in the same area from year to year depending upon the percentage of plants infected and the stage of growth of the crop at the time of infection. Yield loss estimates have been made as high as 28% in rapeseed fields of Alberta, Canada (Morrall et al. 1976).

The causative pathogen *Sclerotinia sclerotiorum* has a vast host range, which, coupled with a lack of host specificity, makes the breeding of resistant varieties a difficult task. *B. napus* var. Isuzu from Japan (Iwata and Igita 1972) and var. Jetneuf from France (Renard and Brun 1982) have been reported to have a high degree of resistance. However, according to Bailey (1987), no current oilseed rape variety has resistance to *Sclerotinia* rot.

The literature reveals a diversity of screening methods for resistance to *Sclerotinia* rot in oilseed brassicas. In China, an artificial inoculation method is the spraying of mycelia and ascospores grown on PDA medium. Hu Baoching and Rimmer (1989) used two methods of inoculation in the greenhouse, one being to use parafilm to bind the mycelial inoculum onto the leaf stalk, and the other being a toothpick, on which mycelium was grown, penetrating the stem.

12.6 Blackleg or Stem Canker

The occurrence of stem canker on *Brassica* crops has been recorded from 11 countries in the world (Table 1). The disease is known to be widespread in Australia, Canada, UK and Germany, causing yield losses ranging from 13-50%, depending upon the virulence of the pathogenic strain (Kolte 1987).

12.6.1 Germplasm Screening Techniques for Testing Resistance

A number of methods for germplasm screening for resistance have been used by different workers. The germplasm screening can be done by planting seed in soil in which diseased *Brassica* plants have been grown or by scattering plant debris containing perithecia and pycnidia over young plants. Under greenhouse conditions, young seedlings can be inoculated by use of infected debris, by pipetting suspensions of ascospores or conidia onto cotyledons (Barbetti 1975; 1976; Wood and Barbetti 1977), by placing suspensions of conidia on the wound, by spraying conidia on the wound, or by spraying conidia on the whole plants (Alabouvette et al. 1974). The advantage of using conidia as inoculum rather than ascospores is that conidia can be produced from pycnidia formed in culture, whereas ascospores can be produced only from perithecia formed in the field, requiring more time. The relative response of a particular variety is the same, irrespective of whether the inoculum consists of conidia or ascospores. Helms and Cruickshank (1979) described a germination-inoculation technique where inoculum is put on the seeds, which has better chances of infection during the process of germination and the emergence of seedlings.

12.6.2 Physiological Specialization

Two strains of *L. maculans* which differ in virulence on host plants and in appearance on culture media have been described (Buddin 1934; Humperson-Jones and Ainsworth 1982). The more virulent form has caused outbreaks of dry rot in the UK and Cabbage blackleg in the USA (Petrie 1975). The virulent strain differs from the avirulent strain in pathogenicity and by formation of pycnidia in culture. The avirulent strain forms only superficial stem lesions and does not form pycnidia, and it produces distinct brownish yellow pigment in culture (McGee and Petrie 1978). According to Petrie (1975), out of three strains found on cruciferous weeds, i.e. *Thlaspi arvensis* L., *Sisymbrium* spp. and *Descurainia* and *Lepidium* sp., the first two are highly pathogenic on rape seedlings. Recently, Newman (1984) identified three virulent strains of *L. maculans* by their differential interaction on 13 cultivars of *B. napus* in Cambridge. *L. maculans* is highly variable. Isolates which vary in pathogenicity occur in western Australia, and new combinations of virulence genes have been obtained by crossing these isolates by Cargeeg (1980).

12.6.3 Sources of Resistance

Amongst the *Brassica* species, *B.juncea* appears to be more resistant to blackleg disease than *B. campestris*, *B. carinata* and *B. napus* (Helms and Cruickshank 1979; Roy and Wesreo 1978; Roy 1978). Seedling and adult plant resistance has been detected in the Novoski, Ceska and Zollerngold varieties of spring-type *B. napus* in Australia through greenhouse and field screening tests (Cargeeg and Thurling 1980). Roy (1978) reported seedling and adult plant resistance in the BJ 168 variety of *B. juncea*. A high level of resistance to blackleg has been reported in the French variety Ramses by Boudart and Lacoste (1972). However, Renard and Brun (1982) found the *B. napus* variety Jetneuf to be the most resistant and reported a *B. juncea* variety called Brown (Chinese) mustard as having good seedling resistance to this disease.

12.6.4 The Genetics of Host-Parasite Interaction

The variation in seedling resistance to blackleg between and within *B. napus* varieties is clearly continuous, suggesting that this resistance is determined by a polygenic system of stable horizontal type (Thurling and Venn 1977). Significant interaction between different *B. napus* varieties, viz. Wesreo, Zollerngold and Ceska and isolates of *L. maculans*, has also been reported (Cargeeg and Thurling 1980). The interaction is particularly significant for major components of resistance such as length of latent period and infection and for partial resistance in the field. According to Wratten and Murray (1982), control of resistance in rape and pathogeniciy in *L. maculans* is polygenic.

12.6.5 Biochemical and Structural Basis of Resistance

Differences of turnip rape varieties in their degree of resistance have been attributed to the production of two fungal toxicants. One appears to be the conversion product of a natural phenol in the host, and the other an induced one in the host by the pathogen after infection (Boudart and Lacoste 1972). Mithen et al. (1986) reported that levels of glucosinolates in leaves of oilseed rape correspond with levels sufficient to generate strongly antifungal hydrolysis products.

A comparative study of the anatomy of resistant and susceptible varieties indicated that in resistant variety an early differentiation of xylem inhibits the fungus (Brunin 1972). Hammond and Lewis (1980) reported that resistance to *L. maculans* in *B. napus* is associated with lignification of walls and rapid, specific accumulation of calcium.

12.6.6 Incorporation of Resistance

In an attempt to transfer resistance from *B. juncea* to *B. napus*, Roy (1978) recovered adult resistant fertile *napus*-type plants in the F_2 cross of *B. juncea* × *B. napus*, indicating the successful transfer of resistance between the two species, and the process suggests that genes for resistance in *B. juncea* must be located in genome A. Some lines of *B. carinata* show seedling as well as adult plant resistance. The genes for seedling resistance in this species lie on the B genome, common to *B. juncea* and *B. carinata*, but absent from *B. napus*, indicating less possibility to transfer seedling resistance from *B. carinata* to *B. napus* (Roy 1978).

Plants resistant or with reduced susceptibility to *L. maculans* could be regenerated from selected callus and embryogenic cultures of haploid rape previously treated with mutagens (Sacristan 1982).

In a population improvement approach for developing resistance to blackleg in rapeseed, Wratten and Murray (1982) established disease nurseries in New South Wales at several locations. At each site single plants were selected for resistance to basal stem canker. By utilizing cross-pollination, the different genes for resistance which these plants are expected to possess together with a recurrent selection system, a highly variable plant population can be developed and manipulated.

12.7 Clubroot

This occurs more frequently in soils which are acidic and poorly drained. Earlier, it was considered as an important general disease of vegetable crops of *Brassica* genus, but now it is known to cause considerable yield losses in eight countries (Table 1).

12.7.1 Germplasm Screening Techniques for Testing Resistance

For resistance breeding to clubroot, a number of inoculation techniques for testing *Brassica* germplasm have been used. For inoculum extraction and growth conditions after inoculation a method was suggested by Buczacki et al. (1975). The different inoculation techniques followed are:

1. The slurry method, in which 400 ml of the spore suspension is mixed with 6000 ml soil and the seeds are sown in the slurry (Toxopeus and Janssen 1975).
2. The modified slurry method, where after mixing of spore suspension and soil, 7-day-old seedlings are planted in the slurry.
3. A root-dip method proposed by Johnston (1968), in which the roots of 7-day-old seedlings are dipped in the spore suspension for 2 h.
4. Dipping of seed in spore suspension for 6 h before sowing was recommended.
5. An injection method was also suggested (cf. Schoeller and Grunewaldt 1987). In this, 7-day-old seedlings are planted in the non-infected soil. Close to each plantlet is a hollow, in which 2 ml of the spore suspension is added so that roots of the plantlets are contaminated with the suspension.

After making a comparison of these inoculation techniques, Schoeller and Grunewaldt (1987) reported that only with the modified slurry method, the root dip method and the injection method could 100 percent of the tested plants be infected.

Rod (1988) described a partly modified inoculation method which is easy and rapid to use under glasshouse or phytotron conditions.

12.7.2 Physiological Specialization

Variability in *P. brassicae* was first demonstrated by Honig (1931). Crute et al. (1980) have critically reviewed the information on the variation of the fungus. A uniform set of international differential hosts is now proposed for identification of physiological races of *P. brassicae*, which is referred to as the European Clubroot Differential (ECD) set (Buczacki et al. 1975). The set consists of 15 different host varieties, five each of *B. campestris*, *B. napus* and *B. oleracea*. Using the ECD set, 34 physiological races have been identified in Europe. Johnston (1968) described a technique which is useful for race surveys of the pathogen and in studies on the genetics of host resistance.

12.7.3 Sources of Resistance

Because of the highly variable nature of the pathogen-resistant sources used for breeding, a clubroot-resistant variety in one country may be completely susceptible to the strains of the pathogen derived from another country. Differential hosts used in the ECD set are resistant to some races and susceptible to others. Black mustard (*B. nigra*) is reported to be resistant due to the volatile mustard oil (Walker et al. 1937).

12.7.4 Incorporation of Resistance

Resistance to *P. brassicae* race 3 has been successfully transferred from the turnip rape (*B. campestris*) variety Wasslander to rape (*B. napus*) variety Nevin by synthesis of the fertile sp. *B. napocampestris* followed by two generations of backcrossing of Nevin (Johnston 1974). Gowers (1979) crossed four oilseed rape cultivars (Norde, Rapol, Lesire, Primor) with a clubroot-resistant *B. campestris* line from the ECD set. The allotriploid hybrids were backcrossed to the rape lines to incorporate resistance into oilseed rape. Using a combination of screening for resistance and chromosome number, a high proportion of 38-chromosome resistant selection has been obtained.

12.8 Conclusions

Oilseed brassicas are susceptible to attack by a large number of pathogenic diseases. Unfortunately, however, satisfactory genetic control is not yet available for most of the diseases. This is mainly due to the absence of adequate sources of resistance. Thus there is an urgent need to develop methods for identifying the resistant genes present in the wild species. Where such genes are not available, attempts must be made to utilize non-host resistance. This is as yet largely unexplored.

References

Alabouvette C, Brunin B, Louvet J (1974) Studies on rape disease caused by *Leptosphaeria maculans* (Desm.) Ces and De Not. III. *Pycnidiospores* infectivity and varietal susceptibility. Ann Phytopathol 64: 265

Anonymous (1980-1989) Ann Prog Rep, 1980-1989, AICORPO, ICAR, Direct Oilseeds Res, Hyderabad, India

Bailey DJ (1987) Screening for resistance to *Sclerotinia sclerotiorum* in oilseed rape using detached leaves. Tests Agrochem Cultiv 8: 152–153

Bains SS, Jhooty JS (1979) Mixed infection by *Albugo candida* and *Peronospora parasitica* on *Brassica juncea* inflorescence and their control. Indian Phytopathol 32: 268

Banga SS (1988) C-genome chromosome substitution lines in *Brassica juncea* (L.) Coss. Genetica 77: 81–84

Bansal VK, Seguin-Swartz G, Rakow GFW, Petrie GA (1990) Reaction of *Brassica* species to infection by *Alternaria brassicae*. Can J Plant Sci 70: 1159–1162

Barbetti MJ (1975) Effect of temperature on development and progression in rape of crown canker caused by *Leptosphaeria maculans*. Aust J Exp Agric Anim Husb 15: 705

Barbetti MJ (1976) Role of pycnidiospores of *Leptosphaeria maculans* in the spread of blackleg disease of rape. Aust J Exp Agric Anim Husb 16: 911

Bhander DS, Maini NS (1965) Studies on the resistance of oleiferous *Brassica* to *Alternaria* blight. Indian Oilseeds J 9: 58

Biga MLB (1955) Review of the species of the genus *Albugo* based on the morphology of the conidia. Sydowia 9: 339–358 (Italian)

Bonnet A (1981) Resistance to white rust in radish. (*Raphanus sativus* L.). Cruciferae Newslett 6: 60

Boudart G, Lacoste L (1972) Presence de substances antifongiques au cours de 1 infection du colza par *Leptosphaeria maculans* (Ces. et de Not). C R Acad Sci Paris 275

Brun H, Plessis J, Renard M (1987) Resistance of some crucifers to *Alternaria brassicae* (Berk) Sacc. In: Proc 7th Rapeseed Conf, Paris, pp 1222–1227

Brunin B (1972) Research on diseases of rape caused by *Leptosphaeria maculans* (Desm) Ces. and de Not III. Anatomical aspects of collar necrosis. Ann Phytopathol 4: 87

Buczacki ST, Toxopeus H, Mattusch P, Johnston TD, Dixon GR, Hobolth LA (1975) Study of physiologic specialization in *Plasmodiphora brassicae* proposals for attempted rationalization through international approach. Trans Br Mycol Soc 65: 295–303

Buddin W (1934) The canker and dry rot diseases of swedes. Minist Agric Fish London Bull 74

Burdyukova LI (1980) Albuginacea fungi. Taxonomy, morphology, biology, and specialization. Ukr Bot Zh 37: 65–74 (Russian)

Cargeeg LA (1980) Host-pathogen relationships of the blackleg disease in rape (*Brassica napus* and *Brassica campestris*) caused by *Leptosphaeria maculans*. Ph D Theses, Univ West Aust

Cargeeg LA, Thurling N (1980) Contribution of host pathogen interactions to the expression of the blackleg disease by *Leptosphaeria maculans* (Desm.) Ces. and De Not. Euphytica 29: 465

Chang IH, Shih NL, Chiu WF (1964) Preliminary studies on physiological differentiation of *Peronospora parasitica* affecting *Brassica pekinensis* and other cruciferous plants in Peking-Tientsing area. Acta Phytopathol Sin 7: 33–44

Conn KL, Tewari JP, Awasthi RP (1990) A disease assessment key for *Alternaria* blackspot in rapeseed and mustard. Can Plant Dis Surv 70: 19–22

Crute IR, Gray AR, Crip P, Buczacki ST (1980) Variation in *Plasmodiphora brassicae* and resistance to club root disease in *Brassicas* and allied crops - critical review. Plant Breed Abstr 50: 91

Delwiche PA, Williams PH (1974) Resistance to *Albugo candida* race 2 in *Brassica* sp. Proc Am Phytopathol Soc 1: 66 (Abstr)

Delwiche PA, Williams PH (1976) Identification of marker genes in *Brassica nigra*. Proc Am Phytopathol Soc 3: 234 (Abstr)

Delwiche PA, Williams PH (1977) Genetic studies in *Brassica nigra* (L.) Koch. Cruciferae Newslett 2: 39

Dickinson CH, Greenhalgh JR (1977) Host range and taxonomy of *Peronospora* on crucifers. Trans Br Mycol Soc 69: 111

Dixon GR (1975) The reaction of some oil rape cultivars to some fungal pathogens. Proc 8th Br Insect Fung Conf, 503–506

Dzhanuzakov A (1962) Specialization and variability in some *Peronosporaceous* fungi. Bot Zh Leningrad 47: 862–866

Eberhardt A (1904) Contribution à l'étude de *Cystopus candidus*. Lev Zentr Bakteriol Parasitenk 12: 235–249; 426–439; 614–631; 714–725

Ebrahimi AG, Delwiche PA, Williams PH (1976) Resistance in *Brassica juncea* to *Peronospora parasitica* and *Albugo candida* Race 2. Proc Am Phytopathol Soc 3: 273 (Abstract)

Edwards MD, Williams PH (1982) Selection for quantitatively inherited resistance to *Albugo candida* race 2 in *Brassica campestris*, CGS-1. Cruciferae Newslett 7: 66–67

Edwards MD, Williams PH (1987) Selection of minor gene resistance to *Albugo candida* in rapid-cycling population of *Brassica campestris*. Phytopathol 77: 527–532

Fan Z, Rimmer SR, Stefansson BR (1983) Inheritance of resistance to *Albugo candida* in rape (*Brassica napus* L). Can J Genet Cytol 25: 420–424

Fox DT, Williams PH (1984) Correlation of spore production by *Albugo candida* and a visual white rust rating scale. Can J Plant Pathol 6: 175–178

Gaumann E (1926) On the specialization of downy mildew *Peronospora brassicae* Gm on cabbage and related species. Landwirtsch Jahrb Schweiz 40: 463–468

Gowers S (1979) Clubroot-resistant oilseed rape. Cruciferae Newslett 4: 26–27

Grontoft M, O'Connor D (1990) Greenhouse method for testing of resistance of young *Brassica* plants to *Alternaria brassicae*. Plant Breed 105: 160–164

Gugel RG, Seguin-Swartz G, Petrie GA (1990) Pathogenecity of three isolates of *Phoma lingam* (Tode ex Fr.) Desm. on *Brassica* species and other curcifers. Can J Plant Pathol 12: 75–82

Gupta SK, Kumar P, Yadav TP, Saharan GS (1984) Changes in phenolic compounds, sugars and total nitrogen in relation to *Alternaria* leaf blight in Indian mustard. Haryana Agric Univ J Res 14: 535–537

Hammond KE, Lewis BG (1980) Ultrastructural studies of the limitation of lesions caused by *Leptosphaeria maculans* in stems of *Brassica napus*. Physiological Mol Plant Pathol 20: 251–265

Helms K, Cruickshank IAM (1979) Germination-inoculation technique for screening cultivars of oilseed rape and mustard for resistance of *Leptosphaeria maculans*. Phytopathol Z 95: 97

Hiura M (1930) Biologic forms of *Albugo candida* (Pers.) Kuntze on some cruciferous plants. Jpn J Bot 5: 1-20

Hiura M, Kanegae H (1934) Studies on the downey mildew of cruciferous vegetables in Japan. Trans Sapporo Nat Hist Soc 13: 125–133

Honig F (1931) Der Kohlkronpferreger (*Plasmodiophora brassicae*) Woronin Edna Monographic. Gartenbauwissenschaft 5: 116–229

Hougas RW, Rieman GH, Stokes GW (1952) Resistance to white rust in horseradish seedlings. Phytopathol 42: 109–110

Hu Baoching, Rimmer SR (1989) Preliminary study of artificial inoculation for resistance (tolerance) to *Sclerotinia sclerotiorum* in rapeseed using detached leaves. Anhui Agric Sci 11: 56–58

Humaydan HS, Williams PH (1976) Inheritance of seven characters in *Raphanus sativus* L. Hortic Sci 11: 146–147

Humperson-Jones FM, Ainsworth IF (1982) Canker of *Brassicas*. Ann Rep 1981 Plant Breed Inst Cambridge, p 66

Iwata I, Igita K (1972) On the growth characteristics of direct sowing rape on upland field. Bull Kyushu Agric Exp Stn 16: 207

Johnston TD (1968) Club root in *Brassica*: a standard inoculation technique and the specification of races. Plant Pathol 17: 184–187

Johnston TD (1974) Transfer of disease resistance from *Brassica campestris* L. to rape (*B. napus*). Euphytica 23: 681

Jönsson R (1966) Peronospora on oil-yielding brassicas: methods for testing resistance in winter rape and their results. Sver Utsaedesfoeren Tidskr 76: 54

Kadian AK, Saharan GS (1983) Symptomathology, host range and assessment of yield losses due to *Alternaria brassicae* infection in rapeseed and mustard. Indian J Mycol Plant Pathol 13: 319–323

Kolte SJ (1985) Diseases of annual edible oilseed crops. Vol II Rapeseed-mustard and sesame diseases. CRC, Boca Raton Florida, 135 pp

Kolte SJ (1987) Important diseases of rapeseed and mustard in India—present research progress and future research needs. Proc 3rd oil crops Network Workshop, Ethiopia, pp 91–106

Kolte SJ, Tewari AN (1980) Note on the susceptibility of certain oleiferous Brassicaceae to downey mildew and white blister diseases. Indian J Mycol Plant Pathol 10: 191–192

Lakra BS, Saharan GS (1988a) Morphological and pathological variations in *Albugo candida* associated with *Brassica* species. Indian J Mycol Plant Pathol 18: 149–156

Lakra BS, Saharan GS (1988b) Correlation of leaf and staghead infection intensities of white rust with yield and yield components of mustard. Indian J Mycol Plant Pathol 18: 157–163

Lakra BS, Saharan GS (1989) Sources of resistance and effective screening techniques in *Brassica-Albugo* system. Indian Phytopathol 42: 293

Lucas JA, Crute IR, Sheriff C, Gordon PL (1988) The identification of a gene for race specific resistance to *Peronospora parasitica* (downy mildew) in *Brassica napus* var. *oleifera* (oilseed rape). Plant Pathol 37: 538–545

Mayee CD, Datar VV (1986) Host gene and disease assessment scales. In: Phytopathometry. Marathwada Agric Univ Parbhani, India, pp 110–112

McGee DC, Petrie GA (1978) Variability of *Leptosphaeria maculans* in relation to blackleg of oilseed rape. Phytopathol 68: 625
Melhus IE (1911) Experiments on spore germination and infection in certain species of oomycetes. Wis Agric Ex Stn Res Bull 15: 25–91
Mithen RF, Lewis BG, Fenwick GR (1986) In-vitro activity of glucosinolates and their products against *Leptosphaeria maculans*. Trans Br Mycol Soc 87: 433–440
Morrall RAA, Dueck J, McKenzie DI, McGee DC (1976) Some aspects of *Sclerotinia sclerotiorum* in Saskatechewan 1970-75. Can Plant Dis Surv 56: 56
Moss NA, Crute IR, Lucas JA, Gordon PL (1988) Requirements for analysis of host-species specificity in *Peronospora parasitica* (downy mildew). Cruciferae Newslett 13: 114–116
Napper ME (1933) Observations on spore germination and specialization of parasitism in *Cystopus candidus*. J Pomol Hortic Sci 11: 81–100
Natti JJ, Dickson MH, Atkin JD (1967) Resistance of *Brassica oleracea* varieties to downy mildew. Phytopathol 57: 144–147
Newman PL (1984) Differential host-parasite interactions between oilseed rape and *Leptosphaeria maculans* the causal fungus of stem canker. Plant Pathol 33: 205–210
Parui MR, Bandyopadhyay D (1973) A note on screening of *rai* (*Brassica juncea* L. Coss). Curr Sci 42: 798
Petrie GA (1975) Diseases of rapeseed and mustard. In: Harapiak J (ed) Oilseed and pulse crops in Western Canada. West Coop Fert, Calgary, pp 399–413
Petrie GA, Vanterpool TC (1974) Fungi associated with hypertrophies caused by infection of Cruciferae by *Albugo cruciferarum*. Can Plant Dis Surv 54: 37–42
Pidskalny RS, Rimmer SR (1985) Virulence of *Albugo candida* from turnip rape (*Brassica campestris*) and mustard (*Brassica juncea*) on various crucifers. Can J Plant Pathol 7: 283–286
Pound GS, Williams PH (1963) Biological races of *Albugo candida*. Phytopathol 53: 1146–1149
Primard C, Vedel F, Mathieu C, Pelletier G, Chevre AM (1988) Interspecific somatic hybridization between *Brassica napus* and *Brassica hirta* (*Sinapis alba* L). Theor Appl Genet 75: 546–552
Rai B, Kolte SJ, Tewari AN (1977) Evaluation of oleiferous *Brassica* germplasm for resistance to *Alternaria* leaf blight. Indian J Mycol Plant Pathol 6: 76
Renard M, Brun H (1982) Screening for resistance to *Phoma lingam* and *Sclerotinia sclerotiorum* in *Brassica napus*. St Amelioration des Plantes (INRA) Le Rheu France, pp 137–147
Rod J (1988) New inoculation method for the testing of Brassicaceae for resistance to *Plasmodiophora brassicae*. Cruciferae Newslett 13: 102–103
Romero-Munoz F, Jimencz-Diaz RM (1980) Black spot: a disease of turnip rape recently recorded in Spain. Rev Plant Pathol 59: 2399
Roy N (1978) Study on disease variation in the populations of an interspecific cross of *Brassica juncea* L. × *Brassica napus* L. Euphytica 27: 145
Roy N, Wesreo A (1978) A blackleg-resistant rapeseed. J Agric West Aust 19: 12
Rozej A (1976) The susceptibility of different varieties of winter rape (*B. napus* var. *oleifera*) to infection by *Alternaria brassicae*. Rev Plant Pathol 55: 4336
Sacristan MD (1982) Resistance responses to *Phoma lingam* of plants regenerated from selected cell and embryogenic cultures of haploid *Brassica napus*. Theor Appl Genet 61: 193–200
Saharan GS (1984) A review of research on rapeseed-mustard pathology in India. Ann Workshop AICORPO, ICAR, Jaipur 6–10 Aug 1984
Saharan GS, Chand JN (1988) Diseases of oilseed crops. Haryana Agric Univ Hissar 268 pp
Saharan GS, Kadian AK (1983a) Physiologic specialization in *Alternaria brassicae*. Cruciferae Newslett 8: 32-33
Saharan GS, Kadian AK (1983b) Analysis of components of horizontal resistance in rapeseed and mustard cultivars against *Alternaria brassicae*. Indian Phytopathol 36: 503–507
Saharan GS, Kaushik JC, Kaushik CD (1981) Progress of *Alternaria* blight on *raya* cultivars in relation to environmental conditions. Abstr 3rd Int Symp Plant Pathol New Delhi, Dec 14-18 1981, p 136
Saharan GS, Kaushik CD, Gupta PP, Tripathi NN (1984) Assessment of losses and control of white rust of mustard. Indian Phytopathol 37: 397 (Abstr)
Saharan GS, Kaushik CD, Kaushik JC (1988) Sources of resistance and epidemiology of white rust of mustard. Indian Phytopathol 41: 96–99

Savulescu O (1946) Study of the species of *Cystopus* (Pers) Lev Bucharest, 1946 Anal Acad Rous Mem Sect Stirntiface Soc 21: 13
Savulescu O, Rayas T (1930) Contribution to the knowledge of the *Peronosporaceae* of Romania. Ann Mycol 28: 297–320
Schoeller M, Grunewaldt J (1987) Comparison of inoculation techniques for testing *Brassica* seedling resistance against *Plasmodiophora brassicae* Wor. Cruciferae Newslett 12: 86–87
Shaw FJF, Ajrekar SL (1915) The genus *Rhizoctonia* in India. Mem Dep Agric India Bot Ser 7: 177–194
Sheriff C, Lucas JA (1987) Variation in host specificity in the *Brassica* population of *Peronospora parasiticia*. In: Day PR, Jellis GJ (eds) Genetics and plant pathogenesis. Sci publ., pp 333–335
Singh BM, Bhardwaj CL (1984) Physiologic races of *Albugo candida* on crucifers in Himachal Pradesh. Indian J Mycol Pl Pathol 14: 25
Singh D, Singh H (1988) Inheritance of white rust resistance in interspecific crosses of *Brassica juncea* L × *B carinata* L. Crop Res 1: 189–193
Singh D, Singh H, Yadava TP (1988) Performance of white rust (*Albugo candida*) resistance genotypes developed from interspecific crosses of *B. juncea* × *B carinata*. Curciferae Newslett 13: 110–111
Singh D, Singh J (1983) A technique for inoculating *Albugo candida* on Lahi (Indian mustard). Indian Phytopathol 36: 139–140
Singh H, Singh D (1987) A note on transfer of resistance to white rust from Ethiopian mustard to Indian mustard. Cruciferae Newslett 12: 95
Skoropad WP, Tewari JP (1977) Field evaluation of the role of epicuticular wax in rapeseed and mustard in resistance to *Alternaria* black spot. Can J Plant Sci 57: 1001
Stankova J (1975) Varietal variability of winter rape in its susceptibility to dark leaf spot and the factors influencing the development of the disease. Rev Plant Pathol 54: 370
Tewari JP, Skoropad WP (1976) Relationship between epicuticular wax and black spot caused by *Alternaria brassicae* in three lines of rapeseed. Can J Plant Sci 56: 781
Tiwari AS, Petrie GA, Downey RK (1988) Inheritance of resistance to *Albugo candida* race 2 in mustard (*Brassica juncea* (L) Czern. Can J Plant Sci 68: 297–300
Thukral SK, Singh H (1986) Inheritance of white rust resistance in *Brassica juncea*. Plant Breed 97: 75–77
Thurling N, Venn IA (1977) Variation in the response of rapeseed (*Brassica napus* and *B. campestris*) cultivars to blackleg (*Leptosphaeria maculans*) infection. Aust J Exp Agric Anim Husb 17: 445
Togashi K, Shibasaki K (1934) Biometrical and biological studies of *Albugo candida* (pers) O Kuntze in connection with its specialization. Bull Imp Coll Agric For (Morioka Japan), pp 18: 88
Toxopeus H, Janssen AMP (1975) Clubroot resistance in turnip. II. The slurry screening method and clubroot races in the Netherlands. Euphytica 24: 751–755
Tripathi NN, Kaushik CD, Yadav TP, Yadav AK (1980) *Alternaria* leaf spot resistance in *raya*. Haryana Agric Univ J Res (India) 10 (2): 166
Verma PR, Harding H, Petrie GA, Williams PH (1975) Infection and temporal development of mycelium of *Albugo candida* in cotyledons of four *Brassica* spp. Can J Bot 53: 1016–1020
Verma PR, Petrie GA (1978) A detached-leaf culture technique for the study of white rust disease of *Brassica* species. Can J Plant Sci 58: 69–73
Verma PR, Petrie GA (1980) Effect of seed infestation and flower bud inoculation on systemic infection of turnip rape by *Albugo candida*. Can J Plant Sci 60: 267–271
Verma VD, Rai B (1980) Note on induced mutagenesis for spotting out the sources of resistance to *Alternaria* leaf spot in Indian mustard. Indian J Agric Sci 50: 278
Walker JC, Morell S, Foster HH (1937) Toxicity of mustard oil and related sulphur compounds to certain fungi. Am J Bot 24: 536
Wang CM (1944) Physiological specialization in *Peronospora parasitica* and reaction of hosts. Chin J Sci Agric 1: 249–257
Williams PH (1985) Resource Book. CrGC Crucifer Genet Coop Univ Wis, Madison

Williams PH, Pound GS (1963) Nature and inheritance of resistance to *Albugo candida* in radish. Phytopathol 53: 1150–1154

Wood PR, Barbetti MJ (1977) A study on the inoculation of rape seedlings with ascospores and pycnidiospores of the blackleg disease causal agent *Leptosphaeria maculans*. J Aust Inst Agric Sci 43: 79

Wratten N, Murray GM (1982) A population improvement approach for developing resistance to blackleg in rapeseed. Cruciferae Newslett 7: 62

CHAPTER 13

Insect Resistance with Special Reference to Mustard Aphid

B.S. Sekhon and *Inger Åhman*

13.1 Introduction

Cruciferous oilseeds are attacked by a large number of insect pests (Lamb 1989). Blossom beetles, *Meligethes* spp., and other species of Coleoptera like *Psylliodes chrysocephala* and *Ceutorhynchus* spp. are severe pests in Europe. Chrysomelid beetles, *Phyllotreta* spp. and *Entomoscelis americana*, are the important pests in Canada. The Indian mustard aphid, *Lipaphis erysimi* (Fig. 1) is by far the most devastating insect on *Brassica* oilseeds in India. The aphid is reported to cause an average loss of about 50% in seed yield (Bakhetia 1983). *Bagrada cruciferarum* and *Athalia proxima* on germinating and young crop, and *Agrotis* sp. on rainfed crop, are other sporadic pests in India (Sekhon, unpubl.). Out of the 38 insect pests attacking rapeseed and mustard crops in India, ten are of particular economic significance (Bakhetia 1987). The cabbage aphid, *Brevicoryne brassicae*, a pest on *Brassica* vegetables in temperate and warm temperate parts of the world, attacks *Brassica* oilseeds only rarely (Blackman and Eastop, 1985). The detailed list of the important pests of brassicas is presented in Table 1.

Differential resistance to some of these pests has been reported and, in general, *Brassica* species with the 'B' genome, namely *B. nigra* (BB), *B. juncea* (AABB) and *B. carinata* (BBCC), have a comparatively greater degree of resistance to some of the pests. For example, these *Brassica* species are less attacked by ovipositing females of cecidomyid *Dasineura brassicae*, a pest on *Brassica* oilseeds in Europe. These species are also less suitable for larval growth of *Dasineura brassicae* than *B. napus* and *B. campestris* (Åhman 1986a). *B. nigra* was also found inferior to *B. campestris* as a larval host for the red turnip beetle (*E. americana*) in Canada, whereas *B. juncea*, *B. napus* and *B. campestris* were equally suitable (Gerber and Obadofin 1981a, b). In India, *B. nigra*, *B. juncea* and *B. carinata* are less infested by the aphid *L. erysimi* than *B. campestris* (Rai and Sehgal, 1975; Prasad and Phadke 1980; Malik 1981; Bakhetia 1987; Ronquist and Åhman 1990). The majority of insect pests can be controlled satisfactorily by insecticides and thus no great efforts have been made to breed varieties resistant to insects (Rogers 1980). However, the severity of *L. erysimi* as a pest, the relatively high cost of insecticides, the concern for pesticidal hazards and some reports regarding pesticide resistance in aphids have promoted such efforts in India. Similarly, attempts are underway in Canada to refine techniques for evaluating resistance to the economically important flea beetle in crucifer seedlings (Lamb 1988). The use of pest-resistant varieties is the best approach to tackle the menace of crop pests. Their use is environmentally safe, cost-effective and is generally compatible with other pest management tactics such as chemical, biological and cultural control.

Fig. 1. Heavy infestation of mustard aphid on *Brassica juncea* shoot

Above all, it is an easy and rapid mean to transfer the gains of development of resistant varieties to farmers, especially in the third world countries. The development of resistant varieties is based on (1) identifying the source of resistance, (2) understanding the genetic basis of resistance, and (3) incorporating the heritable and stable resistance factor(s) into a good agronomic base.

13.2 Screening Techniques

The evaluation of germplasm for insect resistance is generally based upon the following factors:

1) size of insect population and/or
2) assessment of insect damage to plants.

Only for the mustard aphid have major attempts been made to devise suitable screening techniques. Differential developmental patterns of various *Brassica* species significantly affect the assessment of their pest susceptibility. For example, *toria* (*B. campestris*) matures early and normally escapes the attack of *L. erysimi*. On the other

Table 1. List of important pests of crop brassicas (after Lamb 1989)

Pest	Distribution	Reference
Coleoptera		
Meligethes aeneus	Europe	Nilsson (1987)
Phyllotreta cruciferae	Canada, India	Rai (1976) Burgess (1977)
Phyllotreta striolata	Canada, Europe	Augustin et al. (1986) Burgess (1977)
Phyllotreta undulata	Europe	Augustin et al. (1986)
Psylliodes chrysocephala	Europe	Alford (1979) Gerber
Entomoscelis americana	Canada	and Obadofin (1981a)
Ceutorhynchus assimilis	Europe	Free and Williams (1978)
Ceutorhynchus quadridens	Europe	Winfield (1963)
Diptera		
Dasineura brassicae	Europe	Gould (1975)
Phytomyza horticola	India	Rai (1976)
Delia radicum	Canada	Griffiths (1986)
Heteroptera		
Lygus lineolaris	Canada	Lamb (1989)
Lygus elisus	Canada	Lamb (1989)
Bagrada cruciferarum	India	Rai (1976)
Homoptera		
Brevicoryne brassicae	NZ/Australia	Lammerink (1968)
Lipaphis erysimi	India	Rai (1976)
Hymenoptera		
Athalia lugens proxima	India	Rai (1976)
Lepidoptera		
Pieris brassicae	India	Rai (1976)
Diacrisia obliqua	India	Rai (1976)
Mamestra configurata	Canada	Turnock and Philip (1977)
Plutella xylostella	India, Canada, NZ/Australia	Rai (1976) Lee (1968) Palaniswamy et al. (1986)
Agrotis flammatra	India	Sekhon (1990 unpubl.)

Fig. 2. **a** Crucifer flea beetles on *Brassica napus* pods. **b** Red turnip beetle larva feeding on *B. napus* leaves

hand, *B. napus* develops slowly and late in the season, so that hot weather more than host suitability may perhaps put constraints on the performance of the aphid (Bakhetia et al. 1986; Landin and Wennergren 1987). Similarly, the asynchrony in the development of *Brassica* cultivars may be the reason for the lack of correspondence between field and laboratory data on host performance of *Meligethes aeneus* (Charpentier 1985). The differences in the phenologies between annual and biennual overwintering forms of *B. napus* and *B. campestris* are even more extreme and, due to this, the composition of associated pests also differs (Lamb 1989). The literature on the screening techniques for aphid resistance has been reviewed extensively by Bakhetia and Sekhon (1989). Various criteria used in screening include the extent of aphid injury, aphid population, fecundity and longevity of aphids and yields of the cultivars. The important techniques are discussed here.

13.2.1 Screening at the Seedling Stage

From the point of view of breeding for aphid resistance, the seedling screening technique is a welcome substitute for the more laborious and time-consuming adult plant screening. Initial attempts to devise such a technique were made by Jarvis (1969, 1970). Unfortunately, no attempt was made to correlate seedling resistance with adult plant resistance. In a more elaborate study, Bakhetia and Bindra (1977) tried to develop a seedling screening methodology which is compatible with adult plant evaluation. This is based on seedling/plant mortality at a defined aphid population. To further confirm these results the survival and fecundity of the foraging mustard aphids were studied. Population levels of 11,20,20,30 apterae and 1 ml and 3 ml aphids (1 ml = about 600 nymphs + apterae) per plant proved optimal for resistance screening at the cotyledonary, two-leaf, four-leaf, six-leaf, flower bud initiation and flowering stages, respectively. Significantly, the results of screening obtained at all test stages were comparable when screening was conducted under the optimum level of aphid population per plant. The effect on the survival and fecundity of aphids was also similar at all the stages studied. Despite the apparent advantages, no published evidence is available regarding the use of this technique so far.

13.2.2 Screening at the Adult Plant Stage

Screening at the adult plant stage is generally based on easily observable aphid injury symptoms on plants. These include: yellowing, curling and crinkling of the leaves and flower bud abortion. Although different grading systems have been adopted by different workers, the system suggested by Bakhetia and Sandhu (1973) was found most practical for screening at the adult plant stage under field conditions. In this system, the varieties/strains are classified into six grades as described below:

Infestation score	Description
0	Completely free from aphid infestation
1	Normal growth, no visible symptoms of injury like curling or yellowing of the leaves, only a few aphids
2	Average growth, curling and yellowing of a few leaves, average flowering and pod setting on almost all the branches
3	Growth below average, curling and yellowing of the leaves on some branches, poor flowering and pod setting
4	Very poor growth, heavy curling and yellowing of the leaves, little or no flowering with very few well formed pods, heavy aphid colonies on plants
5	Severe stunting of the plants, curling, crinkling and yellowing of almost all the leaves, no flowering and pod formation, plants full of aphids.

Every plant observed is given a specific grade and an infestation index is worked out by multiplying the number of plants falling under each grade with the respective grade number. Indices are calculated for pre-flowering, flowering and pod-formation stages.

The aphid infestation index is calculated as:

$$\text{Aphid infestation index} = \frac{0\times a + 1\times b + 2\times c + 3\times d + 4\times e + 5\times f}{a+b+c+d+e+f},$$

where *a*, *b*, *c*, *d*, *e*, *f* are the number of plants falling in each grade.

The aphid infestation index is taken as the estimate of aphid resistance in the test entry, higher the index number, lower the host resistance.

Although the technique has been frequently used to screen large germplasm collections, it has a limited application to screen a segregating population. Grading based on host morphological symptoms resulting from starvation stress or mechanical damage is also likely to be influenced by both macro and micro-environmental conditions. The level of heterozygosity of the host plant will also modulate the plant response to the external stresses. The majority of the heterozygous plants, primarily due to hybrid vigour, tend to show a greater tolerance and may be wrongly categorized as resistant. This is more so in allogamous or partially allogamous oilseed brassicas.

13.2.3 Other Screening Methods

Aphid population at a defined stage and an increase in aphid population during a given period have also been extensively used to screen large germplasm collections (Bakhetia and Sekhon 1989). Despite the significance of identifying sources of resistance based on aphid biology in relation to the host resistance, only sporadic attempts have been made to develop a technique essentially based on survival and fecundity of the mustard aphid. That it is possible to develop such a criterion is evident from the fact that the survival of nymphs, fecundity, longevity and reproduction of the mustard aphid is similar on all the plant growth stages (Bakhetia and Bindra 1977). Similarly, fecundity has been found to be inversely related to resistance (Singh et al. 1965; Malik 1981). With the recent advances in plant biotechnology, it might now be possible to screen for aphid resistance (based on antibiosis) straight away at plant cell colony (callus) level, followed by regeneration of only the resistant colonies.

13.3 Sources of Resistance

During the last three decades, a number of *Brassica* oilseed cultivars with moderate resistance to *L. erysimi* have been identified (Table 2).

These results are based on the aphid build up under field conditions. Thus the results tend to differ from place to place and year to year. Unfortunately, in none of these instances have the genetics of resistance been studied. The sawfly *Athalia proxima* is another serious pest of young cruciferous plants. Although the primary host (*B. campestris* var. *toria*) is highly susceptible, a few distantly related genera (*Eruca sativa*, *Iberis amara*, *Cheiranthus cheiri* etc.) are fairly resistant (Sehgal and Ujagir 1977). *B. campestris* and *B. juncea*, in general are susceptible to leaf miner, while *B. napus* and *tournefortii* are resistant (Singh et al. 1987).

Table 2. Strains/cultivars of cruciferous oilseeds reported to show resistance/tolerance to the mustard aphid

Strains of cultivars showing resistance/tolerance	Reference
Brassica juncea	
C 294, Laha 101	Teotia and Lal (1970)
T 6342	Bakhetia and Sandhu (1973)
RL 18, RLM 198	Bakhetia and Sandhu (1973)
Rai T 3	Brar et al. (1976)
P 26/21, RCU 10, RH 7326, RK 2, RLM 84, RLM 18, RLM 198	Bakhetia et al. (1984)
Rai 75-1, Pant Rai 34	Singh et al. (1982)
IB 680	Prasad (1983)
B-85, RH 785, RK 8, RLC 1010, KRV Tall, UVR 751, RLM 84, RLM 198, RLM 528	Bakhetia (1985)
T 6342, RH 7846, RH 7847	Singh et al. (1990)
Brassica napus	
Gullivar, GSL-1, Regent, Karat, GSB, GSC, GS-47	Bakhetia (1985)
Brassica campestris	
Pusa Kalyani	Singh et al. (1982)
IB 787	Prasad (1983)
Torpe, CDA-Span, Sariahi	Bakhetia (1985)
Shogoin	Barnes and Cuthbert (1975)
Roots, Charlestowne	Kennedy (1978)
Others	
Eruca sativa (unspecified strain)	Kundu and Pant (1967) Jarvis (1970)
E. sativa var. ITSA	Bakhetia and Bindra (1977)
Crambe abyssinica PI 247310	Jarvis (1969)
C. abyssinica (unspecified strain)	Bakhetia and Sandhu (1973)
B. carinata, *B. alba*, *B. napus*	Bakhetia and Sandhu (1973)
B. nigra, *B. integrifolia*	Singh et al. (1965)

13.4 Plant Characters Associated with Resistance

A good number of researchers have attempted to correlate morphological, anatomical and biochemical plant characteristics with insect resistance.

13.4.1 Morphological Characters

Plant surface characteristics may influence the suitability of plants as a host to the insect (Southwood 1986). The waxy coating on *B. napus* and *B. campestris* leaves may constitute a mechanical barrier to the entrance of leaf-mining larvae of *Phyllotreta nemorum* (Nielsen 1977), and hairs on the pods of *B. hirta* hinder adults of *P. cruciferae* from feeding (Lamb 1980). Furthermore, waxiness but not hairiness hinders *L. erysimi* from reaching the underside of the leaves, where it normally feeds if plants are in the vegetative stage (Ahman, 1990).

At the Asian Vegetable Research and Development Centre in Taiwan, four *B. juncea* accessions had moderate resistance to the mustard aphid (presumably *M. persicae* and *L. erysimi*). The leaves of these accessions are thicker and smoother than those of the susceptible ones, and their surfaces had higher concentrations of long carbon chain waxes, which seem to be associated with resistance to the aphid (Talekar 1980).

Contrarily on B 85, a *B. juncea* strain with glossy surface, the colony formation is disturbed compared to that on the normal waxy B 85 (Chatterjee and Sengupta 1987). The resistance to *L. erysimi* is further improved in a recombinant where the two traits, glossiness and white flower, were combined (Chatterjee, pers. commun.). Contradictory reports about the relationship of some other physical characters with aphid resistance are also available (Pathak 1961; Narang 1982). Plant morphological characters associated with aphid resistance are summarized in Table 3.

Table 3. Plant morphological characters associated with *L. erysimi* resistance in crucifers

Plant character	Plant species	Aphid reaction	Reference
Smaller and hardy inflorescence with loosely packed buds	*B. juncea*	Resistant	Rai and Sehgal (1975)
Darker leaves	*B. juncea* *B. campestris* var. *brown sarson*	Resistant	Pathak (1961)
Non-waxy plant	*B. rapa*	Highly resistant	Srinivasachar and Malik (1972)
More branches with wider angle of orientation	*B. juncea* *B. campestris*	Negative correlation with aphid population	Narang (1982)
Hairiness of plant	*Eruca sativa*	Non-preference	Narang (1982)

The leaf miner resistance of *B. tournefortii* was attributed to leaf hairiness, which results in non-preference for oviposition by flies (Singh et al. 1987).

13.4.2 Anatomical Characters

Phloem sap is the source of nutrition for the aphids which is drawn through their stylets put-forth intercellularly and intracellularly (Miles 1958). The stylets have to pass through epidermis, cortex, endodermis and pericycle. Densely packed and hard cell layers may prove a hindrance to the probing stylets. Such plant species may be rejected or less preferred as hosts (Henning 1966). Moderate resistance to aphids in some species like *B. carinata*, *B. alba* and *Eruca sativa* has been attributed to this factor (Malik 1981). The depth of the vascular bundles is perhaps the most important anatomical character. For feeding on the plant species with deeply localized vascular bundles, the aphids must have a longer stylet (Gibson 1972). While aphids with longer stylets will require more energy for probing deep into the plants, the aphids with short stylets will starve and die (Berlinski 1965). It has also been observed that leaf miner-resistant strains have thick leaves with compactly arranged parenchymatous cells and isolated intercellular spaces (Singh et al. 1983). Susceptible varieties, on the other hand, showed loose mesophyll with large air spaces.

13.4.3 Biochemical Characters

The presence or absence of certain chemical constituents has often been linked to insect resistance in plants (Hsiao 1969; Maxwell 1972). The effect of these chemicals could be stimulatory, antibiotic or toxic (Beland et al. 1970; Virtanen 1965; Tjallingii 1976) for growth and development of the insects. Host preference may depend on the nutritional status of the plant (Finch 1971). Differences in the availability of chemical substances can be both quantitative as well as qualitative. A number of studies aimed at understanding the biochemical factors involved in aphid resistance are summarized in Table 4.

Glucosinolates are the compounds characteristic of cruciferae and a few other plant families (Underhill 1980). These substances and their breakdown products have been demonstrated to stimulate plant approach, feeding and reproduction in a number of insects specialized for feeding on crucifers, whereas other herbivores are often adversely affected by them (Feeny 1977; Lerin 1980). Accordingly the two aphids, *L. erysimi* and *B. brassicae*, feeding primarily on crucifers, were induced to feed and reproduce on non-host leaves systemically treated with allylglucosinolate (Nault and Styer 1972). Feeding by the polyphagus aphid *Myzus persicae*, which includes the cruciferae in its host range, was stimulated to a lesser extent. Three other aphids which do not normally feed on crucifers were deterred by host leaves treated with allylglucosinolate. However, a few studies have suggested that even specialist crucifer feeders may suffer from high glucosinolate levels (Nayar and Thorsteinson 1963; Sömme and Rygg 1972; Louda and Rodman 1983; Åhman 1986b). A positive correlation between a high glucosinolate level and plant resistance to the mustard aphid has also been suggested (Malik 1981). This was further corroborated by a study where the food uptake and survival rate of the *L. erysimi* were reduced on an artificial diet containing 0.25 to 1% allylglucosinolate compared to 0.125% and plain diet (Dilawari and Atwal 1987). Unfortunately, it is difficult to compare these results with the natural situation, since the composition and concentration of glucosinolates encountered by aphids during plant probing and phloem sap feeding are still not known. Elaborate studies conducted by Malik (1981) indicate that the total glucosinolate content in general and sinigrin in particular have the ability to repel the aphids, and gluconapin to attract them. Contrarily, only in one out of several

Table 4. Biochemical factors in plants associated with *L. erysimi* resistance in cruciferous species

Biochemical constituents	Species	Reaction/response to aphid	Reference
Total sugar and sulphur contents	*B. alba* *B. juncea* *B. campestris*	Negative correlation with aphid population	Bakhetia et al. (1982)
Total sugar, phenolic compounds, potassium, sulphur and iron	*E. sativa*	Higher concentration imparted high resistance	Narang (1982)
Protein, free amino acid, reducing sugars and total sugars	20 cruciferous species	Higher concentration responsible for high susceptibility	Malik (1981)
Flavonoids, total sugars and reducing sugars	*B. napus*	Higher contents gave higher resistance	Gill and Bakhetia (1985)
Total ash, nitrogen and phosphorus	*B. napus*	Contents low in resistant varieties	Gill and Bakhetia (1985)
Glucosinolates, especially sinigrin	*B. juncea* *B. carinata* *B. oleracea*	Higher concentration responsible for resistance	Anand (1976)
Total glucosinolates; sinigrin as repellent and gluconapin as attractant	*Brassica* spp.	Negative correlation with aphid fecundity	Malik (1981)
Ascorbic acid and glucosinolate content	*Brassica* spp.	Negative correlation with aphid population	Labana et al. (1983)
Allylisothiocynate (1.0×10^4mg/min as the threshold level of activity)	*Brassica* spp.	Higher release rate act as repellent and lower rate as stimulant (olfactory stimuli)	Dilawari (1985)

studies did the insect responses to newly developed low glucosinolate cultivars of *B. napus* differ from higher glucosinolate cultivar types (Jonasson 1982; Lamb 1989). *Phyllotreta* sp. made more feeding marks on the cotyledons of low than of a high-glucosinolate cultivar of *B. napus*. The feeding rates on true leaves did not differ between the two cultivars (Jonasson 1982). This result was consistent with chemical analysis; the concentration of aliphatic glucosinolates differed between cotyledons but not between true leaves of low and high glucosinolate cultivars (Uppström 1983). Glucosinolates have also been implicated in resistance to leaf miner in *B. tournefortii* (Singh et al. 1985).

The availability of nitrogen may influence growth in insects feeding on plants (McNeill and Southwood 1978). However, to some extent the insects can compensate for a low nitrogen content in the host by increasing its feeding rate. *L. erysimi* suffered from poor growth and survival rate only at very reduced amino acid concentration, i.e. 25% of that in the syntheic standard diet (Sekhon, unpubl.). When the amino acid concentration was 50% of that in the standard diet, the aphids could still compensate by increasing their feeding rate. However, the situation may be different for aphids feeding on plants, since the phloem sap might also contain substances which are harmful at higher intake. There are even certain amino acids which are negatively correlated with the performance of *B. brassicae* and *M. persicae* (Emden 1973). Significant negative correlation between the contents of proline and γ-amino butyric acid (GABA) and relative growth rates of *Myzus persicae* were demonstrated in Brussels sprouts (Emden and Bashford 1971). Thus the composition of amino acids in *Brassica* plants might also be important for the performance of *L. erysimi*. The amino acid content in phloem sap of *Brassica* species and varieties has been analyzed, but the differences in composition and concentrations were small and did not correspond to the differences found between the cultivars in aphid performance.

Some workers have also tried to correlate resistance to aphids with the concentration of various other elements like copper, zinc, sulphur and total reducing sugars. Samples were collected from whole leaves and inflorescence. Results, however, are contradictory. Some other compounds such as cucurbitacin and cardenolides have been shown to be feeding inhibitors to *Phyllotreta* spp., while flavonol glycosides stimulated their feeding (Nielsen 1977; Nielsen et al. 1979).

The notion that the nutrient status of the host plant influences its suitability to insects is well documented, but reports on the differential response of insects to varying amounts of amino acids or glucosinolate concentrations are at the best only circumstantial. None of the studies was based on genotypes isogenic for resistant gene.

13.5 Breeding Methodology

Sporadic attempts, with little success, have been made to exploit intervarietal hybridization, induced mutagenesis or autotetraploidy for insect resistance. Rajan (1961) developed autotetraploids of toria (*B. campestris* var. *toria*). These autotetraploids were reported to be more resistant than diploids and the resistance was attributed to antibiosis. Autotetraploids of *toria* have little breeding value, as they are cytogenetically unstable. Artificially synthesized alloploids, *B. napus* (Prakash and Raut 1983), *B. campestris* × *Eruca sativa* (Agnihotri et al. 1990) and many others (Malik 1981) have not provided any single source for insect resistance.

Lammerink (1968) developed aphid (*B. brassicae*) resistant variety rangi of rape following selection in the F_3 generation of the cross (Broad Leaf Essex rape × Colder Swede) × giant rape. Recurrent selection was also attempted in the crosses involving purple top white Globe and Sjodin turnip for breeding an aphid (*L. erysimi*)-resistant variety. Aphid resistance was associated with non-preference, antibiosis and tolerance. Attempts to induce mutations for aphid resistance in *Brassica juncea* through chemical (Srinivasachar and Verma 1971) and physical mutagens (Srinivasachar and Malik 1972; Labana 1976) were largely inconclusive.

13.6 Conclusions

Even though some species and varieties among *Brassica* oilseeds are less susceptible to their major insect pests, their resistance is probably not sufficient to prevent significant yield losses due to insect attack. Recombinational breeding involving few of the resistant stocks has not been successful. Thus the scope of screening will have to be extended to include other cruciferous plants, which can be utilized for wide hybridization. Interestingly, in late 1987, a group in the United Kingdom reported success in transferring a gene from cowpea to tobacco (Hilder et al. 1987). This gene encodes a serine protease inhibitor. In tobacco the trypsin inhibitor gene imparted resistance to *Heliothis virescens*. Transferring such a gene from legumes to brassicas through the techniques of molecular gene transfer could be an attractive proposition. Similarly, the use of the Bt (*Bacillus thuringiensis*) gene will be an interesting avenue, especially for the chewing type of insects. Currently, transgenic plants with active Bt gene are being evaluated for resistance to a number of *Brassica* insects (Sen, pers. commun.). Development of suitable and rapid screening techniques must also receive due attention. Once truly resistant gene(s) become available, it will be possible to link them with RFLP (restriction fragment length polymorphism) markers. This itself will provide an easy and reliable seedling stage selection criterion.

References

Agnihotri A, Gupta V, Lakshmikumaran MS, Shivanna KR, Prakash S, Jagannathan V (1990). Production of *Eruca-Brassica* hybrid by embryo rescue. Plant Breed 104: 281–289

Åhman I (1986a) Oviposition in *Dasineura brassicae* Winn (Dipt: Cecidomyiidae) adaptive, mechanistic and applied aspects. Dep Plant Prot Swed Univ Agric Sci, Plant Prot Rep Diss 9: 1–25

Åhman I (1986b) Toxicities of host secondary compounds to eggs of the *Brassica* specialist *Dasineura brassicae*. J Chem Ecol 12: 1481–1488

Åhman I (1990) Plant surface characteristics and movements of two *Brassica*-feeding aphids, *Lipaphis erysimi* and *Brevicoryne brassicae*. Sym Biol, Hung 39: 119–125

Alford DV (1979) Observations on the cabbage stem flea beetle, *Psylliodes chrysocephala*, on winter oilseed rape in Cambridgeshire. Ann Appl Biol 93: 117–123

Anand IJ (1976) Oil quality vis a vis aphid resistance in Brassicae. Pap Semin Workshop Oilseed Res India; PKV Nagpur, April 5–9,

Augustin A, Tulisalo U, Korpela S (1986) Flea beetles (Coleoptera, Chrysomelidae, Halticinae) on rapeseed and sugarbeet in Finland. J Agric Sci. Finland 58: 69–82

Bakhetia DRC (1983) Losses in rapeseed-mustard due to *Lipaphis erysimi* (Kalt.) in India: Literature study. Proc 6th Int Rapeseed Conf, Paris, pp 1142–1147

Bakhetia DRC (1985) Summary report 1979-85 and invited paper. Joint rev Meet, Indo-Swed Collab Res Prog Rapeseed-mustard Improv New Delhi Feb 28- March, pp 83–90

Bakhetia DRC (1987) Insect pests of rapeseed-mustard and their management. In: Rao MV, Sithananthans (eds) Plant Protection in Field Crops pp 249–259

Bakhetia DRC, Bindra OS (1977) Screening techniques for aphid resistance in *Brassica* crops. SABRAO J 9: 91–107

Bakhetia DRC, Brar KS, Sekhon BS (1984) Screening of some *Brassica* spp. and their strains for resistance to mustard aphid *Lipaphis erysimi* (Kalt). J Oilseeds Res 1: 81–82

Bakhetia DRC, Brar KS, Sekhon BS (1986) Seasonal incidence of *Lipaphis erysimi* (Kalt.) on the *Brassica* species in the Punjab. In: Agarwala BK (ed) Aphidology in India, pp 20–36

Bakhetia DRC, Rani S, Ahuja KL (1982) Effect of sulphur nutrition on some *Brassica* plants on their resistance response to mustard aphid, *Lipaphis erysimi* (Kalt). Calicut Univ Res J, Special Conf No. May 3-5, p 36

Bakhetia DRC, Sandhu RS (1973) Differential response of *Brassica* species/varieties to the aphid, *Lipaphis erysimi* (Kalt.) infestation. J Res Punjab Agric Univ 10: 272–279

Bakhetia DRC, Sandhu RS (1977) Susceptibility of some *Brassica* species and *Crambe abyssinica* to the leaf miner *Phytomyza horticola* (Meigen). Crop Improv 4: 221–223

Bakhetia DRC, Sekhon BS (1989) Insect pests and their management in rapeseed-mustard. J Oilseeds Res 6: 269–299

Barnes WC, Cuthbert Jr FP (1975) Breeding turnip for resistance to turnip rape. Hortic Sci 10: 59–60

Beland GL, Akason, WR, Manglitz GR (1970) Influence of plant maturity and plant part on nitrate content of the sweet clover weevil-resistant species *Melilotus infesta.* J Econ Entomol 63: 1037–1039

Berlinski K (1965) Studies on food intake and the effects of food plants on the beet aphid—*A. fabae.* (Russian) Pol Pismo Entomol (B) (1-2): 163–168

Blackman RL, Eastop VF (1985) Aphids on the world's crops. An identification guide. John Wiley, New York, pp 247–248

Brar KS, Ratoul HS, Labana KS (1976) Differential reaction of mustad aphid *Lipaphis erysimi* (Kalt.) to different rapeseed and mustard varieties under natural and artificial infestation. J Res Punjab Agric Univ 13: 14–18

Burgess L (1977) Flea beetles (Coleoptera: Chrysomelidae) attacking rape crops in the Canadian prairie provinces. Can Entomol 109: 21–32

Charpentier R (1985) Host plant selection by the pollen beetle *Meligethes aeneus*. Entomol Exp Appl 38: 277–285

Chatterjee SD, Sengupta K (1987) Observations on reaction of mustard aphid to white petal and glossy plants of Indian mustard. J Oilseeds Res 4: 125–127

Dilawari VK (1985) Role of glucosinolates in host selection, development and multiplication of mustard aphid, Ph D, Theses Punjab Agric Univ Ludhiana, p 111

Dilawari VK, Atwal AS (1987) Effect of cruciferous glucosinolates on probing pattern and feed uptake by mustard aphid, *Lipaphis erysimi* (Kalt.). Proc Indian Acad Sci (Anim Sci) 96: 695–703

Emden HF Van (1973) Aphid host plant relationships. In: Lowe HJB (ed) Perspectives in aphid biology. Entomol Soc NZ Bull, pp 54–64

Emden HF Van, Bashford MA (1971) The performance of *Brevicoryne brassicae* and *Myzus persicae* in relation to plant age and leaf amino acids. Entomol Exp Appl 14: 349–369

Feeny P (1977) Defensive ecology of the Cruciferae. Ann M Bot Gon 64: 221–234

Finch S (1971) Comparison of the nutritive values of proteins and related compounds to adult cabbage root fly *Erioischia brassicae*. Entomol Exp Appl 14: 115–124

Free JB, Williams IH (1978) A survey of the damage caused to crops of oilseeds rape (*Brassica napus* L.) by insect pests in South Central England and their effect on seed yield. J Agric Sci Camb 90: 417–424

Gerber GH, Obadofin AA (1981a) Growth, development, and survival of the larvae of the red turnip beetle, *Entomoscelis americana* (Coleoptera: Chrysomelidae), on *Brassica campestris* and *B. napus* (Cruciferae). Can Entomol 113: 395–406

Gerber GH, Obadofin AA (1981b) The suitability of nine species of Cruciferae as hosts for the larvae of the red turnip beetle, *Entomoscelis americana* (Coleoptera: Chrysomelidae). Can Entomol 113: 407–413

Gibson RW (1972) The distribution of aphids on potato leaves in relation to vein size. Entomol Exp Appl 15: 213–223

Gill RS, Bakhetia DRC (1985) Resistance of some *Brassica napus* and *B. campestris* strains to *Lipaphis erysimi* (Kalt.). J Oilseeds Res 2: 227–239

Gould HL (1975) Surveys of pest incidence on oilseed rape in South Central England. Ann Appl Biol 79: 19–26

Griffiths GCD (1986) Relative abundance of the root maggots *Delia radicum* (L.) and *D. floralis* (Fallen) (Diptera: Anthomyiidae) as pests of Canola in Alberta. Quaest Entomol 22: 253–260

Henning E (1966) Zür Histologie und Funktion von Einstichen der Schwarzen Bohnenlaus (*Aphis fabae* scop.) in *Vicia faba* Pflanzen. J Insect physiol 12: 67–76

Hilder VA, Gatehouse AMR, Sheerman SE, Barker RF, Boulter D (1987) A novel mechanism of insect resistance engineered into tobacco. Nature (London) 330: 160–163

Hsiao TH (1969) Chemical basis of host selection and plant resistance in oligophagous insects. Entomol Exp Appl 12: 777–778

Jarvis JL (1969) Differential reaction of introductions of *Crambe* to the turnip aphid and the green peach aphid. J Econ Entomol 62: 697–698

Jarvis JL (1970) Relative injury to some cruciferous oilseeds by the turnip aphid. J Econ Entomol 63: 1498–1502

Jonasson T (1982) Feeding of flea beetles (*Phyllotreta* spp.) on two summer rape cultivars with different glucosinolate content. Entomol Tidskr 103: 140–142

Kennedy GG (1978) Recent advances in insect resistance of vegetable and fruit crops in North America: 1966-1977. Bull Entomol Soc Am 24: 375–384

Kundu GG, Pant NC (1967) Studies on *Lipaphis erysimi* (Kalt.) with special reference to insect plant relationship. I. Susceptibility of different varieties of *Brassica* and *Eruca sativa* to the mustard aphid infestation. Indian J Entomol 29: 241–251

Labana KS (1976) Release of mutant variety of *raya* (*Brassica juncea*). Mutat Breed Newslett J FAO/IAEA, Div At Energ Food Agri 7: 11

Labana KS, Ahuja KL, Gupta ML, Brar KS (1983) Preliminary studies on the chemical basis of resistance in *Brassica* species to mustard aphid (*Lipaphis erysimi*). Proc 6th Int Rapeseed Conf, Paris, pp 1132–1142

Lamb RJ (1980) Hairs protect pods of mustard (*Brassica hirta* Gisilba) from flea beetle feeding damage. Can J Plant Sci 60: 1439–1440

Lamb RJ (1988) Assessing the susceptibility of crucifer seedlings to flea beetles (*Phyllotreta* spp.) damage. Can J Plant Sci 68: 85–93

Lamb RJ (1989) Entomology of oilseed *Brassica* crops. Ann Rev Entomol 34: 211–229

Lammerink J (1968) Rangi: new rape that resists aphids. N Z J Agric 117: 61

Landin J, Wennergren U (1987) Temperature effects on population growth of mustard aphids. Swed J Agric Res 17: 13–18

Lee HS (1968) Evaluation of some granulated insecticides against the aphids and diamond back moth on rapeseed by soil treatment. Plant Prot Bull 10: 69–70

Lerin J (1980) Influence des substances allélochemiques des crucifères sur les insectes. Ecol Gene 1: 215–235

Louda SM, Rodman JE (1983) Ecological patterns in the glucosinolate content of a native mustard, *Cardamine cordifolia* in the Rocky Mountains. J Chem Ecol 9: 397–422

Malik RS (1981) Morphological, anatomical and biochemical basis of aphid, *Lipaphis erysimi* Kalt, resistance in cruciferous species. Sver Utsaedesfoeren Tidskr 91: 25–35

Maxwell FG (1972) Host plant resistance to insect—nutritional and pest management relationship. In: Rodriquez J G (ed) Insect and mite nutrition. North Holland, Amsterdam London, pp 158–161

McNeill S, Southwood TRE (1978) The role of nitrogen in the development of insect/plant relationships. In: Harborne J B (ed) Biochemical aspects of plant and animal coevolution. Academic Press, London, pp 77–98

Miles RW (1958) The stylet movements of a plant sucking bug *Oncopeltus fasciatus* (Dill) (Heteroptera, Lygaeidae). Proc R Entomol Soc London 33: 15–20

Narang DD (1982) Studies on the basis of resistance in *Brassica campestris* var. *yellow sarson*, *B. juncea* and *Eruca sativa* to *Lipaphis erysimi* (Kalt). Ph D Theses Punjab Agric Univ, p 103

Nault LR, Styer WE (1972) Effects of sinigrin on host selection by aphids. Entomol Exp Appl 15: 423–437

Nayar JK, Thorsteinson AJ (1963) Further investigations into the chemical basis of insect-host plant relationships in an oligophagous insect, *Plutella maculipennis* (Curtis) (Lepidoptera: Plutellidae). Can J Zool 41: 923–929

Nielsen JK (1977) Host plant relationships of *Phyllotreta nemorum* L (Coleoptera: Chrysomelidae). I Field studies. Z angew Entomol 84: 396–407

Nielsen JK, Larsen LM, Sørensen H (1979) Host plant selection of the horseradish flea beetle *Phyllotreta armoraciae* (Coleoptera: Chrysomelidae). Identification of two flavonol glycosides stimulating feeding in combination with glucosinolates. Entomol Exp Appl 26: 40–48

Nilsson C (1987) Yield losses in summer rape caused by pollen beetles (*Meligethes* spp.). Swed J Agric Res 17: 105–111

Palaniswamy P, Gillott C, Slater GP (1986) Attraction of diamond back moths, *Plutella xylostella* (L.). (Lepidoptera: Plutellidae) by volatile compounds of canola, white mustard and *faba* bean. Can Entomol 118: 1279–85

Pathak MD (1961) Preliminary notes on the differential response of yellow and brown sarson and rai to mustard aphid (*Lipaphis erysimi* Kalt.). Indian Oilseeds J 5: 39–43

Prakash S, Raut RN (1983) Artificial synthesis of *Brassica napus* and its prospects as an oilseeds crop in India. J Genet 43: 282–290

Prasad SK (1983) Varietal susceptibility of rapeseed and mustard cultivars to the aphid (*Lipaphis erysimi* Kalt). Indian J Entomol 45: 501–503

Prasad SK, Phadke KG (1980) Population dynamics of *Lipaphis erysimi* (Kalt.) on different varieties of *Brassica* species. Indian J Entomol 42: 54–63

Rai BK (1976) Pests of oilseed crops in India and their control. ICAR, New Delhi, pp 121

Rai B, Sehgal VK (1975) Field resistance of *Brassica* germplasm to mustard aphid *Lipaphis erysimi* (Kalt). Sci Cult 41: 444–445

Rajan SS (1961) Aphid resistance of autotetraploid *toria*. Indian Oilseeds J 8: 251–255

Rogers CE (1980) Biology and breeding for insect and disease resistance in oilseed crops. In: Harris MK (ed) Biology and breeding for resistance to arthropods and pathogens in agricultural plants. Tex A M Univ Coll Stn, Texas, pp 359–389

Ronquist F, Åhman I (1990) Reproductive rate of the Indian mustard aphid (*Lipaphis erysimi pseudobrassicae*) on different *Brassica* oilseeds: comparisons with Swedish strains of mustard aphid (*Lipaphis erysimi erysimi*) and cabbage aphid (*Brevicoryne brassicae*). Ann Appl Biol 116: 425–430

Sehgal VK, Ujagir R (1977) Plants' natural defence in Cruciferae and Tropaeolaceae against mustard sawfly, *Athalia proxima* Klug. Indian J Ecol 4: 199–205

Singh H, Singh D, Singh H, Kumar V (1990) Basis of aphid tolerance and combining ability analysis in Indian mustard. Abstr, Nat Semin Genet *Brassica*, p 15

Singh P, Bakhetia DRC, Sidhu HS, Setia RC (1985) Plant apparency and physical defences in *Brassica* species; an index for resistance against the leaf miner *Chromatomiya horticola* Goureau. In: Namopeth B and Subhadra Bandhu J (eds). New frontiers in breeding researches. Kasetsart Univ Bangkok Thailand, pp 755–772

Singh P, Bakhetia DRC, Sidhu HS (1987) Plant apparency and chemical defences in *Brassica* against the leaf miner. 1st Symp Crop Improv Feb 23-27 1987, Ludhiana, India, pp 102–103

Singh P, Mavi GS, Bakhetia DRC (1983) Influence of *Brassica* host plants on the biology of leaf miner, *Chromatomiya horticola*. Colemania 4: 45–55

Singh RN, Dass R, Saran G, Singh RK (1982) Differential response of mustard varieties to the aphid, *Lipaphis erysimi* (Kalt.). Indian J Entomol 44: 408

Singh SR, Narain A, Srivastava KP, Siddiqui RA (1965) Fecundity of mustard aphid on different rape and mustard species. Indian Oilseeds J 9: 215–219

Sömme L, Rygg T (1972) The effect of physical and chemical stimuli on oviposition in *Hylemya floralis* (Fallen) Dpt Anthomyiidae. Norsk Entomol Tidsskr 10: 19–24

Southwood TRE (1986) Plant surfaces and insects—an overview. In: Juniper B and Southwood TRE (eds) Insects and the plant surface. Arnold, London, pp 1–22

Srinivasachar D, Malik RS (1972) An induced aphid-resistant, nonwaxy mutant in turnip, *Brassica rapa*. Curr Sci 41: 820–821

Srinivasachar D, Verma PK (1971) Induced aphid resistance in *Brassica juncea* (L.). Coss Curr Sci 49: 311–313

Talekar NS (1980) Search for host plant resistance to major insect pests in chinese cabbage. Proc Symp Production Insect Control Crucif Vegetables Taiwan, April 1980, Plant Prot Cent Taiwan pp 164–173

Teotia TPS, Lal OP (1970) Differential response of different varieties and strains of oleiferous Brassicaceae to the aphid, *Lipaphis erysimi* Kalt. Labdev J Sci Tech 8-B 219–224

Tjallingii WF (1976) A preliminary study of host selection and acceptance behaviour in the cabbage aphid, *Brevicoryne brassicae* L. Symp Biol Hung 16: 283–285

Turnock WJ, Phillip HG (1977) The outbreak of bertha armyworm *Mamestra configurata* (Noctuidae: Lepidoptera) in Alberta 1971 to 1975. Manit Entomol 11: 10–21

Underhill EW (1980) Glucosinolates. In: Bell EA Charlwood BV (eds) Encyclopedia of plant physiology 8: 493–511

Uppström B (1983) Glucosinolate pattern in different growth stages of high and low glucosinolate varieties of *Brassica napus.* Sver Utsaedesfoeren Tidskr 93: 331–336

Virtanen AL (1965) Studies on organic sulphur compounds and other labile substances in plants. Phytochem 4: 207–268

Winfield AL (1963) Pests of *Brassica* seed crops. Agric London 70: 228–232

CHAPTER 14

Freezing Injury, Resistance and Responses

A.K. Dhawan

14.1 Introduction

The fact that winter frosts cause extensive damage to fruits, vegetables and crop plants is well known. A single night of frost can cause irreversible damage to crops. Sensitivity to this environmental stress depends upon many factors, such as severity of temperature drop and its duration, the species and the genotype involved, the developmental phase of plant, the weather pattern preceding the frost and immediately following. Visible injury may, therefore, range from winter burn to blossom kill, death of buds, damage to seeds and crown kill or outright death of the plant. Considerable physiological research has been directed at the metabolism and membrane function in relation to the chilling stress. The majority of the crucifers are frost-sensitive. Recent research suggests that the injury occurs during freezing at the moment of passing through a critical temperature and, significantly, the primary site of injury is the plasma membrane (Rajashekar et al. 1979). Frost tolerance is an important factor in successful cultivation of *Brassica* crops in many countries, especially in India.

The present chapter aims at providing in brief the conceptualized knowledge of the intricate processes involved in freezing injury and plant responses with special reference to oilseed brassicas. For theoretical details on plant freezing, the reader is referred to Larcher et al. (1973), Burke et al. (1976), Levitt (1980) and Steponkus (1984).

14.2 Frosts in North India

The term "frost" is defined as "the condition that exists when the temperature of the earth's surface and earthbound objects falls below freezing, 0 °C" (Rosenberg et al. 1983). Winter frosts in India are a result of two processes:

1. Advection frost occurs when cold, dry waves enter across the northwest frontier and move towards the northeast, through the northern Indian planes. Air is the coldest element and frost damage occurs mostly in areas adjoining the northwest frontier (northwest India).
2. Radiation frost is caused by a deficit in the radiation balance between the earth and the atmosphere, particularly on clear nights without clouds or high levels of air humidity (Figueiredo 1985). Heat is lost from the earth through longwave emission to the atmosphere and in the absence of wind, the earth's plane or the top of the plant canopy becomes the coldest point, leading to frost damage. Radiation frost is more localized in time and space; advection frost, on the other hand, affects large

areas of land and is more permanent in duration. Both types of frosts occur frequently on clear nights in January and rarely in the second fortnight of December and in the first fortnight of February. On a typical winter day, the temperature begins to drop after 16.00 h, and it reaches minimum just before sunrise. Frost damage occurs if the air temperature is below 10 °C at sunset (17.00h) and the night is without clouds.

14.3 Constitutional Differences in Temperature Resistance

Plants vary in their response to various biotic and abiotic stresses. The following constitutional types may be distinguished with respect to cold resistance.

1. Chilling-Sensitive Plants. These suffer irreversible damage at temperatures between +10 and 0 °C. Many tropical rainforests, fruits and vegetables fall in this category.

2. Freezing-Sensitive Plants. These tolerate low temperature per se, but are damaged immediately when ice begins to form in the tissue. Various woody plants in warm regions and many herbaceous plants remain sensitive to ice formation in tissue. All higher plants are frost-sensitive during the period of active metabolism and growth.

3. Frost-Tolerant Plants. Deep hardening is developed by some algae, lichens and various woody plants of winter regions. The distribution of plants in various regions, along with the temperature at which low temperature injury occurs, is shown in Table 1.

Table 1. Low temperature tolerance of plants in various climatic zones

Distribution	Plants	Low temperature injury to leaves (°C)
Tropical regions	Herbaceous angiosperms, trees of evergreen rain forests	+5 to –2
Subtropical dry regions	Succulents and C_4 plants	–8 to –12
Warm temperate zone	Herbaceous angiosperms, terrestrial halophytes, dwarf shrubs	–15 to –25
Winter cold regions	Herbaceous plants in high altitudes	–20 to –25
	Evergreen conifers	–40 and below

14.4 Ice Formation in the Tissues

Ice formation normally occurs outside the cells either on the surface of the tissues or in the intercellular spaces. Some observations of intracellular ice formation have been made in the laboratory, but in all these cases the rate of temperature fall was rapid (more than 1-2 °C/h), which is uncommon in nature.

Under natural conditions, ice crystallization normally begins from large vessels (Levitt 1980). This is because a dilute sap in these vessels has a higher freezing point, compared to cell sap in other plant parts. External ice, such as hoar frost or soil ice, can readily nucleate plants via entry sites such as stomates and wounds (Burke et al. 1976). Once initiated, ice spreads readily throughout the plants's body, but it stays only in the parts of the plant external to the living cells, because the plasma membrane of the cell bars ice crystals from inoculating the cell contents. Ice crystals in the extra-cellular spaces grow at the expense of water vapour in the extracellular spaces and on the suface of cell walls. Further, the growth of ice crystals in these spaces encourages outward movement of water from the cells because of the concentration gradient. The cell contents thus become concentrated, resulting in freezing point depression and further avoiding the possibility of intracellular ice formation.

The rate of diffusion of water to ice loci outside the cells is limited by the permeability of lipid plasma membrane. Compared to solutes, this membrane is highly permeable to water. If, however, the temperature drop is very rapid and diffusion to extracellular space does not occur with sufficient speed to increase the concentration of cell components as rapidly, the temperature of cells may reach sufficiently below the freezing point and induce spontaneous intracellular ice formation.

Extracellular ice formation is favoured by smaller cell size, resulting in larger specific surface for outward movement of water and high membrane permeability to water. Further, the transition of membrane lipids from the liquid crystalline to solid gel state markedly lowers the permeability of cells. If this transition occurs at a temperature higher than the freezing point of the cell, freezing will occur intracellularly, clearly so because the outward movement of water that favours extracellular ice formation is prevented by this transition.

14.5 Factors Affecting Ice Formation

Burke et al. (1976) have listed factors that affect ice formation in plants. These are:

14.5.1 Freezing Point Depression

An increase in solute concentration in the cells lowers the freezing point of cell sap. However, few species of agricultural importance have freezing point depressions of more than 4 °C. Some halophytes, however, have freezing point depressions as low as –14 °C.

14.5.2 Super Cooling

Freezing point is the temperature at which melting occurs, but not necessarily the temperature at which cell sap will freeze. When the plant is cooled slowly, its temperature generally drops somewhat below the freezing point without ice formation. This process is called supercooling and refers to the lowest subfreezing temperature attained without ice formation. Pure water can super-cool to –38 °C in the absence of nucleating substances.

Herbaceous plants seldom supercool more than –1 or –2 °C, unless the tissue moisture content is very low. Most of the deciduous forest species and some fruit tree cultivars "deep supercool" to as low as –40 °C in mid-winter (George et al. 1974).

14.5.3 Ice Nucleation

One major reason for supercooling of the cell sap is the lack of nucleating substances that are necessary for ice formation. Ice crystals must grow around some nuclei. Indeed, pure water can supercool to –38 °C in the complete absence of nucleating substances.

It has been observed that ice nucleation-active bacteria inhabit plant surfaces and significantly limit the ability of water in herbaceous plant tissues to supercool below –2 °C (Lindow 1983; Lindow et al. 1983; Probsting Jr. and Gross 1988). Contrary to this, chemicals such as fluorophlogopite and acetoacetanilide, which act as ice nucleators, result in lethal injury to *Solanum* leaves (Rajashekar et al. 1983).

14.5.4 Ice Propagation and Growth

Once freezing is initiated, water molecules migrate and freeze at the surface. During extracellular freezing, water becomes frozen in large ice masses at specific sites which accommodate ice. Fast freezing generally results in many small crystals, while fewer large crystals form when freezing is slow. This is because rapid freezing does not permit water migration to favoured sites for growth of large crystals. Single cells or tissues can survive at –196 °C (plunging into liquid nitrogen), when cooling is so rapid that small ice crystals formed are not disruptive (Sakai and Suka 1967).

14.5.5 Tissue Hydration

The optimum water content for overwintering cereal crowns has been reported to be about 65% (Metcalf et al. 1970). At higher moisture contents, water freezes rapidly, while dehydrated cereal crowns supercool significantly. Dehydrated apple buds of the variety Patterson did not freeze until samples were cooled to –20 °C (Tyler et al. 1988).

14.5.6 Membranes and Cell Walls

Cell wall polymers interfere with the freezing process by interacting with ice and modifying the shape of the ice that forms. Cell membrane is a major barrier to the growth of ice crystals into the cell from the extracellular spaces.

14.6 Injury Resulting from Ice Formation

14.6.1 Due to Intracellular Ice Formation

Intracellular freezing occurs with thousands of ice crystals forming throughout the protoplast and vacuoles. This sets in motion several injurious processes such as dehydration imposed on macromolecular cellular structures by ice, membrane destruction and degradation by enzymes released due to breakdown of cellular compartmentalization. Cells in which ice crystals are formed are nearly always killed.

14.6.2 Due to Extracellular Ice Formation

Removal of water from cells to extracellular ice imposes a considerable dehydration stress on the protoplasm. Dehydration of plant cells during freezing is a primary event which leads to death. Apart from the mechanical effects of the presence of ice in the tissues, such dehydration causes damage to the membranes (Olein 1967), protein dissociation into the subunits (Burke et al. 1976) and several other molecular level changes (Olein 1967). Singh and Miller (1982) suggest that extracellular freezing involves irreversible conversion of planar membrane bilayers to structures having less ordered packing.

On thawing, freeze-killed cells characteristically show frost plasmolysis. This is because the cell wall expands back to nearly its original shape, while the dead protoplast contracts. The intercellular ice is converted to water, the cells become flaccid and the tissues give a water-soaked appearance.

14.7 Application of Freezing Treatments

Although natural frosts are of a great value, particularly in experiments of agricultural importance, several problems are encountered in their use as test frosts. These include variations in timing, intensity and frequency of low temperatures, less possibility of applying treatments at the desired stage of plant development, and difficulties in retaining unfrozen control plants for comparison. A primary requirement in frost studies is, therefore, to have a reliable technique of imposing artificial freezing treatments in the field. Artificial freezing cabinets must have defined, reproducible and uniform environmental conditions within the cabinet, sufficient internal space, possibility of varying day length and temperature, and provisions for easy movement. The application of freezing treatments must be such that it mimics the typical natural frosts.

Domestic freezers have been commonly modified into frost chambers (Reid et al. 1976; Wunsche 1966; Dhawan et al. 1983). Specially designed cold rooms have been employed by Paton (1972) and Mather et al. (1980). A movable freezing apparatus designed at the Swedish University of Agricultural Sciences, Uppsala, has several superior features: it distributes cold air uniformly, allows for slow rates of cooling, provides for precise control and continuous monitoring of temperature and eliminates canopy effects (Dhawan and Ohlsson 1989).

One serious drawback in artificial freezing is that lethal intracellular ice formation usually accompanies it due to rapid freezing. Therefore, the application of freezing treatments must invariably be made during the early morning hours (Dhawan et al. 1983; Dhawan and Ohlsson 1989). Apart from the ease in obtaining slow rates of cooling and thawing, this mimics the timing of natural frosts and does not disturb the biological rhythm in plants.

14.7.1 Assessment of Injury

A test for predicting freeze tolerance should be simple, rapid, repeatable, non-destructive and highly correlated to field survival.

Methods available for assessment of cold injury include reduction of triphenyl tetrazolium chloride (Steponkus and Lanphear 1967), measurement of electrolytes (Krasnuk et al. 1978), amino acid concentration (Siminovitch et al. 1964) or potassium efflux (Shcherbackova and Kacperska-Palacz 1983), chlorophyll fluorescence emission (Melcarek and Brown 1977), release of HCN gas (Stout 1983), respiration rates (Siminovitch et al. 1964) and crown water content (Fowler et al. 1981). In spite of promising results in some plant species and the advantage of quicker determinations, such methods cannot be trusted as a true estimate of freeze tolerance in all plants (Levitt 1980).

Direct quantitative assessment of injury to economically important plant parts is, however, a more relevant criteria of assessment of injury in experiments of applied value. Thus in crop plants such as *Brassica*, reduction in seed yield (Dhawan et al. 1983) and percent frost-killed seeds (Dhawan 1985) have been applied as true estimates of frost injury. In crops where vegetative parts are of economic importance, a field survival index (Fowler and Gusta 1979) or the ratio of leaf dry weight to fresh weight (Paton 1972) may be useful. Criteria for assessment of injury would thus depend upon the type of experiment. Use of a field survival index for *Brassica* crops (Dhillon and Larsson 1985) would be of little value.

14.8 Low-Temperature Response of *Brassica* Plants

14.8.1 Tolerance of Various Genotypes

The response of 12 varieties belonging to *B. juncea*, *B. campestris*, *B. chinensis* and *B. napus* species was observed under natural frost conditions in the field (Table 2).

Table 2. Percent undeveloped seeds in various *Brassica* genotypes. Observations recorded 7 days after a natural frost affected the 90-day-old crop. Figures are mean ± S.E.

Species	Variety	Undeveloped seeds (%)
B. chinensis	Local	21.1 ± 1.2
B. napus	Norin-20	84.3 ± 4.8
B. campestris	BSH-1	27.2 ± 2.3
	DBS-1	44.6 ± 3.1
	Span	4.9 ± 0.4
	Torch	7.2 ± 0.4
	Bell	4.8 ± 0.8
B. juncea	RH-30	66.2 ± 4.9
	RH-781	22.3 ± 3.2
	Varuna	39.7 ± 3.1
	Prakash[a]	10.9 ± 0.9
	RH-7513[a]	10.1 ± 1.0

[a] Late-flowering genotypes.

All except *B.juncea* varieties Prakash and RH-7513 showed similar maturity duration. While *B. campestris* varieties Span, Torch and Bell were only slightly susceptible (<10% undeveloped seeds), *B. chinensis* cv. Local, *B. campestris* var. BSH-1 and *B. juncea* var. RH-781 were moderately susceptible and *B. campestris* var. DBS-1 and *B. juncea* var. Varuna and RH-30 were highly susceptible. *B. napus* var. Norin-20 was, however, the most susceptible, with 84.3% seeds affected by frost.

In another study, Banga et al. (1987) evaluated 43 genotypes of Indian mustard for frost tolerance under natural conditions. Initial visual observations were correlated with yield stability and electrolyte leakage. Entries FS Sel 3 and IS 99 were most resistant. IS99 is a chromosome substitution line of *B. juncea* with one C genome chromosome possibly substituting a homoeologus B genome chromosome.

14.8.2 Tolerance of Various Stages of Development

Both very young and nearly mature seeds are less affected by frost, compared to seeds at an intermediate stage of development (Dhawan 1985; Banga et al. 1987). Thus, the crop sown at normal time of planting showed 50% siliquae formation when frost occurred. This showed the highest frost damage (66.2% undeveloped seeds) because of the high proportion of developing (susceptible) seeds at this stage (Table 3).

Studies conducted under artificial frost induction at Uppsala showed the flowering stage to be most susceptible to frost damage in *B. juncea* cv. Varuna (Elf and Ohlsson 1987). A 50% reduction in yield was obtained at –10 °C for 3 h for the flowering stage (70% buds had flowered) and at –2 °C for the pod-filling stage.

Table 3. Percent undeveloped seeds in *B. juncea* var. RH-30 plants sown on different dates. Observations recorded 7 days after a natural frost affected the 90-day-old crop. Figures are mean ± S.E.

Date of sowing	State of development at the time of frost	Undeveloped seeds (%)
Sept. 27	Seed filling nearly complete	23.9 ± 1.3
Oct. 10	Siliquae formation complete	31.7 ± 4.2
Oct. 21[a]	50% siliquae formation	66.2 ± 6.9
Nov. 1	Peak flowering	39.8 ± 2.0
Nov. 17	Early flowering	13.9 ± 1.3

[a] Normal time of sowing.

14.8.3 Effect on Morpho-Physiological Characters

The effect of freezing on morphological characters, seed yield and its components and oil content in *B. juncea* var. Prakash was studied. The number of leaves on main shoot, number of primary branches, length of siliquae and seed number/siliqua are not affected by freezing either at the vegetative or reproductive stages. Plant height, though reduced initially, was regained at maturity. Characters that were adversely affected in unhardened, but not in hardened plants, include seed yield, oil content and harvest index.

Table 4. Effect of freezing on photosynthesis and respiration in *B. juncea* plants. Carbon dioxide measurements were carried out on pot plants using infrared gas analyzer. Values ($g/dm^2/ha$) are mean ± S.E.

Treatment	Photosynthesis	Respiration
Unfrozen	17.02 ± 0.87	4.08 ± 0.24
Frozen	7.91 ± 0.62	5.22 ± 0.11

14.8.4 Effect on Carbon Fixation, Utilization and Assimilate Partitioning Rate

The photosynthetic rate showed a 60% reduction on freezing in *B. juncea* plants. The respiratory rate, however, showed a slight enhancement on freezing (Table 4). Further, freezing had no effect on $^{14}CO_2$ fixation rates of plants that had been previously hardened at 3 °C for 6 days (Table 5).

The Hill activity in chloroplasts isolated from frozen plants was also lower compared to unfrozen plants, but such a decrease did not occur in hardened plants. The photosynthetic apparatus of hardened plants thus appears to be better adapted to survive freezing stress.

Table 5. Effect of freezing on CO_2 fixation, Hill activity and chlorophyll content in control and low temperature-hardened plants of *B. juncea*. $^{14}CO_2$ fixation carried out in assimilation tubes and Hill activity of chlorophyll suspensions was measured as in Hooda and Dhawan (1986). Values are mean ± S.E.

Treatment	$^{14}CO_2$ fixation (cpm/min/g fresh weight)	Hill activity (A_{420}/min/mg chlorophyll)
Control, UF	20757 ± 3397	0.450 ± 0.068
Control, F	5055 ± 1737	0.326 ± 0.031
Hardened, UF	15579 ± 1821	0.297 ± 0.112
Hardened, F	21007 ± 3166	0.337 ± 0.032

UF = Unfrozen; F = Frozen.

The transport of assimilates from source to sink was comparatively less affected by freezing. In control unfrozen, control frozen, low-temperature unfrozen and low-temperature frozen plants, 78, 67, 82 and 63% of the total ^{14}C fixed by the lower leaf was translocated to inflorescence within 24 h.

14.8.5 Effect on Enzyme Activities

Enzyme activities of isocitrate dehydrogenase, malate dehydrogenase and 6-phosphogluconate dehydrogenase increased in the leaves of *B. campestris* L. var. BSH-1 and *B. juncea* var. Prakash after freezing and subsequent thawing. The increase was, generally, pronounced at 1 day after freezing (Gupta et al. 1984). In another study (Dhawan et al. 1986) including several hardening treatments of *B. juncea* var. Prakash, there appeared to be no clear association of malate dehydrogenase and peroxidase activities with freeze tolerance in *Brassica*. The activities of both these enzymes increase on freezing in unhardened plants and seem to be related to freezing injury.

14.8.6 Phospholipids and Sterols

Sikorska and Kacperska-Palacz (1979) reported an increase in phospholipid content during hardening of *B. napus* plants. Further work using ^{32}p incorporation into leaf lipids (Sikorska and Kacperska-Palacz 1980) indicated that an inhibition of phospholipid biosynthesis may actually occur under autumn and winter conditions. The increased total phospholipid level observed at that time may be due to inhibition of phospholipid catabolism.

Sikorska and Forkas (1982) observed a reduction in free sterol content in *B. napus* leaves at the most advanced stage of hardening. The ratio of free sterols to total phospholipids was significantly reduced by hardening due to a decrease in the level of former and an increase in that of latter components.

14.8.7 RNA/Proteins

SDS-PAGE of protein extracts and in vivo labelling with ^{35}S revealed the induction of 15 KD and 27 KD membrane-bound polypeptides during the development of freezing tolerance in *Brassica* and other cruciferous plants. No unique high molecular weight polypeptides were, however, observed in all the species during freezing tolerance (Laroche and Singh 1988).

Weiser (1970) first proposed that cold hardening might involve changes in gene expression. Indeed, cycloheximide has been shown to inhibit cold acclimation in potato (Chen et al. 1983). Studies have shown that the protein synthesis continues even at 0 °C but some polypeptides preferentially accumulate at this temperature. On the other hand, synthesis of several others is repressed while many are insensitive to the cold treatment in *B. napus* (Mezo-Brasso et al. 1986). Similar changes are also observed when mRNA is prepared from cold treated seedlings, translated in vitro in a reticulocyte cell free system and compared with the products of mRNA extracted from control samples. Among repressed genes, a small subunit of ribulose 1,6-biphosphate carboxylase could be identified.

14.9 Freezing Resistance

Freezing resistance may be defined as the ability of an organism to survive the direct action of freezing temperature without suffering permanent damage on its growth or reproduction. Freezing resistance of a plant results from two types of mechanisms. First of these are the mechanisms aimed at delaying or preventing the penetration of stress into the tissues, thereby preventing ice formation. This is freezing avoidance. The second type are the mechanisms that impart the ability to the plant to tolerate freezing, in spite of the penetration of stress into the tissues, i.e. the ice formation does occur, but the plant is able to remain undamaged. This is freezing tolerance.

14.10 Freezing Avoidance

Plants are poikilotherms, i.e. they are unable to maintain a constant tissue temperature different from that of their environment. Thus, plants have no effective insulation against low temperature and, therefore, do not possess true avoidance of low temperature. Plants can, however, avoid ice formation by following different mechanisms (Levitt 1980).

14.10.1 Dehydration

Dehydrated cell sap, as in some halophytes and in seeds and pollen, has little freezable water. In many non-halophytes also, partial freeze avoidance due to solute accumulation does exist. O'Neill (1983) observed that, although solute accumulation is an important factor, it does not function alone to confer freeze avoidance in plants.

14.10.2 Supercooling

Almost all plants supercool to a few degrees, some woody plant stems supercool to –15°C, while wood parenchyma of deciduous woody plants supercool as low as –40 °C (Burke et al. 1976). Deep supercooling found in woody plant tissues has not been observed in herbaceous species. Supercooling is a significant avoidance mechanism for all plants in mild climates and for deciduous woody plants in temperate climates.

Factors that favour supercooling are small cell size, low moisture content, no internal space for nucleation, absence of internal nucleators, barriers against external nucleators and the presence of anti-nucleators.

14.11 Conclusions

Cultivated species of *Brassica* crop plants appear to have little exploitable genetic variability for frost resistance. Available selection methods are not precise enough to allow for identification of small but probably meaningful differences for frost tolerance. A movable frost chamber, with all its limitations, is the only practical screen. However, there is a need to develop rapid and non-destructive techniques which are repeatable and require only a part of a single plant, especially at the juvenile seedling stage, for analysis. Future research should aim at the development of data banks to characterize the germplasm. Plants growing at high altitudes and in the wild state are likely to contain freezing resistance genes as was observed for tomato (Patterson 1988). The search for improved resistance should be expanded to include primary and secondary gene pools. While whole plants show little tolerance, individual cells may have greater variation. Selection at cell level may be an attractive proposition.

References

Banga SK, Banga SS, Labana KS, Dhillon SS (1987) Screening for frost tolerance in Indian mustard (*Brassica juncea* (L.) Coss). In : Proc 7th Int Rapeseed Congr, Poland, pp 771–779

Burke MJ, Gusta LV, Quamme HA, Weiser CJ, Li PH (1976) Freezing and injury of plants. Ann Rev Plant Physiol 27: 507–528

Chen HH, Li PH, Brenner LM (1983) Involvement of abscisic acid in potato cold acclimation. Plant Physiol 71: 362–365

Dhawan AK (1985) Freezing in oilseed *Brassica* ssp: some factors affecting injury. J Agric Sci Camb 104: 513–518

Dhawan AK, Ohlsson I (1989) A movable apparatus for freezing field plants. Indian J Exp Biol 27: 611–614

Dhawan AK, Chhabra ML, Yadava TP (1983) Freezing injury in oilseed *Brassica* species. Ann Bot 51: 673–677

Dhawan AK, Hooda A, Goel RK (1986) Effect of low temperature, short days, water stress and dimethyl sulphoxide on frost tolerance of *Brassica juncea* Coss and Czern var. Prakash. Ann Bot 58: 267–271

Dhillon SS, Larsson S (1985) Screening technique for frost resistance in *Brassica* oilseeds. J Agric Sci Camb 104: 245–248

Elf S, Ohlsson I (1987) Low temperature treatment of *Brassica juncea* grown in field. In: Proc 7th Int Rapeseed Congr, Poland, pp 763–770

Figueiredo P (1985) Frost and freezing injury to crop plants—a literature review. Rep 155, Swed Univ Agric Sci, Uppsala, Sweden, pp 19

Fowler DB, Gusta LV (1979) Selection for winter hardiness in wheat. I. Identification of genotypic variability. Crop Sci. 19: 769–772

Fowler DB, Gusta LV, Tyler NJ (1981) Selection for winter hardiness in wheat. III. Screening methods. Crop Sci 21: 896–901

George MF, Burke MJ, Weiser CJ (1974) Supercooling in overwintering *Azalea* flower buds. Plant Physiol 54: 29–35

Gupta VK, Dendsay JPS, Dhawan AK (1984) Effect of freezing stress on some dehydrogenases in *Brassica* species. Plant Physiol Biochem 11: 103–107

Hooda A, Dhawan AK (1986) Changes in carbon fixation, utilization and assimilate partitioning rates on freezing and low temperature hardening of *Brassica juncea* Czern and Coss plants. Indian J Exp Biol 24: 121–123

Krasnuk M, Witham FH, Jung GA (1978) Hydrolytic enzyme differences in cold tolerant and cold-sensitive alfalfa. Agron J 70: 597–605

Larcher W, Heber U, Santarus KA (1973) Limiting temperatures for life functions. In: Precht H, Christopherson J, Hensel H and Larcher W (eds.) Temperature and life, Springer, Berlin Heidelberg New York, pp 195–263

Laroche A, Singh J (1988) Induction of freezing tolerance in *Arabidopsis thaliana* and other members of the Cruciferae family. Plant Physiol (Suppl) 86: 311

Levitt J (1980) Responses of plants to environmental stresses, Vol I Chilling, freezing and high temperature stresses. Academic Press, New York

Lindow SE (1983) The role of bacteria ice nucleation in frost injury to plants. Ann Rev Phytopathol 21: 363–384

Lindow SE, Arny DC, Upper CD (1983) Biological control of frost injury: establishment of effects of an isolate of *Erwinia herbicola* antagonistic to INA bacteria on corn in the field. Phytopathol 73: 1102–1106

Mather PJC, Modlibowska I, Keep E (1980) Spring frost resistance in black currants (*Ribes nigrum* L.). Euphytica 29: 793–800

Melcarek PK, Brwon GN (1977) Effects of chill stress on prompt and delayed chlorophyll fluorescence from leaves. Plant Physiol 60: 822–825

Metcalf EL, Cress CE, Olien CR, Everson EH (1970) Relationship between crown moisture content and killing temperature for three wheat and three barley cultivars. Crop Sci 10: 362–365

Mezo-Brasso L, Alberd M, Raynal M, Ferrero-Candinanos MD (1986) Changes in protein synthesis in rapeseed (*B. napus*) seedlings during low temperature treatment. Plant Physiol 82: 733–738

Olein CR (1967) Freezing stresses and survival. Ann Rev Plant Physiol 18: 387–408

O'Neill SD (1983) Osmotic adjustment and the development of freezing resistance in *Fragaria virginiana*. Plant Physiol 72: 938–944

Paton DM (1972) Frost resistance in *Euclyptus*: a new method for assessment of frost injury in *E. viminalis*. Aust J Bot 20: 127–139

Patterson BD (1988) Genes for cold resistance from wild tomatoes. Hortic Sci 23: 794

Probsting Jr EL, Gross DC (1988) Field evaluation of frost injury to deciduous fruit trees as influenced by ice nucleation active *Pseudomonas syringae*. J Am Soc Hortic Sci 113: 498–506

Rajashekar CB, Gusta LV, Burke MJ (1979) Frost damage in hardy herbaceous species. In: Lyons JM, Graham D, Raison JK (eds) Low temperature stress in plants. The role of membranes. Academic Press, New York

Rajashekar CB, Li PH, Carter JV (1983) Frost injury and heterogeneous ice nucleation in leaves of tuber-bearing *Solanum* spp.: ice nucleation activity of external source of nucleants. Plant Physiol (Bethesda) 71: 749–755

Reid WS, Harris RE, McKenzie JS (1976) A portable freezing cabinet for cold-stressing plants growing in the field. Can J Plant Sci 56: 623–625

Rosenberg NJ, Blaine LB, Joshi BV (1983) Microclimate-biological environment. John Wiley, New York

Sakai A, Suka O (1967) Survival of plant tissue at super-low temperatures. V. An EM study of ice in cortical cells cooled rapidly. Plant Physiol 42: 1680–1694

Shcherbackova A, Kacperska-Palacz A (1983) Water stress injuries and tolerance as related to potassium efflux from winter rape hypocotyls. Physiol Plant 57: 296–300

Sikorska E, Forkas T (1982) Sterols and frost hardening of winter rape. Physiol Plant 56: 349–352

Sikorska E, Kacperska-Palacz A (1979) Phospholipid involvement in frost tolerance. Physiol Plant 47: 144–150

Sikorska E, Kacperska-Palacz A (1980) Phospholipid biosynthesis in winter rape plants during the winter season. Bull Acad Pol Sci 28: 191–194

Siminovitch D, Therrien H, Gfeller F, Rheamme B (1964) The quantitative estimation of frost injury and resistance in black locust, alfalfa and wheat tissues by determination of amino acids and other ninhydrin-reacting substances released after thawing. Can J Bot 42: 637

Singh J, Miller RW (1982) Spin-probe studies during freezing of cells isolated from cold-hardened and non-hardened winter rye-molecular mechanisms of membrane freezing injury. Plant Physiol 69: 1423–1428

Steponkus PL (1984) Role of the plasma membrane in freezing injury and cold acclimation. Ann Rev Plant Physiol 35: 543–584

Steponkus PL, Lanphear FD (1967) Refinement of the tetrazolium chloride method of determining cold injury. Plant Physiol 42: 1423–1426

Stout DG (1983) Evaluation of freezing injury and freeze-thaw injury to membranes of Saskatoon service berry twigs by measuring HCN release. Physiol Plant 57: 573–578

Tyler N, Stushnoff C, Gusta LV (1988) Freezing of water in dormant vegetative apple buds in relation to cryopreservartion. Plant Physiol 87: 201–205

Weiser CJ (1970) Cold resistance and injury in woody plants. Science 169: 1269–1278

Wunsche U (1966) The use of food freezers to build plant growth cabinets. Lantbrukshoegsk Ann 32: 417–426

Author Index

For authors and co-authors cited in the text, no page numbers of the list of references have been given. For co-authors not cited in the text the relevant page numbers of the list of references are given between brackets only once.

Subject Index

Terms which occur throughout the book with high frequency (e.g. generic names etc.) have been indexed only in case of special significance.

Springer-Verlag and the Environment

We at Springer-Verlag firmly believe that an international science publisher has a special obligation to the environment, and our corporate policies consistently reflect this conviction.

We also expect our business partners – paper mills, printers, packaging manufacturers, etc. – to commit themselves to using environmentally friendly materials and production processes.

The paper in this book is made from low- or no-chlorine pulp and is acid free, in conformance with international standards for paper permanency.

GPSR Compliance

The European Union's (EU) General Product Safety Regulation (GPSR) is a set of rules that requires consumer products to be safe and our obligations to ensure this.

If you have any concerns about our products, you can contact us on ProductSafety@springernature.com

In case Publisher is established outside the EU, the EU authorized representative is:

Springer Nature Customer Service Center GmbH
Europaplatz 3
69115 Heidelberg, Germany

Batch number: 10844669

Printed by Printforce, the Netherlands